AF228991

Nonlinear Problems in Accelerator Physics

Sponsored by

Forschungszentrum Jülich
Hahn-Meitner-Institut
Michigan State University

Nonlinear Problems in Accelerator Physics

Proceedings of the International Workshop
on Nonlinear Problems in Accelerator Physics
held in Berlin, Germany, 30 March–2 April, 1992

Edited by M Berz, S Martin and K Ziegler

Institute of Physics Conference Series Number 131
Institute of Physics, Bristol and Philadelphia

CODEN IPHSAC 131 1–265 (1993)

British Library Cataloguing in Publication Data

A catalogue record for this book is available from the British Library

ISBN 0-7503-0238-0

Library of Congress Cataloging-in-Publication Data are available

Published by IOP Publishing Ltd, a company wholly owned by the Institute of Physics, London, Techno House, Redcliffe Way, Bristol BS1 6NX, England
US Editorial Office: IOP Publishing Inc., The Public Ledger Building, Suite 1035, Independence Square, Philadelphia, PA 19106, USA

Printed in the UK by Galliard (Printers) Ltd, Great Yarmouth, Norfolk

Preface

From March 30 to April 2, 1992, about 80 scientists working on various aspects of accelerator physics and particle optics gathered in a small eastern suburb of Berlin to discuss nonlinear phenomena in beam physics. The setting at Humboldt University's historically important Gosen Science Communication and Conference Center with its gently rolling hills, pristine lakes, and serene landscape provided a beautiful environment for their activities, and the capricious spring weather helped to focus on questions of science rather than beauty of nature.

A variety of topics were covered: There were talks about the foundations of the field and new techniques; several contributions addressed codes that allow their application in practice, and two papers summarized experiments subjecting the fruits of theoretical pursuits to the uncompromising test of reality. Besides the talks, there were many stimulating discussions, and until late at night several enthusiasts studied various new features of codes.

We would like to thank all the attendants for their interest, the speakers for stimulating talks and interesting papers, the referees for valuable comments, and all those who helped with organizational questions for their tireless efforts. We are grateful for financial support from the Hahn-Meitner-Institut, KFA Jülich, and Michigan State University.

M Berz
S Martin
K Ziegler

Contents

Inst. Phys. Conf. Ser. No 131
Paper presented at Int. Workshop Nonlinear Problems in Accelerator Phys. Berlin, 1992

Nonlinear Problems in Accelerator Physics

H Mais

Abstract. Nonlinear problems in accelerator physics are reviewed. Theoretical tools and methods are introduced and discussed, and it is shown how these concepts can be applied to the study of various nonlinearities in storage rings. The first part treats Hamiltonian systems (proton accelerators) whereas the second part is concerned with explicitly stochastic systems (e.g. electron storage rings).

Contents

1. Introduction

As synchrotron radiation sources and as colliders, storage rings have become an important tool in physical research. Colliders are devices which allow two beams of ultrarelativistic charged particles circulating in opposite directions to be accumulated, stored and collided. (see Figure 1)

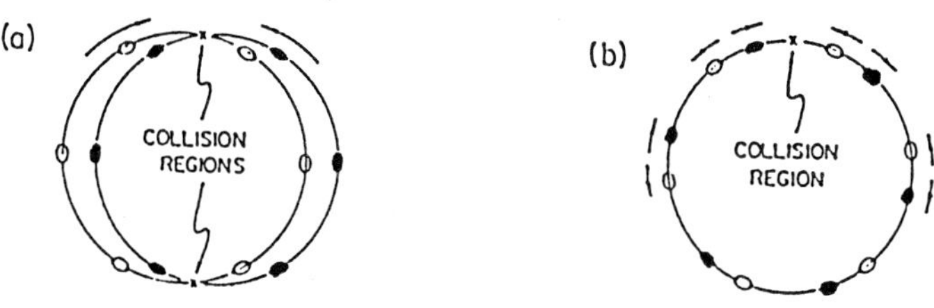

Figure 1. storage rings

The technical components of such an accelerator are magnets, a toroidal vacuum chamber and accelerating rf structures (cavities). Usually the stored beams consist of bunches, each of them containing 10^{10} to 10^{11} particles. The size of these bunches ranges from a tenth of a millimeter to a few centimeters. Some details of the electron-proton storage ring HERA at DESY in Hamburg are listed below:

	$electron-ring$	$proton-ring$
$circumference$	$6.336km$	$6.336km$
$energy$	$26/30GeV$	$820GeV$
$No.\ of\ bunches$	210	210
$particles/bunch$	3×10^{10}	10^{11}
$bunch-width\ at\ interaction\ point$	$0.27mm$	$0.29mm$
$bunch-height\ at\ interaction\ point$	$0,017mm$	$0,095mm$
$bunch-length\ at\ maximum\ energy$	$30mm$	$440mm$

An accelerator constitutes a complex many body system - namely an ensemble of 10^{10} to 10^{11} charged ultrarelativistic particles subject to external electromagnetic fields, radiation fields and various other influences such as restgas scattering, space charge effects and wakefields.

Although collective phenomena, as for example instabilities, are very important for the performance of an accelerator we restrict ourselves in this lecture to the classical single particle dynamics, i.e. we study the equations of motion of a single charged ultrarelativistic particle under the influence of external electromagnetic fields and radiation effects. In general these equations are nonlinear [1] [2]. The main nonlinearities are due to beam-beam interaction, due to nonlinear cavity fields or due to transverse multipole fields. These multipole fields are either introduced artificially, e.g. by sextupoles which compensate the natural chromaticity or they occur naturally as deviations from linear fields due to errors.

Because of these nonlinearities a storage ring acts as a nonlinear device schematically sketched in Figure 2. a_{in} is some initial amplitude (position, momentum given by the injection

Figure 2. storage ring as a nonlinear device

conditions) and a_{fin} is the amplitude after N (10^8 - 10^{10}) revolutions in the ring. In accelerator physics one often tries to define different zones according to a_{in}. For small amplitudes up to a certain boundary a_{lin} - the linear aperture - the storage ring behaves more or less like a linear element (at least for the time scales of interest, i.e. 10 - 20 hours storage time). For larger amplitudes the behaviour becomes more and more nonlinear and eventually at a_{dyn} - the dynamic aperture - the particle motion becomes unbounded. One problem of accelerator physics is to make quantitative predictions of these different zones, or stated in a different way, to calculate quantities such as the linear aperture a_{lin} or the dynamic aperture a_{dyn}. Furthermore one wants to know how these quantities depend on various machine parameters and the type of the nonlinearity. A better and - from a practical point of view - more relevant question is: what is the lifetime of the particle, or what is the probability for the particle to hit the vacuum chamber (first passage time) if it is injected into a certain volume in phase space. In order to solve these problems, various numerical and analytical tools have been developed, some of which will be described in the following.

This survey lecture is organized as follows: In the first part we will consider storage rings where radiation phenomena can be neglected, i.e. accelerators for protons or heavy ions. In HERA for example the radiation losses of a proton are a factor 10^{-7} less than the losses of an electron. Thus these storage rings can be modelled mathematically by nonlinear (in general nonintegrable) Hamiltonians. Nonintegrable means that the corresponding nonlinear equations of motion cannot be solved analytically. As we will see later the phase space dynamics of these systems shows a very rich and complicated structure. The questions we want to answer in the first part are:

- what does the Hamiltonian for the particle dynamics look like?

- what is in principle possible in these systems? (qualitative theory)

- which analytical (i.e. perturbative) tools are available for a quantitative study of these problems?

In the second part of this survey we will treat systems where radiation effects or noise effects are important. Because of the stochastic emission of the radiation, radiative systems can be modelled by explicit stochastic dynamical systems. A staightforward way to extend deterministic systems to include noise effects and explicit stochastic phenomena is to write down stochastic differential equations. In this lecture we will illustrate some of the subtleties related to stochastic differential equations including Gaussian white noise, and we will mention and illustrate some applications in accelerator physics.

This lecture cannot cover the whole subject exhaustively, we can only sketch the basic ideas and illustrate these ideas with simple (sometimes oversimplified) models. For many details we have to refer the reader to the references. Our main aim is to show that the single particle dynamics of storage rings represents an interesting field for nonlinear dynamics with a practical background.

2. Hamiltonian dynamics

2.1. Hamiltonian for coupled synchro-betatron motion

Starting point is the following relativistic Lagrangian for a charged particle under the influence of an electromagnetic field described by a vector potential $\vec{A}$ [3] :

$$\mathcal{L} = -m_0 c^2 \sqrt{1 - \frac{\dot{\vec{r}}^2}{c^2}} + \frac{e}{c}(\dot{\vec{r}}\vec{A}) \tag{1}$$

with

- e=elementary charge
- c=speed of light
- m_0=rest mass of the particle
- $\dot{\vec{r}}$=particle velocity

Usually one changes to a Hamiltonian description of motion and one introduces the curvilinear coordinate system depicted in Figure 3 [4].

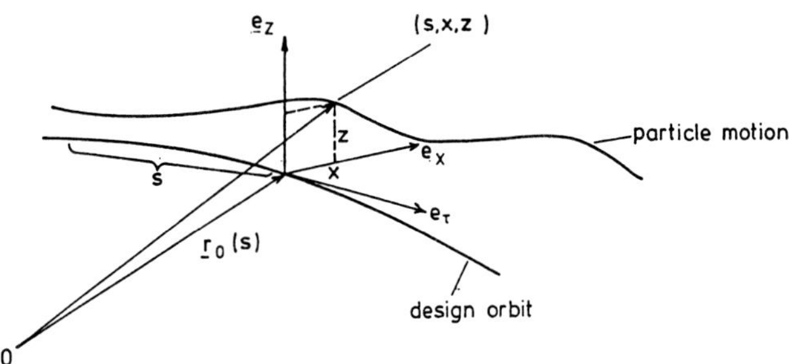

Figure 3. curvilinear coordinate system

It consists of three unit vectors $\vec{e}_\tau, \vec{e}_x, \vec{e}_z$ attached to the design orbit of the storage ring. s is the pathlength along this trajectory. For simplicity we have assumed a plane reference orbit with horizontal curvature κ only. Using s as an independent variable and introducing difference variables with respect to an equilibrium particle on the design orbit one obtains

$$\mathcal{H}(x, z, \tau, p_x, p_z, p_\tau; s) =$$
$$-(1 + \kappa x)\{(1 + p_\tau)^2 - (p_x - \frac{e}{E_0}A_x)^2 - (p_z - \frac{e}{E_0}A_z)^2\}^{1/2} -$$
$$-(1 + \kappa x) \cdot \frac{e}{E_0} \cdot A_\tau + (1 + p_\tau) \tag{2}$$

where we have used

- $v \approx c$ (ultrarelativistic particles)

- $\tau = s - ct$

- $p_\tau = \frac{\Delta E}{E_0}$

- E_0=energy of design particle.

The corresponding equations of motion are:

$$\frac{d}{ds}\, x = +\frac{\partial \mathcal{H}}{\partial p_x}; \qquad \frac{d}{ds}\, p_x = -\frac{\partial \mathcal{H}}{\partial x}$$
$$\frac{d}{ds}\, z = +\frac{\partial \mathcal{H}}{\partial p_z}; \qquad \frac{d}{ds}\, p_z = -\frac{\partial \mathcal{H}}{\partial z} \qquad (3)$$
$$\frac{d}{ds}\, \tau = +\frac{\partial \mathcal{H}}{\partial p_\tau}; \qquad \frac{d}{ds}\, p_\tau = -\frac{\partial \mathcal{H}}{\partial \tau}\,.$$

$\vec{A}(x, z, s) = (A_x(x, z, s), A_z(x, z, s), A_\tau(x, z, s))$ is the vector potential which determines the external electromagnetic fields. The transverse coordinates (x, z, p_x, p_z) describe the *betatron* motion and the longitudinal coordinates (τ, p_τ) describe the *synchrotron* motion. Some examples for the vector potential $\vec{A}(x, z, s)$ are shown below:

rf - cavity:

$$A_\tau = -\frac{L}{2\pi k} \cdot V_0 \cdot \cos(k\frac{2\pi}{L}\tau) \cdot \delta(s - s_0) \qquad (4)$$

with

- V_0=peak voltage of cavity

- L=circumference of storage ring

- k=harmonic number

- $\delta(s - s_0)$=delta function (localized cavity)

bending (dipole) magnet:

$$\frac{e}{E_0} A_\tau = -\frac{1}{2}(1 + \kappa \cdot x) \qquad (5)$$

with

- $\kappa = \frac{e}{E_0} B_z(x = z = 0)$=horizontal curvature of design orbit. B_z=z-component of magnetic field

quadrupole:

$$\frac{e}{E_0} A_\tau = \frac{1}{2}g_0 \cdot (z^2 - x^2) \qquad (6)$$

with

- $g_0 = \frac{e}{E_0} \cdot (\frac{\partial B_z}{\partial x})_{x=z=0}$= focusing strength of quadrupole

multipole (sextupole):

$$\frac{e}{E_0} \cdot A_\tau = -\frac{1}{6} \cdot \lambda_0 \cdot (x^3 - 3xz^2) \tag{7}$$

with

- $\lambda_0 = \frac{e}{E_0} \cdot \left(\frac{\partial^2 B_z}{\partial x^2}\right)_{x=z=0} =$ strength of sextupole

multipole (octupole):

$$\frac{e}{E_0} \cdot A_\tau = \frac{1}{24} \cdot \mu_0 \cdot (z^4 - 6x^2z^2 + x^4) \tag{8}$$

with

- $\mu_0 = \frac{e}{E_0} \cdot \left(\frac{\partial^3 B_x}{\partial z^3}\right)_{x=z=0} =$strength of octupole

Further examples for other types of electromagnetic fields can be found in [4].

Generally, by expanding the square root in equation (2) and the vector potential $\vec{A}(x,z,s)$ into a Taylor series around a reference orbit, various examples of nonlinear motion can be investigated. The linear part of the Hamiltonian is given by [5]:

$$\mathcal{H}_0(x,z,\tau,p_x,p_z,p_\tau;s) =$$
$$\frac{1}{2}p_x^2 + \frac{1}{2}(\kappa^2(s) + g_0(s)) \cdot x^2 + \frac{1}{2}p_z^2 - \frac{1}{2}g_0(s) \cdot z^2 - \frac{1}{2}V(s) \cdot \tau^2 - \kappa(s) \cdot x \cdot p_\tau \tag{9}$$

where $V(s) = V_0 \cdot \delta_p(s - s_0)$ with $\delta_p(s - s_0) = \sum_{n=-\infty}^{n=+\infty} \delta(s - (s_0 + n \cdot L))$ describes a localized cavity at position s_0 and where $g_0(s)$ characterizes the (periodic) focusing strength of the magnet system. $\mathcal{H}_0$ describes three coupled linear Floquet oscillators [5].

Two simple examples of nonlinear motion are given below:

Example 1:Nonlinear Cavity

$$\mathcal{H}(x,z,\tau,p_x,p_z,p_\tau;s) =$$
$$\frac{1}{2}p_x^2 + \frac{1}{2}p_z^2 + \frac{1}{2}g_0(s) \cdot (x^2 - z^2) + \frac{1}{2}\kappa^2(s) \cdot x^2 - \kappa(s) \cdot x \cdot p_\tau + V(s) \cdot \cos(\tau) \tag{10}$$

Introducing the dispersion function D defined by

$$D''(s) = -(\kappa^2(s) + g_0(s)) \cdot D(s) + \kappa(s) \tag{11}$$

with

$$(.)' = \frac{d}{ds}$$

via the canonical transformation [6],[7],[8] (depending on the *old* coordinates x, z, τ and the *new* momenta $\bar{p}_x, \bar{p}_z, \bar{p}_\tau$)

$$F_2(x,z,\tau,\bar{p}_x,\bar{p}_z,\bar{p}_\tau;s) =$$
$$\bar{p}_x \cdot (x - \bar{p}_\tau \cdot D(s)) + \bar{p}_\tau \cdot D'(s) \cdot x + \bar{p}_\tau \cdot \tau + \bar{p}_z \cdot z - \frac{1}{2} \cdot D(s) \cdot D'(s) \cdot \bar{p}_\tau^2 \tag{12}$$

and the corresponding transformation rules,

$$\begin{cases} x = \bar{x} + \bar{p}_\tau \cdot D(s) \\ z = \bar{z} \\ \tau = \bar{\tau} + \bar{p}_x \cdot D(s) - \bar{x} \cdot D'(s) \end{cases} \tag{13}$$

$$\begin{cases} p_x = \bar{p}_x + \bar{p}_\tau \cdot D'(s) \\ p_z = \bar{p}_z \\ p_\tau = \bar{p}_\tau \end{cases} \tag{14}$$

one obtains the Hamiltonian in the *new* variables $(\bar{x}, \bar{z}, \bar{\tau}, \bar{p}_x, \bar{p}_z, \bar{p}_\tau)$ as follows

$$\begin{aligned} \bar{\mathcal{H}}(\bar{x}, \bar{z}, \bar{\tau}, \bar{p}_x, \bar{p}_z, \bar{p}_\tau; s) =& \\ & \frac{1}{2}\bar{p}_x^2 + \frac{1}{2}(g_0(s) + \kappa^2(s)) \cdot \bar{x}^2 + \frac{1}{2}\bar{p}_z^2 - \frac{1}{2}g_0(s) \cdot \bar{z}^2 \\ & -\frac{1}{2}\kappa(s) \cdot D(s) \cdot \bar{p}_\tau^2 + V(s) \cdot \cos(\bar{\tau} + D(s) \cdot \bar{p}_x - D'(s) \cdot \bar{x}) \end{aligned} \tag{15}$$

If there is no dispersion in the cavity region ($V \cdot D = 0$), the synchrotron motion $(\bar{\tau}, \bar{p}_\tau)$ is completely decoupled from the betatron motion $(\bar{x}, \bar{z}, \bar{p}_x, \bar{p}_z)$ [9].

Example 2:multipole

As a second example of nonlinear motion we consider the influence of transverse multipole fields with the following Hamiltonian:

$$\mathcal{H}(x, z, p_x, p_z, s) = \frac{1}{2}p_x^2 + \frac{1}{2}p_z^2 - \frac{e}{E_0} \cdot A_\tau(x, z, s) \tag{16}$$

The equations of motion are given by

$$\begin{cases} \frac{d}{ds} x = p_x \\ \frac{d}{ds} z = p_z \\ \frac{d}{ds} p_x = \frac{e}{E_0} \cdot \frac{\partial A_\tau}{\partial x} = -\frac{e}{E_0} \cdot B_z(x, z, s) \\ \frac{d}{ds} p_z = \frac{e}{E_0} \cdot \frac{\partial A_\tau}{\partial z} = \frac{e}{E_0} \cdot B_x(x, z, s) \end{cases} \tag{17}$$

The magnetic field components B_x and B_z are usually expressed in terms of the skew and normal multipole expansion coefficients a and b according to :

$$(B_z + iB_x) = B_0 \cdot \sum_{n=2}^{\infty} (b_n + ia_n) \cdot (x + iz)^{n-1} \tag{18}$$

It is an easy exercise to verify, that these simple examples (15) and (16) contain the standard map [10] [11]:

$$\begin{cases} \bar{\tau}(n) = \bar{\tau}(n-1) + \bar{p}_\tau(n) \\ \bar{p}_\tau(n) = \bar{p}_\tau(n-1) + V \cdot \sin(\bar{\tau}(n-1)) \end{cases} \tag{19}$$

and the quadratic map of Henon [12]:

$$\left(\begin{array}{c} x(n+1) \\ p_x(n+1) \end{array} \right) = \left(\begin{array}{cc} \cos(\phi) & \sin(\phi) \\ -\sin(\phi) & \cos(\phi) \end{array} \right) \cdot \left(\begin{array}{c} x(n) \\ p_x(n) \end{array} \right) + \left(\begin{array}{c} 0 \\ x^2(n+1) \end{array} \right) \qquad (20)$$

as limiting cases.These maps are extensively studied in nonlinear dynamics and show a very complex behaviour.Regular and chaotic motion is intricately mixed in phase space.For the quadratic map of Henon this is illustrated in Figure 4.

Thus one can expect, that the original system as described by (2) also shows highly nontrivial behaviour.

In order to get a better understanding of this complex dynamical phase space pattern, we will briefly repeat some facts from the *qualitative* theory of nonintegrable Hamiltonian systems.

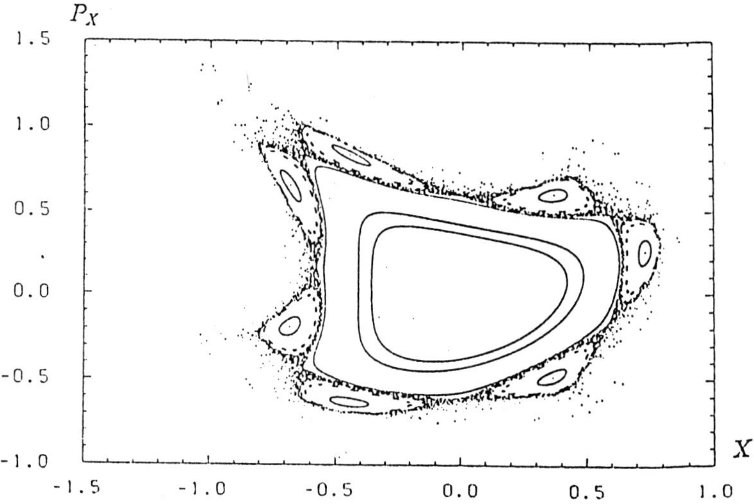

Figure 4. Hénon map (20) for an angle $\phi = 0.7071$

2.2. Qualitative theory of nonlinear Hamiltonian systems

Excellent and detailed reviews can be found in [10], [11],[13],[14],[15],[16].

The easiest way to investigate weakly perturbed nonintegrable Hamiltonian systems is via a map. The reduction of a Hamiltonian system to a nonlinear mapping has been a well-known procedure since Poincaré (1890). Consider for example a two-dimensional Hamiltonian system without explicit time (or s-) dependence $\mathcal{H}(q_1, q_2, p_1, p_2)$. The corresponding phase space is four-dimensional, and since $\mathcal{H}$ itself is a constant of the motion, the physically accessible phase space is three-dimensional. Consider a surface Σ in this three-dimensional -not necessarily Euclidean- space as depicted for example in Figure 5. The bounded particle motion induced by the Hamiltonian $\mathcal{H}$ will generally intersect this surface in different points $(P_0, P_1, ...P_n...)$. If one is not interested in the fine details of the orbit but only in the behaviour over longer time scales it is sufficient to consider the consecutive points $(P_0, P_1....)$ of intersection. These contain complete information on the Hamiltonian system. In this sense one has reduced the Hamiltonian dynamics to a mapping of Σ to itself, which is in general nonlinear (Poincaré surface of section technique). The Hamiltonian character is reflected in the symplectic structure of the map. Symplectic means that the Jacobian $\underline{J}$ of the map is a symplectic matrix with

$$\underline{J}^T \cdot \underline{S} \cdot \underline{J} = \underline{S} \tag{21}$$

where $\underline{J}^T$ is the transpose of $\underline{J}$ and where $\underline{S}$ is the symplectic unity

$$\begin{pmatrix} \underline{0} & \underline{1} \\ -\underline{1} & \underline{0} \end{pmatrix} \tag{22}$$

($\underline{1}$ designates the unit matrix). Similar mappings can also be derived for Hamiltonian systems with explicit periodic time (s-) dependence (this is normally the case in storage rings).

Another important fact and, after the work of Chirikov [11] one of the few beacons among an otherwise still dense mist of diverse phenomena, is the KAM-theorem (Kolmogorov,Arnold,Moser see for example [10]). We will only illustrate this theorem in the two-dimensional case and instead of concentrating on mathematical rigour we will discuss its physical implications.

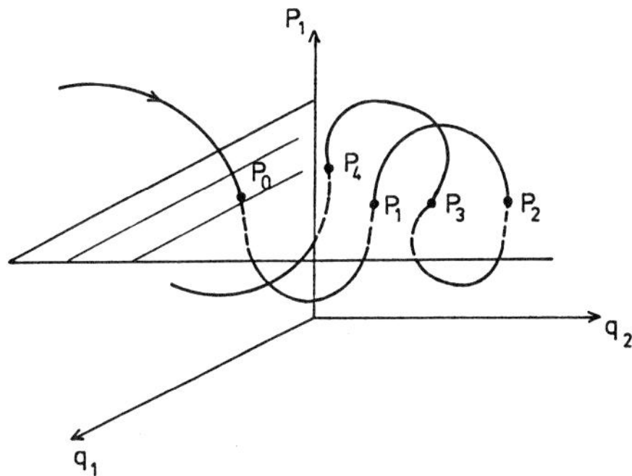

Figure 5. Poincaré surface of section method

Consider first the bounded motion of a two-dimensional autonomous (not explicitly time (s-) dependent) Hamiltonian system which is integrable. Roughly speaking, an n-dimensional system $\mathcal{H}(q_1, q_2, ..., q_n, p_1, p_2, ..., p_n)$ is integrable if there exists a canonical transformation to action-angle variables $(I_1, I_2, ..., I_n, \Theta_1, \Theta_2, ..., \Theta_n)$ such that the transformed Hamiltonian depends only on the n (constant) action variables $(I_1, I_2, ..., I_n)$ alone. For the two-dimensional case under consideration this implies, that $\mathcal{H}(q_1, q_2, p_1, p_2)$ is transformed into $\mathcal{H}(I_1, I_2)$ with the corresponding equations of motion:

$$\begin{cases} \frac{d}{ds} I_1 = 0 \\ \frac{d}{ds} I_2 = 0 \\ \frac{d}{ds} \Theta_1 = \frac{\partial \mathcal{H}}{\partial I_1} = \omega_1(I_1, I_2) = \text{const} \\ \frac{d}{ds} \Theta_2 = \frac{\partial \mathcal{H}}{\partial I_2} = \omega_2(I_1, I_2) = \text{const} \end{cases} \tag{23}$$

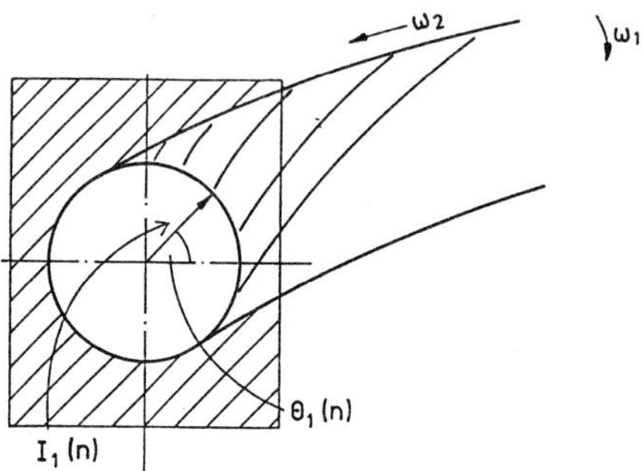

Figure 6. surface of section method for a two-dimensional integrable system

The motion is restricted to a two-torus, parametrized by the two angle variables Θ_1 and Θ_2, as depicted in Figure 6.

As surface of section one can choose the $(I_1 - \Theta_1)$ plane for Θ_2=constant. In this surface of section, which may be chosen to be just the plane of the page, the motion of the integrable two-dimensional system looks very simple. During the motion around the torus from one crossing of the plane to the next the radius of the torus (action variable) does not change (see (23))

$$I_1(n) = I_1(n-1) \tag{24}$$

and the angle Θ_1 changes according to (see (23))

$$\Theta_1(n) = \Theta_1(n-1) + \omega_1 \cdot T \tag{25}$$

where T is just the revolution time in Θ_2-direction from one intersection of the plane to the next, namely

$$T = \frac{2\pi}{\omega_2}. \tag{26}$$

Thus, for an integrable system one obtains the so-called twist-mapping:

$$\begin{cases} I_1(n) = I_1(n-1) \\ \Theta_1(n) = \Theta_1(n-1) + 2\pi \cdot \alpha(I_1(n)) \end{cases} \tag{27}$$

The term $\alpha = \frac{\omega_1}{\omega_2}$ is called the winding number and is the ratio of the two frequencies of the system. In general α will depend on the actions. If α is irrational, the $\Theta_1(n)$ form a dense circle while if α is rational the $\Theta_1(n)$ close after a finite sequence of revolutions (periodic orbit or resonance).Thus, there are invariant curves (circles) under the mapping which belong to rational and irrational winding numbers. What happens now if a perturbation is switched on, i.e. if

$$\begin{cases} I_1(n) = I_1(n-1) + \varepsilon \cdot f(I_1(n), \Theta_1(n-1)) \\ \Theta_1(n) = \Theta_1(n-1) + 2\pi \cdot \alpha(I_1(n)) + \varepsilon \cdot g(I_1(n), \Theta_1(n-1)) \end{cases} ? \tag{28}$$

In particular, can one still find invariant curves? The KAM-theorem says that this is indeed the case if the following conditions are fulfilled

- perturbation must be weak

- $\alpha = \frac{\omega_1}{\omega_2}$ must be sufficiently irrational, i.e. $|\alpha - \frac{p}{q}| \geq \frac{k(\varepsilon)}{q^{2+\delta}}$ with p, q integers, $\delta > 0$ and $k(\varepsilon) \to 0$ for $\varepsilon \to 0$

together with some requirements of differentiability and periodicity for f and g. For further details see for example [10], [16]. Under these assumptions most of the unperturbed tori survive the perturbation although in slightly distorted form.

The rational and some nearby tori, however, are destroyed, only a finite number of fixed points of the rational tori survive - half of them are stable (elliptic orbits around this fixed point), half of them are unstable (hyperbolic orbits).This is a consequence of the Poincaré-Birkhoff fixed point theorem [13].

The hyperbolic fixed points with their stable and unstable branches, which generally intersect in the homoclinic points, see Figure 7, are the source of chaotic motion in phase space, i.e. motion which is extremely sensitive to the variation of initial conditions. From a historical point of view it is interesting to note, that these facts were already known by Poincaré 100 years ago [13].

The motion around the elliptic fixed points can be considered as motion around a torus with smaller radius and the arguments used till now can be repeated on this smaller scale giving rise to the - now well known - schematic picture shown below (see Figure 8 ,"chaos-scenario" of weakly perturbed two-dimensional twist maps).

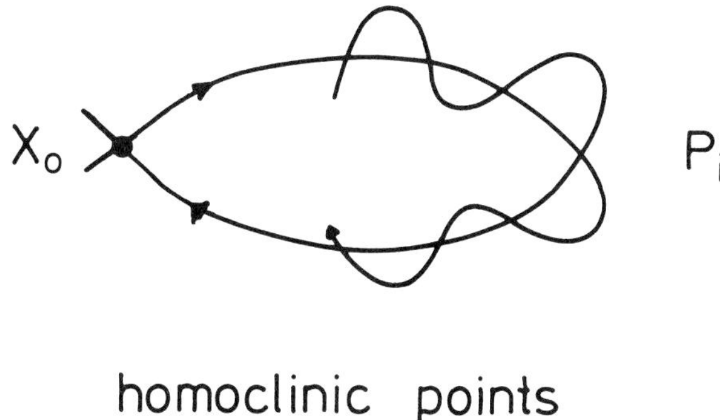

Figure 7. homoclinic intersections of stable and unstable branch of fixed point $\vec{X}_0$

Thus, the phase space pattern of a weakly perturbed integrable two-dimensional system looks extremely complicated. There are regular orbits confined to tori with chaotic trajectories delicately distributed among them. As the strength of the perturbation increases more and more KAM tori break up giving rise to larger and larger chaotic regions. This onset of large scale chaos has been the subject of many studies [10],[17]. For the standard map (see equation (19)) the situation is depicted in Figure 9.

One comment is pertinent at this point - two-dimensional systems are special in that the existence of KAM circles implies exact stability for orbits starting inside such an invariant curve. Since these trajectories cannot escape without intersecting the KAM tori, they are forever trapped inside.

The situation is much more complex and less well-understood for higher-dimensional systems like our storage ring (six-dimensional, explicitly s- dependent Hamiltonian system). The KAM theorem predicts three-tori in six-dimensional phase space, four-tori in eight-dimensional phase space etc. In this case chaotic trajectories can in principle always escape and explore

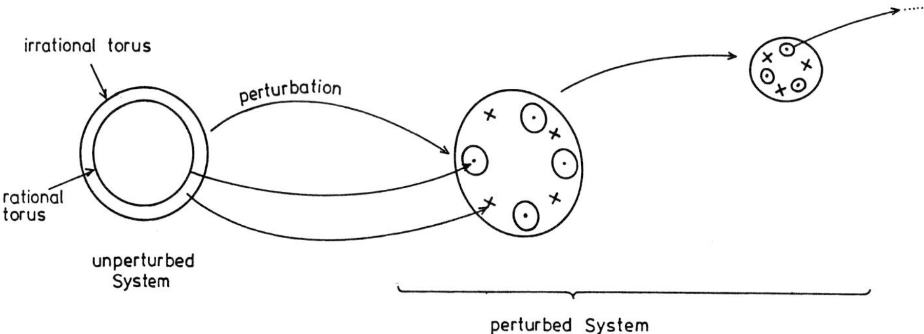

Figure 8. chaos scenario

all the accessible phase space although the motion can be obstructed strongly by existing tori. Chaotic regions can form a connected web along which the particle can diffuse, as has been demonstrated by Arnold (Arnold diffusion see for example [10], [17]).We will come back to this point later.

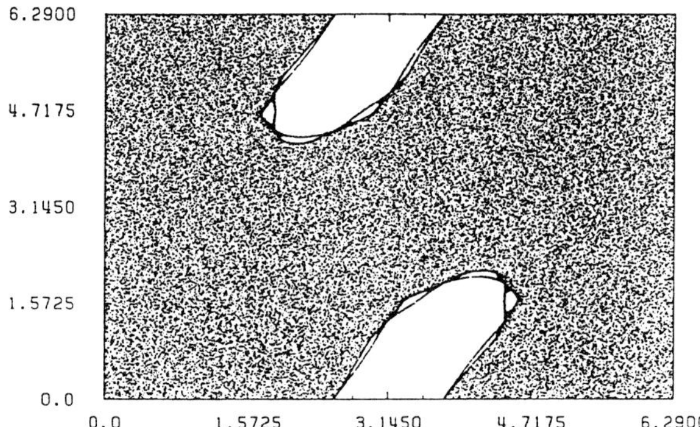

Figure 9. $(\bar{p}_\tau - \bar{\tau})$-phase space plot of the standard map (19) showing global chaos for $V = 3.3$

Figure 10 shows examples of regular and chaotic trajectories in a realistic model of a storage ring [18],[19]. We have used the characteristic Lyapunov exponent λ to distinguish between regular and chaotic motion [10], [20],[21]:

$$\lambda = \lim_{t\to\infty, d(0)\to 0} \frac{1}{t} \cdot \ln\left(\frac{|d(t)|}{|d(0)|}\right) \tag{29}$$

$d(t)$ describes how the (Euclidean) distance between two adjacent phase space points evolves with time and $d(0)$ is the initial distance. In a chaotic region of phase space this distance will grow exponentially fast and a non-zero Lyapunov-exponent λ is a quantitative measure

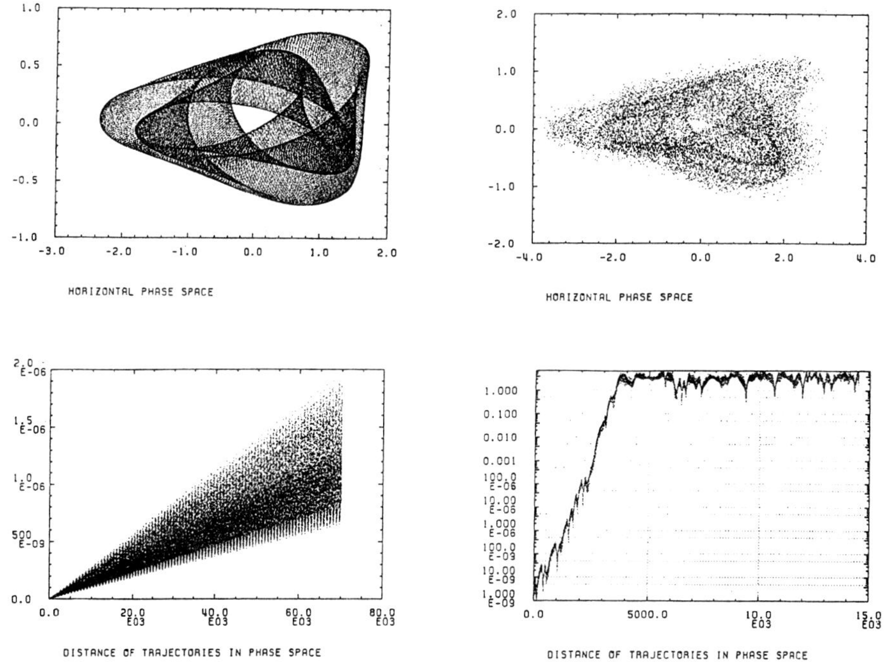

Figure 10. regular and chaotic trajectories in a realistic model of a storage ring and the evolution of the distance of two adjacent initial conditions

for this separation. For the details of an explicit numerical calculation of the characteristic Lyapunov-exponents (for continuous and discrete dynamical systems) the reader is referred to [10], [20],[21].

Another method derived from the homoclinic structure of nearly integrable symplectic mappings is due to Melnikov, and this method belongs to one of the few analytical tools for investigating chaotic behaviour. It is applicable to dynamical systems of the form:

$$\begin{cases} \frac{d}{ds}\,\vec{x}(s) = \vec{F}(\vec{x}) + \varepsilon \cdot \vec{G}(\vec{x},s) \\ \vec{x} = (x_1, x_2)^T \in \mathbf{R}^2 \\ s \in \mathbf{R} \\ \vec{F} = (F_1, F_2)^T \\ \vec{G} = (G_1, G_2)^T \end{cases} \tag{30}$$

where $\vec{F}$ usually describes a Hamiltonian system. The perturbation $\varepsilon\vec{G}$, which may also be weakly dissipative, is periodic in s, and the unperturbed system

$$\frac{d}{ds}\,\vec{x}(s) = \vec{F}(\vec{x}) \tag{31}$$

has a homoclinic orbit belonging to a saddle point $\vec{x}_0$. Homoclinic orbit means smooth joining of the stable and unstable branch of the saddle or hyperbolic fixed point (see Figure 11). The

Melnikov method enables a kind of directed distance between the stable and unstable branch of the perturbed saddle $\vec{x}_0^s$ to be calculated and thus allows the existence of homoclinic points to be predicted, a prerequisite of chaotic dynamics. A derivation of the Melnikov function and further details and applications can be found in [22].

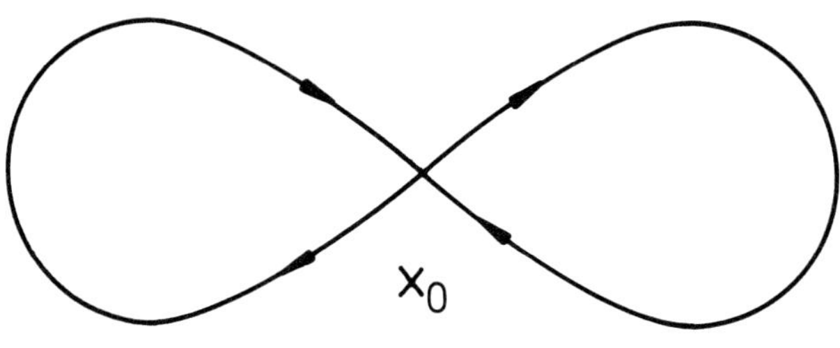

Figure 11. homoclinic orbit belonging to saddle $\vec{x}_0$

Remark:

High resolution 3D-colour graphics can be a very helpful tool for visualizing the dynamics of nonlinear four-dimensional mappings [19] [23]. Toy models like

$$\vec{x}(n+1) = \underline{R} \cdot \vec{x}(n) + \vec{f}(\vec{x}(n)) \tag{32}$$

with

$$\vec{x}(n) = \begin{pmatrix} x(n) \\ p_x(n) \\ z(n) \\ p_z(n) \end{pmatrix} \tag{33}$$

and

$$\underline{R}(\phi,\theta) = \begin{pmatrix} \cos(\phi) & \sin(\phi) & 0 & 0 \\ -\sin(\phi) & \cos(\phi) & 0 & 0 \\ 0 & 0 & \cos(\theta) & \sin(\theta) \\ 0 & 0 & -\sin(\theta) & \cos(\theta) \end{pmatrix} \tag{34}$$

and

$$\begin{cases} \vec{f}(\vec{x}(n)) = \begin{pmatrix} 0 \\ \frac{\partial f}{\partial x}(x(n+1), z(n+1)) \\ 0 \\ \frac{\partial f}{\partial z}(x(n+1), z(n+1)) \end{pmatrix} \\ (f = f(x,z)) \end{cases} \tag{35}$$

can help to get a better understanding of the break-up mechanism of invariant tori and the role periodic orbits play in this process [23],[24]. (($n+1$) periodic orbits are defined by: $\vec{x}(n+1) = \underline{T}(\vec{x}(n)) = \vec{x}(0)$ where $\underline{T}$ is some nonlinear (symplectic) map).

In the last chapter we have seen that the single particle dynamics of a proton in a storage ring can be modelled by nonintegrable Hamiltonians. The qualitative theory we have briefly sketched predicts a very rich and complicated phase space structure - regular and chaotic regions are intricately mixed in phase space.

When applying these concepts to accelerators one is immediately faced with questions such as:

What is the relevance of chaos for the practical performance of a storage ring? How do KAM tori break up as the strength of the nonlinearity increases? Can we somehow estimate the size of the chaotic regions in phase space? What is the character of the particle motion in this region? Can it be described by diffusion-like models? Is it possible to calculate escape rates of the particle if it is in such a chaotic region of phase space?

A quantitative analysis of these and other questions makes extensive use of perturbation theory and numerical simulations of the system.

2.3. Perturbation theory

Perturbation theory for weakly perturbed integrable Hamiltonian systems is a vast field and an active area of research, which we cannot treat exhaustively in this survey lecture. We will only illustrate some of the basic results and ideas. Before we enter into detail we will briefly repeat some facts from the linear theory of particle motion in storage rings (linear theory of synchro-betatron oscillations [5])

In simple cases, as for example pure x- or z-motion without any coupling, the system is described by Floquet oscillators of the form:

$$\mathcal{H}(q, p, s) = \frac{1}{2} \cdot p^2 + \frac{1}{2} \cdot g_0(s) \cdot q^2 \tag{36}$$

with $p = p_x,\ p_z,\ q = x,\ z$ and $g_0(s) = g_0(s + L)$ periodic function of circumference L.

It is well known that these Floquet type systems can be solved exactly. Using the optical functions $\alpha(s), \beta(s)$ and $\gamma(s)$ defined by the following set of differential equations:

$$\frac{d}{ds}\,\alpha(s) = -\gamma(s) + \beta(s) \cdot g_0(s) \tag{37}$$

$$\frac{d}{ds}\,\beta(s) = -2 \cdot \alpha(s) \tag{38}$$

$$\frac{d}{ds}\,\gamma(s) = 2 \cdot \alpha(s) \cdot g_0(s) \tag{39}$$

one can find a canonical transformation to action angle variables I and Θ such that the Hamiltonian in equation (36) is transformed into [25]:

$$\bar{\mathcal{H}}(\Theta, I) = \frac{2\pi \cdot Q}{L} \cdot I \tag{40}$$

with

$$Q = \frac{1}{2\pi} \cdot \int_0^L \frac{ds'}{\beta(s')} \tag{41}$$

(Q is the so-called tune of the machine) and

$$I = \frac{q^2}{2\beta(s)} \cdot \{1 + (\frac{\beta(s) \cdot p}{q} + \alpha(s))^2\} \tag{42}$$

(I is called Courant-Snyder invariant see [26]).

In realistic cases there is always some coupling between the different degrees of freedom and the situation is more complicated. In these cases machine physicists rely on the one-turn matrix $\underline{M}$ relating some initial state phase space vector $\vec{y}(s_{in})$ to the final state vector $\vec{y}(s_{fin})$ after one complete revolution around the ring:

$$\vec{y}(s_{fin} = s_{in} + L) = \underline{M}(s_{in} + L, s_{in}) \cdot \vec{y}(s_{in}) \tag{43}$$

In general $\vec{y}$ is six-dimensional and consists of the phase space coordinates $x, z, \tau, p_x, p_z, p_\tau$. The linear one-turn map $\underline{M}(s_{in} + L, s_{in})$ contains all the information about the system. For example the stability of the particle motion depends on the eigenvalue spectrum of the (symplectic) matrix $\underline{M}$ [5] - stability is only guaranteed if the eigenvalues lie on the complex unit circle (see also Figure 12).

What happens now if we perturb such a linear system with some nonlinear terms? How can we extend the linear analysis to the nonlinear case?

In simple models we can start with a perturbed Hamiltonian

$$\mathcal{H}(q, p, s) = \mathcal{H}_0(q, p, s) + \varepsilon \cdot \mathcal{H}_1(q, p, s) \tag{44}$$

Using the action angle variables of the unperturbed system one can apply conventional Hamiltonian perturbation theory [25],[27], which we will sketch in a moment. The advantages of such an approach are that one easily gets simple analytical expressions for interesting machine parameters of the perturbed system in terms of the unperturbed quantities. The price one has to pay, however, is an over-simplification of the problem. Realistic machines with all their nonlinearities and perturbations are extremely complex and cannot be handled efficiently in such a way. In such a case one should try to extend the concept of the one-turn map to the nonlinear case. This "contemporary" approach of Hamiltonian-free perturbation theory for particle dynamics in storage rings has been strongly advanced by E.Forest [28] and a recent description can be found in [29].

The basic idea of perturbation theory (common to both the direct Hamiltonian approach and the contemporary Hamiltonian-free approach) is to find - in a way to be specified - new variables, such that the system becomes solvable or at least easier to handle in this new set.

Let us first consider the Hamiltonian formalism.Various realizations exist for this kind of perturbation theory:Poincaré-von Zeipel [25],[27],[30], Lie methods [31], [32], and normal form algorithms [33],[34].

Here we will illustrate the Poincaré-von Zeipel method.

Assume our Hamiltonian is of the form

$$\mathcal{H}(\vec{q}, \vec{p}) = \mathcal{H}_0(\vec{q}, \vec{p}) + \varepsilon \cdot \mathcal{H}_1(\vec{q}, \vec{p}) \tag{45}$$

where the vectors for the coordinates and momenta $\vec{q}$ and $\vec{p}$ may have arbitrary dimension (3 in the storage ring case)

$$\vec{q} = \begin{pmatrix} q_1 \\ \cdot \\ \cdot \\ \cdot \\ q_n \end{pmatrix}, \qquad \vec{p} = \begin{pmatrix} p_1 \\ \cdot \\ \cdot \\ \cdot \\ p_n \end{pmatrix}. \tag{46}$$

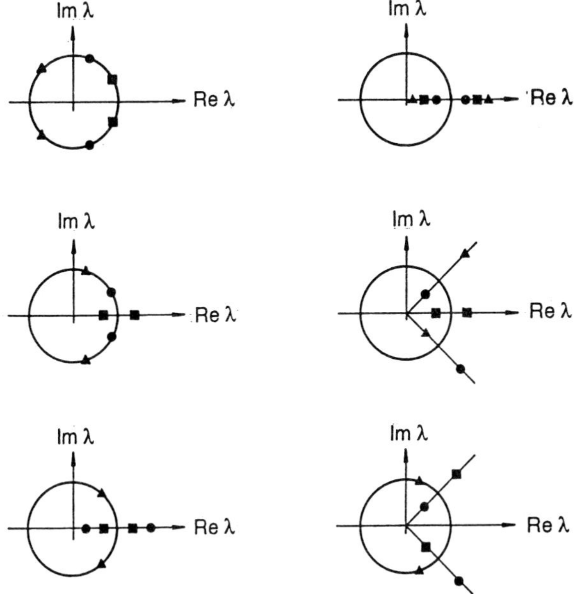

Figure 12. eigenvalue spectrum of a six-dimensional symplectic matrix

Introducing the action angle variable vectors $\vec{I}$ and $\vec{\Theta}$ of the unperturbed system $\mathcal{H}_0$

$$\vec{I} = \begin{pmatrix} I_1 \\ \cdot \\ \cdot \\ \cdot \\ I_n \end{pmatrix}, \qquad \vec{\Theta} = \begin{pmatrix} \Theta_1 \\ \cdot \\ \cdot \\ \cdot \\ \Theta_n \end{pmatrix} \tag{47}$$

the Hamiltonian (45) can be rewritten in the form:

$$\mathcal{H}(\vec{I}, \vec{\Theta}) = \mathcal{H}_0(\vec{I}) + \varepsilon \cdot \mathcal{H}_1(\vec{I}, \vec{\Theta}) \tag{48}$$

The problem would be trivial, if we could find a transformation to new variables $\vec{J} = (J_1...J_n)^T$ and $\vec{\psi} = (\psi_1...\psi_n)^T$ such that the transformed Hamiltonian depends only on the new action variables $J_1...J_n$ alone. Since most Hamiltonian systems are nonintegrable [13],[14] this cannot be done exactly. The best one can achieve is to push the nonlinear perturbation to higher and higher orders in ε i.e. after a sequence of N canonical transformations

$$\mathcal{H}(\vec{I}, \vec{\Theta}) = \mathcal{H}_0(\vec{I}) + \varepsilon \cdot \mathcal{H}_1(\vec{I}, \vec{\Theta})$$

is transformed into a form given by :

$$\begin{cases} \bar{\mathcal{H}}(\vec{J}^{(N)}, \vec{\psi}^{(N)}) = \bar{\mathcal{H}}_0(\vec{J}^{(N)}) + \varepsilon^{N+1} \cdot R_N(\vec{\psi}^{(N)}, \vec{J}^{(N)}) \\ \bar{\mathcal{H}}_0(\vec{J}^{(N)}) = \sum_{i=0}^{N} \varepsilon^i \cdot \mathcal{H}_0^{(i)}(\vec{J}^{(N)}) \\ \bar{\mathcal{H}}_0^{(0)}(\vec{J}^{(N)}) = \mathcal{H}_0(\vec{J}^{(N)}) \end{cases} \tag{49}$$

where

$$
\vec{J}^{(N)} = \begin{pmatrix} J_1^{(N)} \\ \cdot \\ \cdot \\ \cdot \\ J_n^{(N)} \end{pmatrix}, \qquad \vec{\psi}^{(N)} = \begin{pmatrix} \psi_1^{(N)} \\ \cdot \\ \cdot \\ \cdot \\ \psi_n^{(N)} \end{pmatrix} \tag{50}
$$

are the new variables after N transformations. Neglecting the remainder $\varepsilon^{N+1} \cdot R_N$ (which is of order ε^{N+1} i.e. one order higher than the first part in equation (49)) the system is then trivially solvable.

For example in first order of perturbation theory this is achieved by a canonical transformation (depending on the *old* coordinates $\vec{\Theta}$ and the *new* momenta $\vec{J}$)

$$
F_2(\vec{\Theta}, \vec{J}) = \vec{\Theta} \cdot \vec{J} + \varepsilon \cdot S_1(\vec{\Theta}, \vec{J}) \tag{51}
$$

where $S_1(\vec{\Theta}, \vec{J})$ is given by:

$$
S_1(\vec{\Theta}, \vec{J}) = -\frac{1}{i} \cdot \sum_{\vec{n} \neq 0} \frac{\mathcal{H}_{1,\vec{n}}(\vec{J})}{\vec{\omega} \cdot \vec{n}} \cdot \exp(i\vec{n} \cdot \vec{\Theta}) \tag{52}
$$

$\vec{\omega}$ designates the frequency vector

$$
\vec{\omega}(\vec{J}) = \frac{\partial \mathcal{H}_0(\vec{J})}{\partial \vec{J}} \tag{53}
$$

and $\mathcal{H}_{1,\vec{n}}(\vec{J})$ is defined by the Fourier expansion of $\mathcal{H}_1(\vec{\Theta}, \vec{J})$

$$
\mathcal{H}_1(\vec{\Theta}, \vec{J}) = \sum_{\vec{n}} \mathcal{H}_{1,\vec{n}}(\vec{J}) \cdot \exp(i\vec{n} \cdot \vec{\Theta}). \tag{54}
$$

However, there is a serious problem concerning the convergence of our perturbative approach: even if we exclude the *nonlinear resonances* $\vec{n} \cdot \vec{\omega} = 0$ in equation (52) the infinite sum always contains n_i's such that the denominator in equation (52) can become arbitrarily small (small divisor problem) making this whole enterprise very doubtful. Generally these expansions diverge. Nevertheless the hope is that these expansions can be useful as asymptotic series and one hopes that the new invariants $(J_1, ...J_n)$ calculated in this way approximate in some sense our original system. However, for finite perturbations, there is no proof for the accuracy, if any, of such an approximation. So some care is always needed when one applies perturbation theories of this kind. A careful analysis of the convergence properties of (52) leads immediately to the heart of the KAM theory and requires sophisticated mathematical tools, which are far beyond the scope of this survey lecture.

Let us now briefly illustrate the perturbative techniques for mappings [28],[35],[36],[37],[38].

As mentioned already in the introduction, an accelerator acts as a nonlinear device and an initial state phase space vector $\vec{y}(s_{in})$ is nonlinearly related to the final state $\vec{y}(s_{fin})$ by a symplectic map

$$
\vec{y}(s_{fin}) = \mathcal{M}(\vec{y}(s_{in})) \ . \tag{55}
$$

Let us assume this map can be Taylor expanded up to some order N with respect to $\vec{y}(s_{in})$

$$y_i(s_{fin}) = \sum_j A_{ij} \cdot y_j(s_{in}) + \sum_{jk} B_{ijk} \cdot y_j(s_{in}) \cdot y_k(s_{in}) + \dots \tag{56}$$

with the transfer matrix or aberration coefficients A_{ij}, B_{ijk} etc.

Dragt and Finn [39],[40] have shown that Lie algebraic techniques can be very efficient for handling maps like (56). The factorization theorem [37] for example states that $\mathcal{M}$ can be expressed as a product of *Lie transforms*

$$\mathcal{M} = e^{:f_1:} \cdots e^{:f_k:} \cdots \tag{57}$$

where $: f_i :$ denotes a *Lie operator* related to a homogeneous polynomial of degree i in the variables $y_i(s_{in})$ and $: f :$ acts on the space of phase space functions g via the Poisson bracket operation of classical mechanics [27]

$$: f : g \equiv \{f, g\} \tag{58}$$

Example

The map $e^{:\frac{a}{3}x^3:}$ gives the known expression for a sextupole in thin lens (kick) approximation [37], [38]:

$$\begin{pmatrix} x(s_{fin}) \\ p_x(s_{fin}) \end{pmatrix} = \begin{pmatrix} x(s_{in}) \\ p_x(s_{in}) + a \cdot x^2(s_{in}) \end{pmatrix} \tag{59}$$

because

$$\begin{cases} e^{:\frac{a}{3}x^3(s_{in}):}\, x(s_{in}) = x(s_{in}) \\ e^{:\frac{a}{3}x^3(s_{in}):}\, p_x(s_{in}) = p_x(s_{in}) + a \cdot x^2(s_{in}) \end{cases} \tag{60}$$

In principle one could now try to construct the one-turn map for a nonlinear accelerator using these Lie algebraic tools. However, beyond an order $N = 3$ in the Taylor expansion, this becomes incredibly tedious and complicated. So we will discuss a more efficient way of obtaining Taylor expanded maps for one turn later.

The advantage of using maps in the form (57) is formal and lies in the Lie algebraic tools that are available for treating these systems. There is an elegant extension of the normal form theory to such cases [41]. The problem is -roughly stated- that given a map $\mathcal{M}$ one has to determine a map $\mathcal{A}$ such that

$$\mathcal{N} = \mathcal{A} \cdot \mathcal{M} \cdot \mathcal{A}^{-1} \tag{61}$$

is as simple as possible. Simple means that the action of the map is simple. We can easily illustrate this fact with the following map which describes the action of a single multipole in kick approximation [35],[42]

$$\begin{pmatrix} x(n+1) \\ p_x(n+1) \end{pmatrix} = \begin{pmatrix} \cos(2\pi \cdot Q) & \sin(2\pi \cdot Q) \\ -\sin(2\pi \cdot Q) & \cos(2\pi \cdot Q) \end{pmatrix} \cdot \begin{pmatrix} x(n) \\ p_x(n) + \varepsilon \cdot x^p(n) \end{pmatrix} \tag{62}$$

In complex notation $z = x + ip_x$ equation (62) can be rewritten as:

$$z(n+1) = \exp(-i \cdot 2\pi \cdot Q) \cdot \left\{ z(n) + \frac{i\varepsilon}{2^p} \cdot (z(n) + z^\star(n))^p \right\} \tag{63}$$

where $z^\star$ designates the complex conjugate of z.

Finding a map $\mathcal{A}$ implies that one transforms to a new set of variables

$$z \longrightarrow \xi \tag{64}$$

such that equation (63) takes the following form in the new variables:

$$\xi(n+1) = \exp\left\{ -i\Omega(\xi(n) \cdot \xi^\star(n)) \right\} \xi(n) + o(\varepsilon^2) + \ldots \tag{65}$$

Now, up to order ε, the action of the map is very simple - it is just a rotation in the ξ-plane with a frequency (winding number) Ω which depends on the distance from the origin (see Figure 13). This kind of perturbative analysis has been developed in detail in [28], [35] and has led to a powerful strategy for investigating the nonlinear motion of particles in storage rings. We will come back to this point after the description of numerical simulations in the next section.

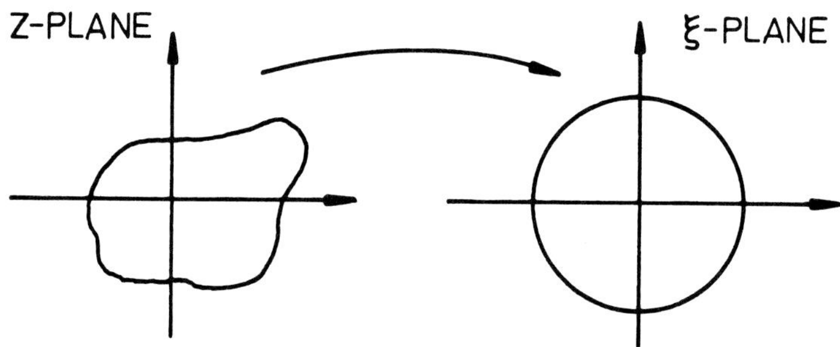

Figure 13. normal form theory of maps

2.4. Numerical simulations and particle tracking

The main idea of numerical simulations is to track particles over many revolutions in realistic models of the storage ring and to observe the amplitude of the particle at a special point s_0 [43], [44]. Given the initial amplitude $\vec{y}(s_0) = (x(s_0), z(s_0), \tau(s_0), p_x(s_0), p_z(s_0), p_\tau(s_0))$ one needs to know $\vec{y}(s_0 + n \cdot L)$ for n of the order of 10^9 (corresponding to a storage time of a particle of about 10 hours in HERA). Different methods and codes have been developed to evaluate $\vec{y}(s_0 + n \cdot L)$. Among others there are COSY INFINITY [45],TEAPOT [46],MARYLIE [47],TRANSPORT [48], RACETRACK [49]. We will not go into the details of these codes - much more will be said about these things in other contributions to this workshop - we restrict ourselves to some general remarks and facts instead.

An ideal code should be fast and accurate. It should allow for six-dimensional phase space calculations, thus allowing all kinds of coupling between the synchrotron and betatron oscillations. Error simulations of the storage ring should be possible as well as the calculation of interesting physical quantities such as tunes, (perturbed) invariants, nonlinear resonance widths etc.

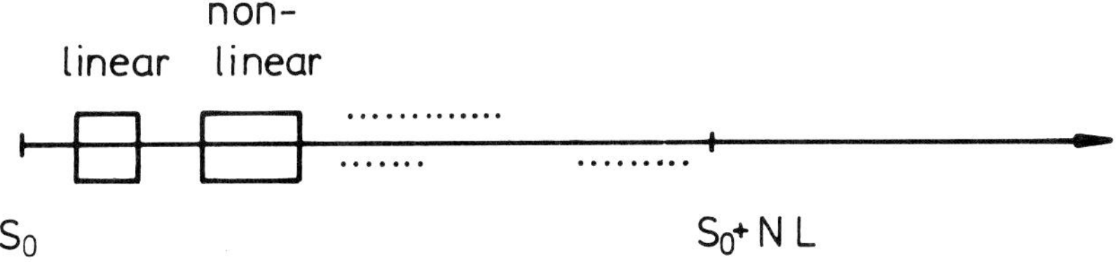

Figure 14. beam line

One way of achieving this is by naive element to element tracking.

One has to solve the corresponding equations of motion for each element (linear or nonlinear). In each case the symplectic structure of the underlying Hamiltonian system has to be preserved by using suitable symplectic integration schemes [50], [51]. Such a code would be accurate but also extremely slow especially for large colliders like HERA and the SSC. A modification of these element-to-element tracking codes is the so-called kick approximation. Nonlinear elements described for example by terms

$$\mathcal{H}_1 = \sum_{n,m} a_{n,m}(s) \cdot x^n \cdot z^m$$

in the Hamiltonian are replaced by

$$\mathcal{H}_1 = \sum_{n,m} \bar{a}_{n,m} \cdot x^n \cdot z^m \cdot \delta_p(s - s_i)$$

where s_i denotes the localization of the nonlinear kick. These codes can speed up the calculations considerably and they also preserve the symplectic structure of the underlying equations automatically. However, one has to check carefully the accuracy of this kind of approximation.

Recent developments in particle tracking and numerical simulations use differential algebra tools as developed by M. Berz and described in other contributions to this workshop.

A typical numerical investigation of particle motion in nonlinear storage rings then comprises the following steps:

1. specification of the storage ring model by a Hamiltonian $\mathcal{H}$

2. numerical integration of the corresponding equations of motion for one complete revolution using symplectic integrators

3. extraction of the Taylor expanded form of the one-turn map (see equation (56)) from this calculation

4. use of this map for long-time tracking and for a perturbative analysis (to get physical quantities of interest such as invariants, perturbed frequencies and tunes of the synchro-betatron oscillations, nonlinear resonance widths etc.).

Step 3 is elegantly solved by using the powerful differential algebra package developed by M. Berz [52], [53] [54], [55], [56]. Figure 15 shows a flow chart for this approach.

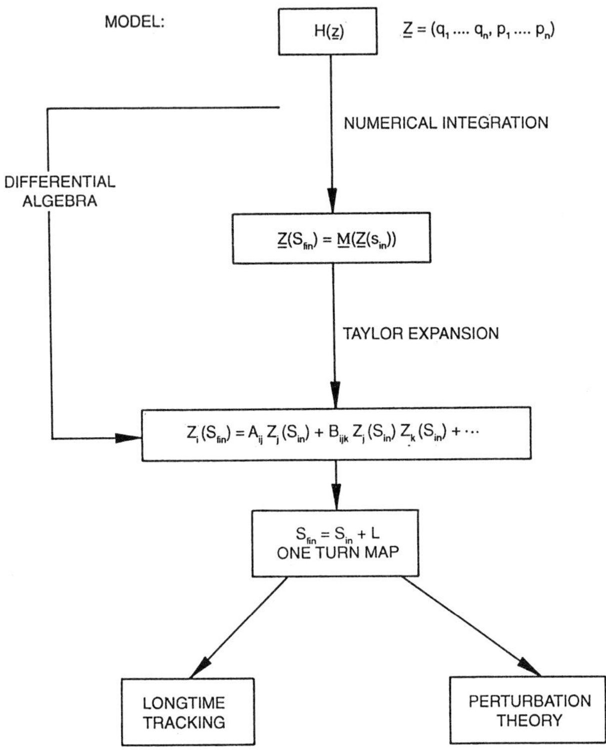

Figure 15. particle tracking

Let us conclude this section by mentioning some problems related to tracking namely the unavoidable rounding errors of the computers and the limited CPU time. The rounding errors depend on the number system used by the compiler and they can destroy the symplectic structure of the nonlinear mappings. Thus these rounding errors can simulate non-physical damping (anti-damping) effects [57]. In order to estimate the order of magnitude of these effects, one can switch to a higher precision structure in the computer hardware or software and observe the differences. Another way is to compare the differences between forward tracking of the particles and backward tracking [58].

The limited CPU time could be improved by developing special tracking processors [59]. Special processors have been successfully used in celestial mechanics for studies of the long-time stability of the solar system [60].

Besides these technical problems there are also some physical problems related to the evaluation and interpretation of the tracking data. For example fast instabilities with an exponential increase of amplitudes beyond a certain boundary (dynamic aperture) can easily be detected, whereas slow, diffusion-like processes which are very important for an understanding of the long-time dynamics are much more difficult to detect.

Nevertheless, tracking is the only way to obtain realistic estimates for the dynamic aperture up to $10^5 - 10^6$ revolutions, but it is very difficult and sometimes dangerous to extrapolate these data to longer times (10^9 revolutions or more). Furthermore, tracking is always very important for checking perturbative calculations because of the divergence problems in perturbation theories as mentioned above. We conclude this chapter on Hamiltonian systems with some final remarks.

2.5. Remarks

As mentioned already in the introduction, in accelerator physics one often tries to define different zones or regions corresponding to the importance of the nonlinearities. For small nonlinearities the accelerator behaves more or less like a linear element. A quantitative measure for this quasilinear behaviour is the so-called smear, a concept developed during the design studies for the SSC [61]. This quantity indicates how much the invariants of the linear machine are changed due to the nonlinear perturbations (see Figure 16). Another measure could be the amplitude dependence of the tunes. In the weakly nonlinear region one would expect that perturbation theory is the adequate theoretical tool.

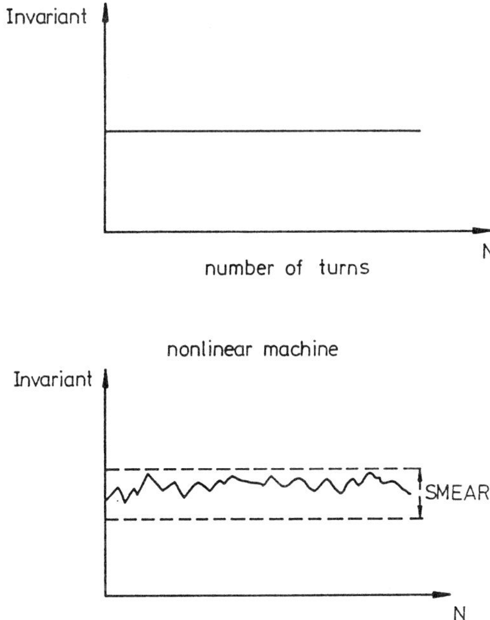

Figure 16. concept of smear

For stronger nonlinearities the dynamics becomes more and more nonlinear and chaotic i.e. sensitively dependent on the initial conditions. A quantitative measure for the onset of large scale chaos can be derived from Chirikov's resonance overlap criterion.One estimates the resonance widths and resonance distances and the criterion roughly states that no KAM tori survive in the region where resonance overlap occurs which leads to a completely chaotic particle motion in this area of phase space. The formal steps for applying this criterion are carefully described in [11] and [62]. Direct application to the standard map (19) for example yields a critical nonlinearity parameter of $V \approx 2.47$, which is in qualitative agreement with numerical simulations (see Figure 9).

The problem of beam-beam interaction in storage rings is an example where this method has been applied extensively by Tennyson et. al. [63], [64], [65].

An interesting and important question is: how does the particle motion look in this extended chaotic region of phase space? Can it be described by a diffusion-like process and can probabilistic concepts be used successfully in this context [66], [67]?

In order to illustrate some of the ideas and techniques used in this case we choose the following simple model [10], [11], [17]:

$$\mathcal{H}(\psi_1, \psi_2, J_1, J_2, t) =$$
$$\frac{1}{2} \cdot (J_1^2 + J_2^2) + \varepsilon \cdot (\cos \psi_1 - 1) \cdot (1 + \mu \cdot \sin \psi_2 + \mu \cdot \cos t) \tag{66}$$

which in extended phase space $(J_1, J_2, p, \psi_1, \psi_2, x = t)$ can be written as :

$$\mathcal{K}(J_1, J_2, p, \psi_1, \psi_2, x) =$$
$$\frac{1}{2} \cdot (J_1^2 + J_2^2) + p + \varepsilon \cdot (\cos \psi_1 - 1) - \mu \cdot \varepsilon \cdot \sin \psi_2 - \mu \cdot \varepsilon \cos x +$$
$$\frac{\mu \cdot \varepsilon}{2}(\sin(\psi_2 - \psi_1) + \sin(\psi_2 + \psi_1) + \cos(\psi_1 - x) + \cos(\psi_1 + x)) \tag{67}$$

$\mathcal{K}$ represents now an autonomous system in six-dimensional phase space. The primary resonances of the system (67) and the corresponding resonance widths are given by:

$$\left\{ \begin{array}{l} \frac{d}{dt}\,\psi_1 \approx J_1 = 0; \ \ \text{width} \sim \sqrt{\varepsilon} \\ \frac{d}{dt}\,\psi_2 \approx J_2 = 0; \ \ \text{width} \sim \sqrt{\varepsilon \cdot \mu} \\ \frac{d}{dt}\,(\psi_1 \pm \psi_2) \approx J_1 \pm J_2 = 0; \ \ \text{width} \sim \sqrt{\varepsilon \cdot \mu} \\ \frac{d}{dt}\,(\psi_1 \pm x) \approx J_1 \pm 1 = 0; \ \ \text{width} \sim \sqrt{\varepsilon \cdot \mu}. \end{array} \right. \tag{68}$$

For small ε, μ the energy surface is approximated by

$$\mathcal{K}_0(J_1, J_2, p) \approx \frac{1}{2}(J_1^2 + J_2^2) + p \tag{69}$$

and the resonance zones are given approximately by the intersection of the resonance surfaces (68) with the unperturbed energy surface (69), see Figures 17,18.

If $\varepsilon \gg \varepsilon \cdot \mu$, $J_1 = 0$ is the dominant resonance (guiding resonance). The motion (transport, diffusion) along the guiding resonance and the resonances which intersect it is called Arnold diffusion, see Figures 17,18.

For $\mu = 0$ the Hamiltonian in equation (67) is integrable (nonlinear pendulum) and a constant of the motion. For $\mu \neq 0$ the Hamiltonian is nonintegrable and the separatrix of the $J_1 = 0$ resonance will be replaced by a chaotic layer. Furthermore, we expect some diffusive variation of the energy in this case. One can calculate this variation approximately [10], [11], [17].

Using

$$\Delta\mathcal{H}(J_1, J_2, \psi_1, \psi_2, t) =$$
$$\int_{-\infty}^{\infty} dt\frac{d\mathcal{H}}{dt} = \int_{-\infty}^{\infty} \varepsilon \cdot \mu \sin t \cdot (1 - \cos \psi_1) dt \tag{70}$$

(see equation (66)) and replacing $\psi_1(t)$ by the unperturbed separatrix expression $\psi_{1sx}(t)$ the evaluation of the resulting Melnikov-Arnold integral [11] gives:

$$< (\Delta\mathcal{H})^2 > \approx 8\pi^2 \cdot \mu^2 \cdot \exp\{-\frac{\pi}{\sqrt{\varepsilon}}\} \tag{71}$$

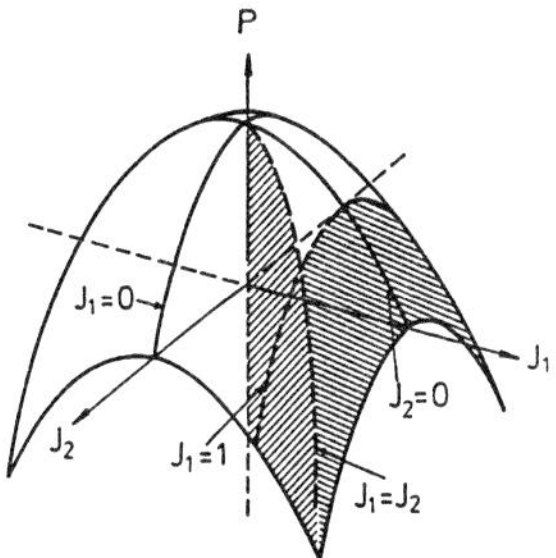

Figure 17. energy surface of unperturbed system (69)

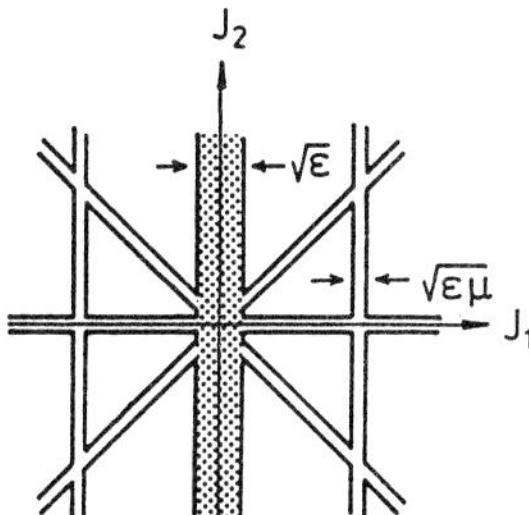

Figure 18. projection of resonance curves on $(J_1 - J_2)$ plane

Equation (71) is an estimate for the short time variation of the energy of the system for trajectories deep inside the chaotic layer of the $J_1 = 0$ resonance.Further details can be found in [10], [11], [17]. This kind of analysis has been applied in accelerator physics by Brüning see [68].

As mentioned already, the outstanding problem of accelerator physics is the long time stability of particle motion under the influence of various nonlinearities such as magnetic multipoles, rf fields and beam-beam forces.

In perturbation theory one usually approximates a nonintegrable system by an integrable (solvable) system. Whether this approximation really reflects the "reality" of the nonintegrable case has to be checked very carefully especially because integrable systems have no chaotic regions in phase space and because the dominant instability mechanisms are related to chaotic diffusion or transport [69]. For two-dimensional systems chaotic transport is in general only possible by breaking KAM tori (but see also [70]).For higher-dimensional systems the chaotic layers can form a connected web along which diffusion like motion is always possible.

To extract information about the long time stability of particle motion from numerical simulations is also a difficult task as mentioned above. So-called survival plots (see Figure 19) [71] can be helpful in getting some insight into the problem.

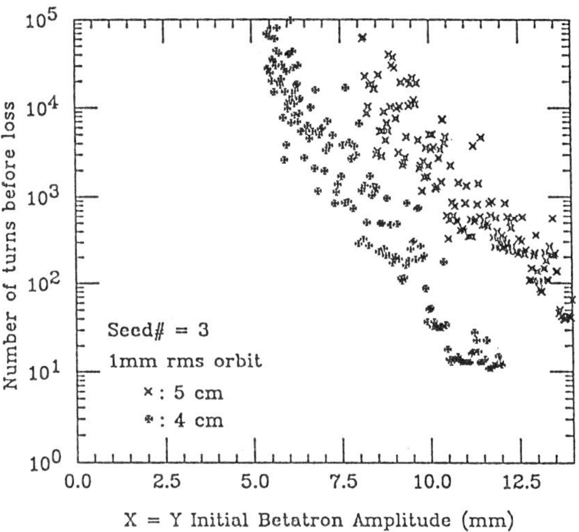

Figure 19. survival plot for the SSC, i.e. number of turns of the particle before loss versus initial betatron amplitude of the particle

An ideal and realistic - but mathematically very complicated - approach would be to consider the perturbed system

$$\mathcal{H}(\vec{I}, \vec{\Theta}) = \mathcal{H}_0(\vec{I}) + \varepsilon \cdot \mathcal{H}_1(\vec{I}, \vec{\Theta})$$

and to find some rigorous estimates for the time variation of the actions $\vec{I}$ - for example to predict a time T for which the variation of $\vec{I}$ is less than some fixed upper limit. This is the spirit of Nekhoroshev's theory [72]. First promising attempts to apply these ideas to accelerator physics problems have been made by Warnock, Ruth and Turchetti see [73],[74], [75] for more details.

Since there are no exact solutions available for the complicated nonlinear dynamics in storage rings and in order to check and test the theoretical concepts and tools described above, existing accelerators at FERMILAB, CERN and at the University of Indiana have been used for experimental investigations of the particle motion. Summaries of these results and further details can be found in [76], [77], [78].

As an example of an experimentally observed phase space plot of a nonlinear machine we show Figure 20 [79].

Figure 21 summarizes the status of the art of the nonlinear particle motion in storage rings.

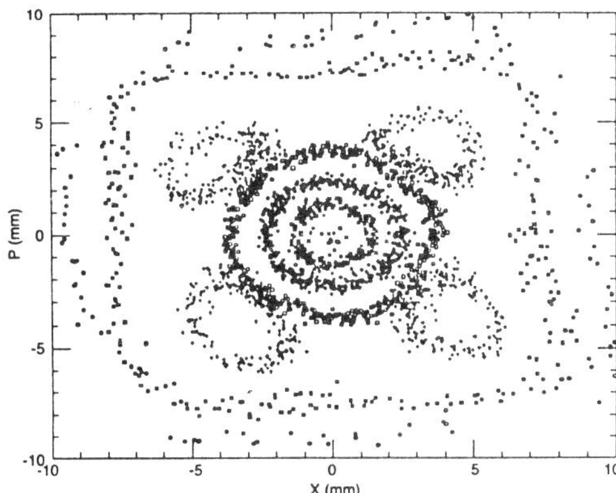

Figure 20. measured transverse phase space plot near a $Q_x = 15/4$ nonlinear resonance in the IUCF

In the next chapter we will investigate explicitly stochastic systems i.e. systems subject to noise or stochastic forces.

3. Stochastic dynamics in storage rings

In the first part of this lecture we have shown how the single particle dynamics of proton storage rings can be described by nonintegrable Hamiltonians. These systems show a very complex dynamics - regular and chaotic motion is intricately mixed in phase space. We have mentioned the concept of Arnold diffusion and chaotic transport and we have asked whether probabilistic methods can be applied successfully in this context.

In the second part of our survey we want to investigate systems where probabilistic tools are necessary, because we want to study the influence of stochastic forces and noise. In this case the equations of motion, which describe the dynamics, take the form :

$$\frac{d}{dt}\,\vec{x}(t) = \vec{f}(\vec{x},t;\vec{\xi}(t)) \tag{72}$$

or in the discrete time (mapping) case

$$\vec{x}(n+1) = \vec{f}(\vec{x}(n),\vec{\xi}(n)) \tag{73}$$

where $\vec{\xi}(t)$ or $\vec{\xi}(n)$ designates some explicit stochastic vector process with known statistical properties. Our aim will be to study the temporal evolution of $\vec{x}(t)$ or $\vec{x}(n)$ under the influence of these explicit stochastic forces. We will call this kind of (probabilistic) dynamics *stochastic dynamics* in contrast to the (deterministic) *chaotic dynamics* investigated in the first part of this review. Questions we want to answer in the following are:

Given the statistical properties of the random forces, what are the statistical properties of $\vec{x}(t)$ or $\vec{x}(n)$? How can we treat these systems mathematically? And how can we calculate, for example, average values $< x_i(t) >$ or correlations $< x_i(t)\,x_j(t') >$?

This part of the review is organized as follows. At first we will summarize some basic results of probability theory and the theory of stochastic processes. Then we will concentrate

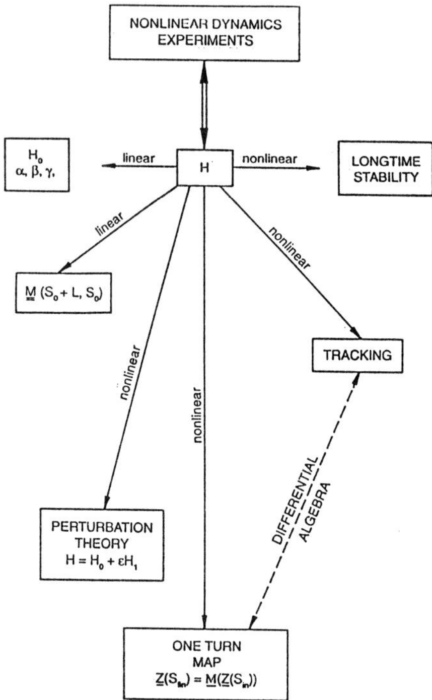

Figure 21. nonlinear dynamics of storage rings

on stochastic differential equations and their use in accelerator physics problems. In the case that the fluctuating random forces are modelled by Gaussian white noise processes (which is quite often a very good approximation) we will illustrate the mathematical subtleties related to these processes.

Examples of stochastic differential equations are

1. Langevin equation approach to Brownian motion

$$\frac{d}{dt}\, v = -\eta \cdot v + \xi(t) \tag{74}$$

with v particle velocity, η friction coefficient and $\xi(t)$ fluctuating random force

2. stochastically driven harmonic oscillator

$$\frac{d}{dt}\begin{pmatrix} x_1(t) \\ x_2(t) \end{pmatrix} = \begin{pmatrix} 0 & 1 \\ -1 & 0 \end{pmatrix} \cdot \begin{pmatrix} x_1(t) \\ x_2(t) \end{pmatrix} + \begin{pmatrix} 0 \\ \xi(t) \end{pmatrix} \tag{75}$$

3. spin diffusion or Brownian motion on the unit sphere [83] (see Figure 22)

$$\frac{d}{dt}\, \vec{S}(t) = \vec{H}(t) \times \vec{S}(t) \tag{76}$$

where $\vec{H}(t)$ denotes a fluctuating field.

As we will see later, single particle dynamics in accelerators is a rich source for stochastic differential equations. Before we start with a systematic study of these systems we have to repeat some basic facts.

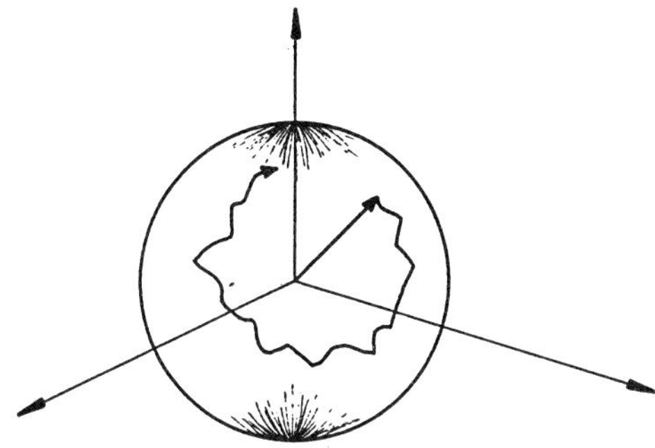

Figure 22. Brownian motion on the unit sphere

3.1. Summary of mathematical facts

Fundamental concepts of probability theory are the *random experiment* (e.g. throwing dice) and the *probability space*. This space consists of the sample space Ω of outcomes ω of the random experiment, a sigma algebra of events $\mathcal{A}$ i.e. a family of sets defined over Ω such that

1. $\Omega \in \mathcal{A}$

2. for every $A_j \in \mathcal{A}, \ \bar{A}_j \in \mathcal{A}$

3. for $A_j, j = 1, 2, \dots$ with $A_j \in \mathcal{A}$

$$\bigcup_{j=1}^{\infty} A_j \in \mathcal{A}$$

(where $\bar{A}$ denotes the complement of A with respect to Ω) and a probability measure Pr defined over $\mathcal{A}$:

$$Pr : \mathcal{A} \longrightarrow [0,1].$$

Pr is a measure for the frequency of the occurance of an event in $\mathcal{A}$ and it satisfies the following axioms:

1. $Pr(\phi) = 0$

2. $Pr(\Omega) = 1$

3. $Pr(A_i \cup A_j) = Pr(A_i) + Pr(A_j)$ for $A_i \cap A_j = \phi$

(ϕ is the empty set and designates the impossible event whereas Ω is the certain event).

The aim of probability theory is not the calculation of the probability measure of the underlying sample space Ω, but it is concerned with the calculation of new probabilities from given ones [80]. A rigorous treatment needs sophisticated measure theoretic concepts and is beyond the scope of this review. We will restrict ourselves to some basic facts and results which will be needed later. Detailed presentations of probability theory can be found in the references [80], [81], [82], [84], [85], [86], [87], [88]. In the following summary we will closely follow the book of Horsthemke and Lefever [80].

3.1.1. Random variables (r.v.) The first notion we need is that of a random variable (r.v.) $\mathcal{X}$. A random variable is a function from the sample space Ω to $\mathbf{R}$ i.e. $\mathcal{X} : \Omega \longrightarrow \mathbf{R}$ with the property that

$$A = \{\omega | \mathcal{X}(\omega) \le x\} \in \mathcal{A} \tag{77}$$

for all $x \in \mathbf{R}$. This means that A is an event which belongs to $\mathcal{A}$ for all x.

Remark: One should always distinguish carefully between the random variable $\mathcal{X}$ (calligrahic letter) and the realization x of the r.v., i.e. the value $\mathcal{X}$ takes on $\mathbf{R}$.

In the following we will only consider continuous r.v. which can be characterized by probability density functions $p_\mathcal{X}(x)dx$ which - roughly stated - give the probability of finding $\mathcal{X}$ between x and $x + dx$ i.e.

$$p_\mathcal{X}(x)dx = Pr(\{\omega | x \le \mathcal{X}(\omega) \le x + dx\}) \tag{78}$$

or in shorthand notation

$$p_\mathcal{X}(x)dx = Pr(x \le \mathcal{X} \le x + dx)$$

Given this probability density one can define expectation value, moments and mean square deviation or variance of a r.v.:

1. expectation value of a random variable $\mathcal{X}$

$$E\{\mathcal{X}\} \equiv\ <\mathcal{X}> \equiv m_\mathcal{X} \stackrel{\text{def}}{=} \int_{-\infty}^{\infty} x p_\mathcal{X}(x)dx. \tag{79}$$

2. moment of order r

$$E\{\mathcal{X}^r\} \equiv\ <\mathcal{X}^r> \stackrel{\text{def}}{=} \int_{-\infty}^{\infty} x^r p_\mathcal{X}(x)dx. \tag{80}$$

3. mean square deviation or variance

$$E\{(\delta\mathcal{X})^2\} \equiv E\{(\mathcal{X}- <\mathcal{X}>)^2\} \equiv \sigma^2 \stackrel{\text{def}}{=}$$
$$\stackrel{\text{def}}{=} \int_{-\infty}^{\infty} (x - m_\mathcal{X})^2 p_\mathcal{X}(x)dx \tag{81}$$

As an example we consider the Gaussian distribution, which because of the central limit theorem (see the references), plays an important role in statistics. This distribution is defined by (see also Figure 23):

$$p_\mathcal{X}(x) = [(2\pi)^{1/2}\sigma]^{-1} \cdot \exp\{-\frac{(x - m)^2}{2\sigma^2}\} \tag{82}$$

In this case equations (79), (80) and (81) yield

$$E\{\mathcal{X}\} = m \tag{83}$$

$$E\{(\delta\mathcal{X})^2\} = \sigma^2 \tag{84}$$

$$E\{(\delta\mathcal{X})^r\} = \begin{cases} 0, & \text{for } r \ge 1 \text{ odd} \\ (r-1)!! \cdot \sigma^r, & r \text{ even} \end{cases} \tag{85}$$

where we have used the following definition $(r-1)!! = 1 \cdot 3 \cdot 5 \cdot .... \cdot (r-1)$.

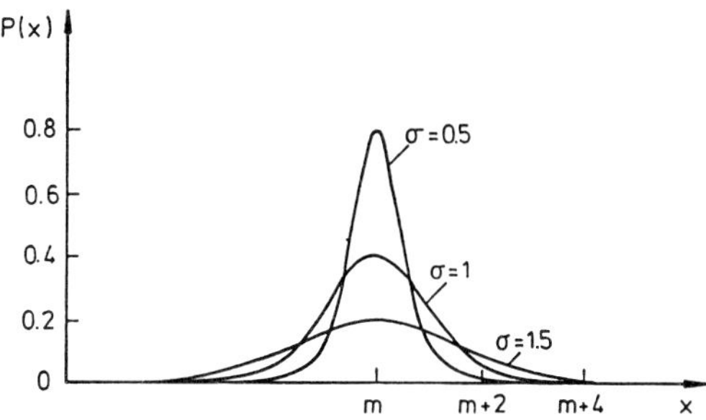

Figure 23. Gaussian distribution

A Gaussian variable is thus completely specified by its first two moments.

Extending these considerations to the multivariable or random vector case $\mathcal{X}_1, ... \mathcal{X}_n$ requires the notion of joint probability densities i.e.

$$p_{\mathcal{X}_1 ... \mathcal{X}_n}(x_1, ... x_n)dx_1 ... dx_n =$$
$$Pr(x_1 \leq \mathcal{X}_1 \leq x_1 + dx_1, \ ..., \ x_n \leq \mathcal{X}_n \leq x_n + dx_n). \tag{86}$$

Moments, cross correlations, covariance matrix etc can then be defined. For example in the two-dimensional case $\vec{\mathcal{X}} = (\mathcal{X}, \mathcal{Y})^T$ mixed moments are defined by:

$$E\{\mathcal{X}^r \cdot \mathcal{Y}^p\} = \int_{-\infty}^{\infty} \int_{-\infty}^{\infty} x^r \cdot y^p p_{\mathcal{X},\mathcal{Y}}(x,y)dxdy \tag{87}$$

Another important definition we will need is the conditional density function $p_{\mathcal{X},\mathcal{Y}}(x|y)$. Conditional probability $p(A|B)$ means the probability that event A will take place, knowing with certainty that another event B has occured. For random variables the corresponding density is given by:

$$p_{\mathcal{X},\mathcal{Y}}(x|y) = \frac{p_{\mathcal{X},\mathcal{Y}}(x,y)}{p_{\mathcal{Y}}(y)} \tag{88}$$

or similarly

$$p_{\mathcal{X},\mathcal{Y}}(x,y) = p_{\mathcal{X},\mathcal{Y}}(x|y) \cdot p_{\mathcal{Y}}(y). \tag{89}$$

In the multivariable case one has accordingly:

$$p_{\mathcal{X}_1 ... \mathcal{X}_n}(x_1 ... x_n) =$$
$$p_{\mathcal{X}_1 ... \mathcal{X}_n}(x_1|x_2 ... x_n) \cdot p_{\mathcal{X}_2 ... \mathcal{X}_n}(x_2|x_3 ... x_n) \cdot$$
$$... p_{\mathcal{X}_{n-1}, \mathcal{X}_n}(x_{n-1}|x_n) \cdot p_{\mathcal{X}_n}(x_n). \tag{90}$$

3.1.2. stochastic processes (s.p.) Next we introduce stochastic processes (s.p.) with the following definition: a family of random variables indexed by a parameter t, $\mathcal{X}_t$, is called a random or stochastic process (t may be continuous or discrete). A stochastic process thus depends on two arguments (t,ω). For fixed t, $\mathcal{X}_t$ is a random variable whereas for fixed ω and continuous

t , $\mathcal{X}_{(.)}(\omega)$ is a real valued function of t which is called realization or sample path of the s.p. (realizations are designated by $x(t)$).

Generally, stochastic processes $\mathcal{X}_t$ are defined by an infinite hierarchy of joint distribution density functions

$$
\begin{cases}
\quad p(x_1, t_1)dx_1 \\
\quad p(x_1, t_1; x_2, t_2)dx_1 dx_2 \\
\qquad\qquad \cdot \\
\qquad\qquad \cdot \\
\qquad\qquad \cdot \\
\quad p(x_1, t_1; ...x_n, t_n)dx_1...dx_n \\
\qquad\qquad \cdot \\
\qquad\qquad \cdot
\end{cases}
\tag{91}
$$

and are a complicated mathematical object. A proper treatment requires the so-called stochastic calculus see for example [82].

In a similar way as for random variables we can define moments and correlation functions. For example

$$
\begin{aligned}
E\{\mathcal{X}_{t_1}\mathcal{X}_{t_2}\} = &< x(t_1)x(t_2) >= \\
= &\int_{-\infty}^{\infty}\int_{-\infty}^{\infty} x_1 x_2 p(x_1, t_1; x_2, t_2)dx_1 dx_2
\end{aligned}
\tag{92}
$$

is called the two-time correlation function of the stochastic process $\mathcal{X}_t$. Higher order correlations are obtained analogously

$$
\begin{aligned}
E\{\mathcal{X}_{t_1}...\mathcal{X}_{t_n}\} = &< x(t_1)...x(t_n) >= \\
= &\int_{-\infty}^{\infty} ... \int_{-\infty}^{\infty} x_1...x_n \cdot p(x_1, t_1; ...; x_n, t_n)dx_1...dx_n
\end{aligned}
\tag{93}
$$

One way of characterizing a stochastic process is by looking at its history or memory. The completely independent process is defined by

$$
p(x_1, t_1; ...; x_n, t_n) = \prod_{i=1}^{n} p(x_i, t_i)
\tag{94}
$$

i.e. only the one-time distribution density is needed to classify and determine this process.

The next simplest case is the so-called *Markov process*. It is defined by

$$
p(x_n, t_n | x_{n-1}, t_{n-1}; ..; x_1, t_1) = p(x_n, t_n | x_{n-1}, t_{n-1})
\tag{95}
$$

with

$$
t_1 \leq t_2 \leq ... \leq t_n
\tag{96}
$$

Equation (95) implies that if the present state is known, any additional information on the past history is totally irrelevant for predicting the future evolution.

Markov processes are completely specified by the transition probability density $p(x_n, t_n | x_{n-1}, t_{n-1})$ and the one-time probability density $p(x, t)$. Because of (95) and (90) we have:

$$p(x_n, t_n; ...; x_1, t_1) =$$
$$= p(x_n, t_n | x_{n-1}, t_{n-1}) \cdot p(x_{n-1}, t_{n-1} | x_{n-2}, t_{n-2}) \cdot ... \cdot p(x_2, t_2 | x_1, t_1) \cdot p(x_1, t_1). \quad (97)$$

The transition probability densities fulfill the following nonlinear functional relation (Chapman-Kolmogorov equation):

$$p(x_3, t_3 | x_1, t_1) =$$
$$= \int_{-\infty}^{\infty} p(x_3, t_3 | x_2, t_2) \cdot p(x_2, t_2 | x_1, t_1) dx_2 \quad (98)$$

Examples of stochastic processes are:

1. Gaussian stochastic processes $\mathcal{X}_t$
   $\mathcal{X}_t$ is specified by

$$\left\{ \begin{array}{c} p(x, t) \\ p(x_1, t_1; x_2, t_2) \\ \cdot \\ \cdot \\ \cdot \\ p(x_1, t_1; ...; x_n, t_n) \\ \cdot \\ \cdot \end{array} \right.$$

If all the m-th order distributions are Gaussian i.e.

$$p(x_1, t_1; ...; x_m, t_m) =$$
$$(2\pi)^{-m/2} \cdot (\det \underline{\Lambda})^{-1/2} \cdot \exp\{-\frac{1}{2}(\vec{x} - \vec{m})^T \cdot \underline{\Lambda}^{-1} \cdot (\vec{x} - \vec{m})\} \quad (99)$$

with $\vec{m}^T = (m_{\mathcal{X}}(t_1), ..., m_{\mathcal{X}}(t_m))$ and $\underline{\Lambda} = \Lambda_{ij} = E\{(\mathcal{X}_{t_i} - m_{\mathcal{X}}(t_i))(\mathcal{X}_{t_j} - m_{\mathcal{X}}(t_j))\}$, $\mathcal{X}_t$ is called a Gaussian stochastic process.

2. The Wiener process $\mathcal{W}_t$ which plays an important role in probability theory and which is defined by:

$$p(w_n, t_n; ...; w_0, t_0) = \prod_{i=0}^{n-1} p(w_{i+1}, t_{i+1} | w_i, t_i) \cdot p(w_0, t_0) =$$
$$= \prod_{i=0}^{n-1} [2\pi(t_{i+1} - t_i)]^{-1/2} \cdot \exp\{-\frac{(w_{i+1} - w_i)^2}{2(t_{i+1} - t_i)}\} \cdot p(w_0, t_0). \quad (100)$$

The Wiener process is an example for an independent increment process. This can be seen as follows: defining the increments

$$\Delta \mathcal{W}_i = \mathcal{W}_i - \mathcal{W}_{i-1} \quad (101)$$

and

$$\Delta t_i = t_i - t_{i-1} \quad (102)$$

we obtain

$$p(\Delta w_n, t_n; ...; \Delta w_1, t_1; w_0, t_0) =$$
$$= \prod_{i=1}^{n} (2\pi \Delta t_i)^{-1/2} \cdot \exp\{-\frac{(\Delta w_i)^2}{2\Delta t_i}\} \cdot p(w_0, t_0) \tag{103}$$

i.e. the random variables $\Delta \mathcal{W}_{t_i}$ are statistically independent. Furthermore one calculates in this case

$$E\{\mathcal{W}_t\} = 0 \tag{104}$$
$$E\{\mathcal{W}_t \mathcal{W}_s\} = \min(t, s) \tag{105}$$
$$E\{\mathcal{W}_t^2\} = t. \tag{106}$$

A typical path of a Wiener process is shown in Figure 24. These paths are continuous but nowhere differentiable.

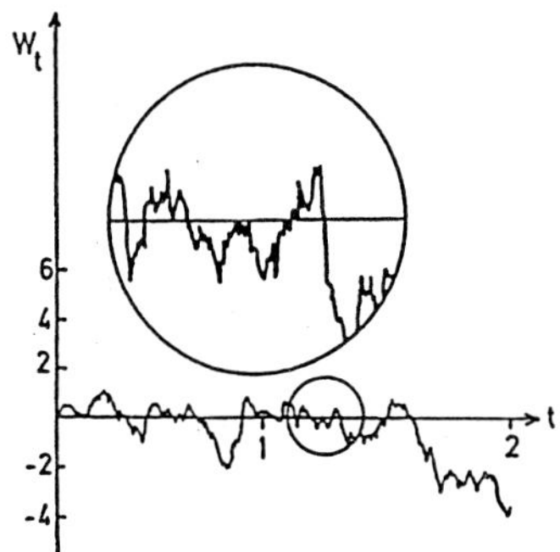

Figure 24. typical path of a Wiener process

3. Gaussian white noise process $\mathcal{Z}_t$. It is a completely random process with

$$p(z_1, t_1; ...; z_n, t_n) = \prod_{i=1}^{n} p(z_i) \tag{107}$$

i.e. independent values at every instant of time. It has

$$E\{\mathcal{Z}_t\} = 0 \tag{108}$$

and the two-time correlation function is given by

$$E\{\mathcal{Z}_t \mathcal{Z}_s\} = \delta(t - s) \ . \tag{109}$$

Since $\mathcal{Z}_t$ is Gaussian all odd correlations vanish automatically (see equation (85)) and the even correlations are given by [84]

$$E\{\mathcal{Z}_{t_1} \cdot \ ... \ \cdot \mathcal{Z}_{t_{2n}}\} = \sum_{P_i} \delta(t_{i_1} - t_{i_2}) \cdot \ ... \ \cdot \delta(t_{i_{2n-1}} - t_{i_{2n}}) \tag{110}$$

where the sum is taken over $(2n)!/(2^n \ n!)$ permutations. Gaussian white noise is a mathematical idealization and does not occur in nature. It plays a similar role in the theory of stochastic processes as the Dirac δ-function in functional analysis. One can show, that in a generalized sense [80] Gaussian white noise is the derivative of the Wiener process.

The last concept we need is that of a Markovian diffusion process. This is a Markov process with continuous sample paths. Diffusion processes play an important role in physics and in the context of stochastic differential equations with Gaussian white noise, as we will see in the next section. The temporal evolution of these diffusion processes is described by the so-called Fokker-Planck equation. This is a linear partial differential equation for the transition density $p(x, t|x_0, t_0)$ or the one-time density $p(x, t)$.

3.1.3. stochastic differential equations (s.d.e.) As mentioned already stochastic differential equations are the natural extension of deterministic systems, if one wants to include noise effects

$$\frac{d}{dt}\, \vec{x}(t) = \vec{f}(\vec{x}, t; \vec{\xi}(t)) \ .$$

Often $\vec{\xi}(t)$ is modelled by a Gaussian white noise process. This approximation is well justified if the fluctuating forces show only short-time correlations compared to other typical time-scales of the system. Introducing Gaussian white noise in dynamical systems is related to some mathematical problems which we want to illustrate now. In order to keep the notation as simple as possible we will restrict ourselves for the moment to scalar equations of motion (multiplicative stochastic processes [2]) of the form

$$\frac{d}{dt}\, \mathcal{X}_t = f(\mathcal{X}_t) + g(\mathcal{X}_t) \cdot \mathcal{Z}_t \tag{111}$$

where $\mathcal{Z}_t$ designates Gaussian white noise and where we have switched to the notation introduced above in order to make clear that we are treating stochastic processes.

Before we start our investigation let us repeat what Horsthemke and Lefever have written in this context:

The transition to Gaussian white noise sounds rather harmless but it is actually at this point that dangerous territory is entered, which contains hidden pitfalls and traps to ensnare the unwary theoretician ... if one succeeds in avoiding the various traps, either by luck, intuition or whatever, one captures a treasure which might be bane or boon: the white noise

The mathematical problems are related to the irregular behaviour of white noise and one has to ask which sense can be given to equations like (111). Remember, the sample paths of a Wiener process are continuous but nowhere differentiable, so what is the meaning of $\frac{d}{dt}$ in (111)?

We will not go through all the mathematical details, we only want to illustrate the subtleties, so that readers who are confronted with this problem are reminded of being careful when using stochastic differential equations with white noise. Excellent presentations of this problem can be found in [89],[90].

After these remarks we try to give a sense to the stochastic differential equation by rewriting it as an integral equation

$$\mathcal{X}_t = \mathcal{X}_0 + \int_0^t f(\mathcal{X}_s)ds + \int_0^t g(\mathcal{X}_s)\mathcal{Z}_s ds \tag{112}$$

which is eqivalent to

$$\mathcal{X}_t = \mathcal{X}_0 + \int_0^t f(\mathcal{X}_s)ds + \int_0^t g(\mathcal{X}_s)d\mathcal{W}_s. \tag{113}$$

---

[2] the external noise is coupled in a multiplicative manner to $\mathcal{X}_t$ the statistical properties of which are sought

As before $\mathcal{W}_s$ denotes the Wiener process. The second integral on the right hand side of equation (113) - a kind of a stochastic Stieltjes integral - is the main reason for the mathematical problems. Let us quote again Horsthemke and Lefever :

The problem is though a sense can be given to this integral and thus to the stochastic differential equation in spite of the extremely irregular nature of the white noise, there is no unique way to define it, precisely because white noise is so irregular. This has nothing to do with the different definitions of ordinary integrals by Riemann and Lebesgue. After all for the class of functions for which the Riemann integral as well as the Lebesgue integral can be defined, both integrals yield the same answer. The difference between the definitions for the above stochastic integral, connected with the names of Ito and Stratonovich , is much deeper; they give different results.

This difficulty can be illustrated as follows:

Consider a stochastic integral of the form

$$\mathcal{S}_t = \int_{t_0}^{t} \mathcal{W}_s d\mathcal{W}_s \tag{114}$$

If (114) would be Riemann integrable the result would be

$$\mathcal{S}_t = \frac{1}{2}(\mathcal{W}_t^2 - \mathcal{W}_{t_0}^2) \tag{115}$$

As in the Riemann case we try to evaluate (114) by a limit of approximating sums of the form

$$\mathcal{S}_n = \sum_{i=1}^{n} \mathcal{W}_{\tau_i^{(n)}} \cdot (\mathcal{W}_{t_i^{(n)}} - \mathcal{W}_{t_{i-1}^{(n)}}) \tag{116}$$

with a partition of the interval $[t_0, t]$

$$t_0 \le t_1^{(n)} \le t_2^{(n)} \le \dots \le t_{n-1}^{(n)} \le t$$

and

$$\tau_i^{(n)} \in [t_{i-1}^{(n)}, t_i^{(n)}]$$

or

$$\tau_i^{(n)} = (1 - \alpha)t_{i-1}^{(n)} + \alpha t_i^{(n)}$$

with $0 \le \alpha \le 1$.

Using the stochastic calculus (the proper calculus to treat stochastic processes as mentioned above) one can show that the limit of $\mathcal{S}_n$ for $n \to \infty$ depends on the evaluation points $\tau_i^{(n)}$ or α [80], [81]

$$\lim_{n \to \infty} \mathcal{S}_n = \frac{1}{2}(\mathcal{W}_t^2 - \mathcal{W}_{t_0}^2) + (\alpha - \frac{1}{2})(t - t_0) \ . \tag{117}$$

Thus, the stochastic integral is no ordinary Riemann integral. However, an unambiguous definition of the integral can be given - and thus a consistent calculus is possible - if $\tau_i^{(n)}$ is fixed once and forever. Two choices are convenient

- $\alpha = 0$, Ito definition

- $\alpha = \frac{1}{2}$, Stratonovich definition.

Thus, a stochastic differential equation has always to be supplemented by a kind of interpretation rule for the stochastic integral. In both cases mentioned above one can show, that the solutions of the corresponding equations are Markovian diffusion processes. In the Ito case

$$(I) \quad d\mathcal{X}_t = f(\mathcal{X}_t, t)dt + g(\mathcal{X}_t, t)d\mathcal{W}_t \tag{118}$$

the corresponding Fokker-Planck equation for the transition probability density $p(x, t|x_0, t_0)$ reads

$$\frac{\partial}{\partial t}p(x, t|x_0, t_0) =$$
$$= -\frac{\partial}{\partial x}[f(x, t) \cdot p(x, t|x_0, t_0)] + \frac{1}{2} \cdot \frac{\partial^2}{\partial x^2}[g^2(x, t) \cdot p(x, t|x_0, t_0)] \tag{119}$$

whereas in the Stratonovich case

$$(S) \quad d\mathcal{X}_t = f(\mathcal{X}_t, t)dt + g(\mathcal{X}_t, t)d\mathcal{W}_t \tag{120}$$

the Fokker-Planck equation is given by

$$\frac{\partial}{\partial t}p(x, t|x_0, t_0) =$$
$$= -\frac{\partial}{\partial x}[f(x, t) \cdot p(x, t|x_0, t_0)] + \frac{1}{2} \cdot \frac{\partial}{\partial x}[g(x, t) \cdot \frac{\partial}{\partial x}(g(x, t)p(x, t|x_0, t_0))] \tag{121}$$

or equivalently by

$$\frac{\partial}{\partial t}p(x, t|x_0, t_0) =$$
$$= -\frac{\partial}{\partial x}([f(x, t) + \frac{1}{2}\frac{\partial g(x, t)}{\partial x} \cdot g(x, t)] \cdot p(x, t|x_0, t_0)) +$$
$$+ \frac{1}{2} \cdot \frac{\partial^2}{\partial x^2}[g^2(x, t) \cdot p(x, t|x_0, t_0)] \ . \tag{122}$$

Equations (119) and (121) have to be supplemented with the initial conditions

$$p(x, t|x_0, t_0)_{t=t_0} = \delta(x - x_0)$$

and suitable boundary conditions for x.

The Ito calculus is mathematically more general but leads to unusual rules such as

$$\int_{t_0}^{t} \mathcal{W}_s d\mathcal{W}_s = \frac{1}{2}(\mathcal{W}_t^2 - \mathcal{W}_{t_0}^2) - \frac{1}{2}(t - t_0)$$

and some care is needed when one transforms from one process $\mathcal{X}_t$ to $\mathcal{R}_t = h(\mathcal{X}_t)$. For further details and a discussion of the relationship between Ito and Stratonovich approach (which preserves the "normal" rules of calculus) the reader is referred to the references.

Remarks:

1. The one-time probability density $p(x,t)$ of a Markovian diffusion process $\mathcal{X}_t$ also satisfies the Fokker-Planck equation (119) or (121).

2. In the case of purely additive noise where g does not depend on $\mathcal{X}_t$ there is no difference between the Ito and Stratonovich approach, so both stochastic differential equations define the same Markovian diffusion process.

3. Since Gaussian white noise is a mathematical idealization and can only approximately model real stochastic processes in nature, there is always the question how to interpret equation (111) in practical problems. In most physical cases one will rely on the Stratonovich interpretation as is suggested by a theorem due to Wong and Zakai [80] which roughly states :

   if we start with a phenomenological equation containing realistic noise $\mathcal{W}_t^{(n)}$ of the form

$$\frac{d}{dt}\,\mathcal{X}_t = f(\mathcal{X}_t) + g(\mathcal{X}_t)\cdot \frac{d}{dt}\,\mathcal{W}_t^{(n)} \tag{123}$$

   where all the integrals can be interpreted in the usual (e.g. Riemann) sense and if we pass to the white noise limit

$$\mathcal{W}_t^{(n)} \longrightarrow \mathcal{W}_t \tag{124}$$

   so that a stochastic differential equation of the form

$$\frac{d}{dt}\,\mathcal{X}_t = f(\mathcal{X}_t) + g(\mathcal{X}_t)\cdot \frac{d}{dt}\,\mathcal{W}_t \tag{125}$$

   is obtained (remember that Gaussian white noise is the derivative of the Wiener process $\mathcal{Z}_t = \frac{d}{dt}\,\mathcal{W}_t$) the latter has to be interpreted as a Stratonovich equation.

4. The above considerations can be extended to the multivariable case where $\mathcal{X}_t$, $f(\mathcal{X}_t)$ and $\mathcal{W}_t$ have to be replaced by vector quantities and where $g(\mathcal{X}_t)$ has to be replaced by a matrix. Now,the stochastic differential equation takes the form

$$d\vec{\mathcal{X}}_t = \vec{f}(\vec{\mathcal{X}}_t,t)dt + \underline{g}(\vec{\mathcal{X}}_t,t)d\vec{\mathcal{W}}_t \ . \tag{126}$$

   The Ito interpretation leads to a Fokker-Planck equation for the transition density $p(\vec{x},t|\vec{x}_0,t_0)$ of the form

$$\frac{\partial}{\partial t}p(\vec{x},t|\vec{x}_0,t_0) =$$

$$= -\sum_i \frac{\partial}{\partial x_i}[f_i(\vec{x},t)\cdot p(\vec{x},t|\vec{x}_0,t_0)] +$$

$$+\frac{1}{2}\cdot\sum_{i,j} \frac{\partial}{\partial x_i}\cdot\frac{\partial}{\partial x_j}[\{\underline{g}(\vec{x},t)\underline{g}^T(\vec{x},t)\}_{ij}\cdot p(\vec{x},t|\vec{x}_0,t_0)] \tag{127}$$

   whereas the Stratonovich interpretation gives

$$\frac{\partial}{\partial t} p(\vec{x}, t | \vec{x}_0, t_0) =$$

$$= -\sum_i \frac{\partial}{\partial x_i} [f_i(\vec{x}, t) \cdot p(\vec{x}, t | \vec{x}_0, t_0)] +$$

$$+ \frac{1}{2} \cdot \sum_{ijk} \frac{\partial}{\partial x_i} \{ g_{ik}(\vec{x}, t) \frac{\partial}{\partial x_j} [g_{jk}(\vec{x}, t) \cdot p(\vec{x}, t | \vec{x}_0, t_0)] \} \ . \tag{128}$$

Examples

1. Wiener process

$$d\mathcal{X}_t = d\mathcal{W}_t \tag{129}$$

with the corresponding Fokker-Planck (diffusion) equation ($f = 0, g = 1$)

$$\frac{\partial}{\partial t} p(x, t | x_0, t_0) = \frac{1}{2} \cdot \frac{\partial^2}{\partial x^2} \cdot p(x, t | x_0, t_0) \tag{130}$$

2. The Ornstein Uhlenbeck process (see also equation (74))

$$d\mathcal{V}_t = -\eta \mathcal{V}_t dt + \sigma \cdot d\mathcal{W}_t \tag{131}$$

leads to the following Fokker-Planck equation ($f = -\eta, g = \sigma$)

$$\frac{\partial}{\partial t} p(v, t | v_0, t_0) =$$

$$= \frac{\partial}{\partial v} [\eta \cdot v \cdot p(v, t | v_0, t_0)] + \frac{1}{2} \sigma^2 \frac{\partial^2}{\partial v^2} \cdot p(v, t | v_0, t_0) \ . \tag{132}$$

3. Stochastically driven harmonic oscillator as an example of a multivariable system:

$$\begin{pmatrix} d\mathcal{X}_{1_t} \\ d\mathcal{X}_{2_t} \end{pmatrix} = \begin{pmatrix} 0 & 1 \\ -1 & 0 \end{pmatrix} \cdot \begin{pmatrix} \mathcal{X}_{1_t} \, dt \\ \mathcal{X}_{2_t} \, dt \end{pmatrix} + \begin{pmatrix} 0 \\ d\mathcal{W}_t \end{pmatrix} \tag{133}$$

In this case one obtains as Fokker-Planck equation

$$\frac{\partial}{\partial t} p(x_1, x_2, t | x_{1_0}, x_{2_0}, t_0) =$$

$$= -\frac{\partial}{\partial x_1} [x_2 \cdot p(x_1, x_2, t | x_{1_0}, x_{2_0}, t_0)] + \frac{\partial}{\partial x_2} [x_1 \cdot p(x_1, x_2, t | x_{1_0}, x_{2_0}, t_0)] +$$

$$+ \frac{1}{2} \cdot \frac{\partial^2}{\partial x_2^2} p(x_1, x_2, t | x_{1_0}, x_{2_0}, t_0) \ . \tag{134}$$

Summarizing, we can say that stochastic differential equations are the natural extension of deterministic systems if one wants to study the influence of noise. Often these noise processes are approximated by Gaussian white noise, a mathematical idealization, which has to be treated with great care. However, the mathematical subtleties related to stochastic differential equations with white noise are outweighed by the results which are available for these Markovian diffusion processes namely the Fokker-Planck equation [86].

3.2. Stochastic dynamics problems in accelerator physics

After this - admittedly - very sketchy summary of mathematical results we will now investigate where stochastic differential equations arise in the single particle dynamics of accelerators [91].

At first we will study the particle motion in electron storage rings where radiation effects play an important role. Classically, radiation is taken into account by the following modified Lorentz equation [92], [93] :

$$\frac{d}{dt}\left(\frac{E}{c^2}\cdot\dot{\vec{r}}\right) = \frac{e}{c}\cdot\dot{\vec{r}}\times\vec{B}(\vec{r},t) + e\cdot\vec{\epsilon}(\vec{r},t) + \vec{R}^{rad}(\vec{r},t) \tag{135}$$

with

- $\vec{B}(\vec{r},t)$ magnetic field

- $\vec{\epsilon}(\vec{r},t)$ electric field

- $\vec{R}^{rad}(\vec{r},t)$ radiation reaction force

- E energy of particle

Because of the stochastic emission of the radiation, $\vec{R}^{rad}(\vec{r},t)$ is modelled by a stochastic force, which we divide into its average part and its fluctuating part according to

$$\vec{R}^{rad}(\vec{r},t) = <\vec{R}^{rad}(\vec{r},t)> +\delta\vec{R}^{rad}(\vec{r},t) \ . \tag{136}$$

The average part is identified with the classical Lorentz-Dirac radiation reaction force, see for example [94]

$$<\vec{R}^{rad}(\vec{r},t)> = -\frac{2}{3}\cdot\frac{e^2}{c^5}\cdot\gamma^4\cdot\dot{\vec{r}}\cdot\{(\ddot{\vec{r}})^2 + \frac{\gamma^2}{c^2}\cdot(\dot{\vec{r}}\cdot\ddot{\vec{r}})^2\} \tag{137}$$

with $\gamma = \frac{E}{m_0 c^2}$.

$<\vec{R}^{rad}(\vec{r},t)>$ leads in general to the so-called radiation damping. The fluctuating part in equation (136),which mainly effects the energy variation in equation (135) is usually approximated by Gaussian white noise [5],[93],[95].

Thus, in the curvilinear coordinate system of the accelerator one generally obtains a system of six coupled nonlinear stochastic differential equations of the form [5]:

$$\frac{d}{ds}\vec{y}(s) = \vec{f}(\vec{y},s) + \underline{T}(\vec{y},s)\cdot\delta\vec{c}(s) \tag{138}$$

with $\vec{y}^T = (x,z,\tau,p_x,p_z,p_\tau)$. $\delta\vec{c}(s)$ designates a Gaussian white noise vector process.

Interesting physical questions one wants to answer are :

What are the average fluctuations of the particle around the closed orbit (beam emittances)? What is the particle distribution $p(\vec{y},s)$ i.e. what is the probability for finding the particle between $\vec{y}$ and $\vec{y}+d\vec{y}$ at location s ? Is there a stationary solution of this density i.e. what is $\lim_{s\to\infty} p(\vec{y},s)$? What is the particle lifetime in the finite vacuum chambers of the accelerator (time to hit the border) ?

These questions have been extensively studied in the linear case [92], [93], [95],[96], [97]

$$\frac{d}{ds}\,\vec{y}(s) = (\underline{A}(s) + \delta\underline{A}(s))\cdot\vec{y}(s) + \vec{c}(s) + \delta\vec{c}(s) \tag{139}$$

$\underline{A}$ designates the Hamiltonian part of the motion (six-dimensional coupled synchro-betatron oscillations) [5]; $\delta\underline{A}(s)$ describes the radiation damping due to $< \vec{R}^{rad} >$ (see equation (137)); $\delta\vec{c}$ is the fluctuating part of the radiative force and $\vec{c}$ denotes some additional field errors of the system. In this case one obtains compact and easily programmable expressions for the important beam parameters (beam emittances) [95], [96]. Furthermore, the corresponding Fokker-Planck equation for the probability distribution can be solved exactly [98], [99], [100].

An investigation of the nonlinear system is much more complicated and is an active area of research. Nonlinear systems such as an octupole-dipole wiggler or the beam-beam interaction in electron storage rings have been analyzed by various authors [101], [102], [103], [104], [105], [106]. Let us consider the latter case in more detail. The main problem is to understand the motion of a test particle under the influence of the nonlinear electromagnetic fields of the counter rotating beam [107] (see Figure 25).

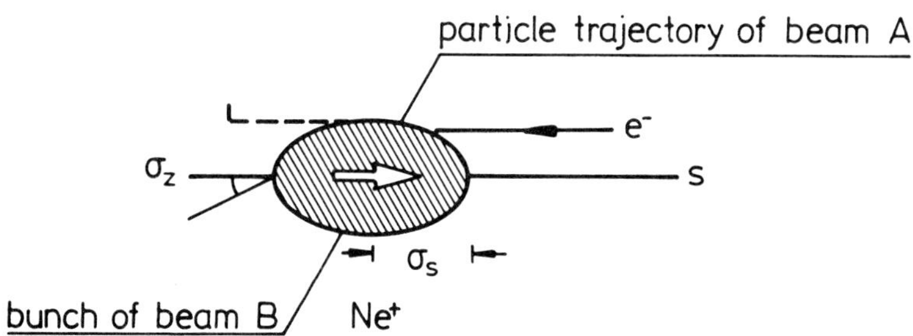

Figure 25. deflection of a test particle in beam A in the field of the counter rotating beam B

This so-called weak-strong model of the beam-beam interaction is mathematically described by the following set of equations [107] (perturbed Hamiltonian system) :

$$\begin{cases} \frac{d}{ds}\,x(s) = \frac{\partial}{\partial p_x}\,\mathcal{H}(x,z,p_x,p_z,s) \\ \frac{d}{ds}\,p_x(s) = -\frac{\partial}{\partial x}\,\mathcal{H}(x,z,p_x,p_z,s) - \gamma_x\cdot p_x + d_x\cdot\Gamma \\ \frac{d}{ds}\,z(s) = \frac{\partial}{\partial p_z}\,\mathcal{H}(x,z,p_x,p_z,s) \\ \frac{d}{ds}\,p_z(s) = -\frac{\partial}{\partial z}\,\mathcal{H}(x,z,p_x,p_z,s) - \gamma_z\cdot p_z + d_z\cdot\Gamma \end{cases} \tag{140}$$

The Hamiltonian $\mathcal{H}(x,z,p_x,p_z,s)$ consists of the linear part and the nonlinear potential due to the beam-beam interaction

$$\mathcal{H}(x,z,p_x,p_z,s) =$$
$$= \frac{p_x^2}{2} + K_x(s)\cdot\frac{x^2}{2} + \frac{p_z^2}{2} + K_z(s)\cdot\frac{z^2}{2} + U(x,z)\cdot\delta_p(s-s_0). \tag{141}$$

This nonlinear term is given by [107]

$$U(x,z) = \frac{N_b \cdot r_e}{\gamma} \cdot \int_0^\infty \frac{1 - \exp\{-\frac{x^2}{2\sigma_x^2+q} - \frac{z^2}{2\sigma_z^2+q}\}}{(2\sigma_x^2 + q)^{1/2} \cdot (2\sigma_z^2 + q)^{1/2}} \cdot dq. \tag{142}$$

In equations (140), (141) and (142) we have used the following definitions : $\delta_p(s - s_0)$ periodic delta function, r_e classical electron radius, N_b number of particles in the counter rotating bunch, σ_x, σ_z rms beam sizes of the strong bunch. Radiation damping is described by the two damping constants γ_x, γ_z and the strength of the stochastic excitation Γ is denoted by d_x, d_z.

These equations have been extensively used in numerical simulations. These simulations are very helpful for understanding the complicated interplay of nonlinearity, damping and stochastic excitation in lepton colliders. Figure 26 [108] shows such a calculation. The combined effect of quantum fluctuations and nonlinearity can move a particle starting near the origin in phase space to a (nonlinear) resonance island before it is damped again and eventually pushed to another resonance nearby.

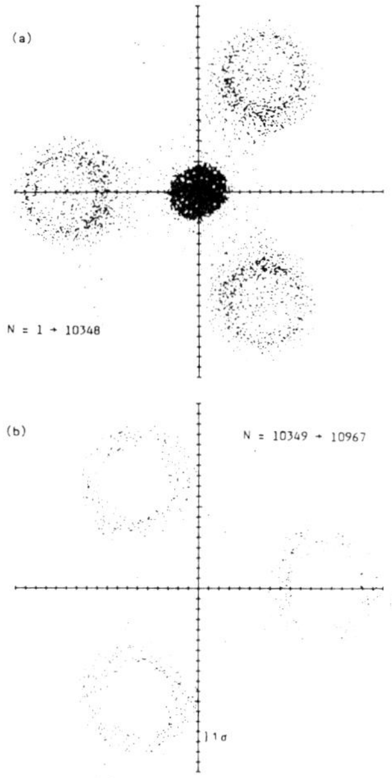

Figure 26. $(p_z - z)$ phase diagram of a simulation

An interesting and important question in this context is :

How does the distribution function $p(x, z, p_x, p_z, s)$ of a stochastic system like (140) evolve with s or time?

The corresponding Fokker-Planck equation would be a five-dimensional partial differential equation, and because of the highly singular behaviour of its coefficients (delta functions), this equation would be extremely complicated to solve. In this case stochastic mappings are more suitable. In the following we will sketch an algorithm for calculating the temporal evolution of the density function for such a stochastic map. This algorithm is based on an idea of Gerasimov and gives much faster results than direct numerical simulations (see also [109]).

We will illustrate this approach with a simple two-dimensional model of the beam-beam interaction. The details are described in [110] and in a PhD thesis of Pauluhn [111].

The considered model is given by (see also [65]):

$$\begin{pmatrix} x(n+1) \\ p_x(n+1) \end{pmatrix} = \begin{pmatrix} \cos(2\pi Q) & \beta \cdot \sin(2\pi Q) \\ -\frac{1}{\beta} \cdot \sin(2\pi Q) & \cos(2\pi Q) \end{pmatrix} \cdot \begin{pmatrix} x(n) \\ p_x(n) - \gamma_x p_x(n) + u(x(n)) + d_x\Gamma \end{pmatrix} \tag{143}$$

Γ is now a random variable, Q is the tune of the storage ring, β is the beta-function at the interaction point and $u(x(n))$ is given by

$$u(x(n)) = -\frac{4\pi\xi}{\beta} \cdot x(n) \cdot \frac{1 - \exp(-\frac{x^2(n)}{2\sigma^2})}{\frac{x^2(n)}{2\sigma^2}} \tag{144}$$

with ξ beam-beam strength parameter [107].

The main steps of this algorithm are:

1. discretization of the two-dimensional phase space

2. use of the microscopic dynamics (see equation (143)) to calculate the transition rates A_{ij} between the discretized bins of the phase space

3. use of this (stochastic) transition matrix A_{ij} as macroscopic propagator for the time evolution of an initial particle distribution

Figure 27 shows how an initially constant and homogeneous distribution evolves with time (after 1000, 3000, 15000, 99000 turns respectively). These results are in excellent agreement with direct numerical simulations [111].

Spin dynamics in electron storage rings constitutes another interesting application of stochastic differential equations in beam dynamics. In this case the orbital equations of motion (138)

$$\frac{d}{ds}\,\vec{y}(s) = \vec{f}(\vec{y},s) + \underline{T}(\vec{y},s) \cdot \delta\vec{c}(s)$$

have to be supplemented by the spin equation of motion, the so-called BMT equation (Bargman, Michel, Telegdi see for example [3])

$$\frac{d}{ds}\,\vec{S} = \vec{\Omega}(\vec{y}) \times \vec{S} \tag{145}$$

where the field $\vec{\Omega}$ depends on the orbital degrees of freedom. Since (138) is a stochastic differential equation, we end up with a kind of spin diffusion or random motion on the unit sphere [83].

The linearized spin-orbit motion in electron storage rings has been studied in [112], [113] and the reader is referred to these references for further details.

Until now we have seen, that radiation is a natural source for noise in electron accelerators. However, there are also other sources for stochastic forces such as rf noise, power supply noise,

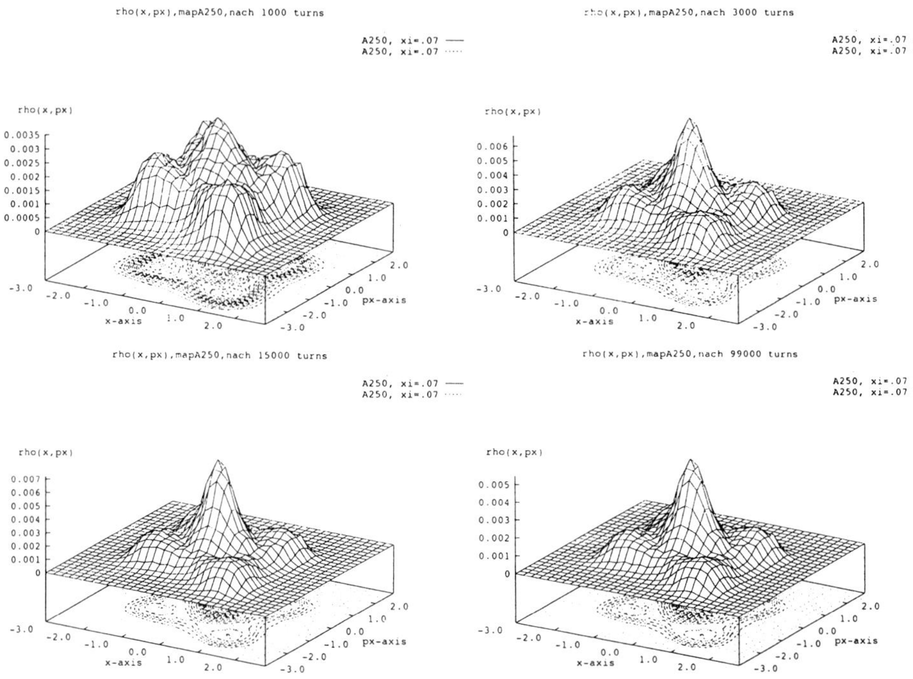

Figure 27. evolution of the density function for the stochastic map (143) (here denoted by $\rho(x, p_x, t)$) as a function of time

random ground motion or restgas scattering. These influences can also be present in proton (Hamiltonian) systems and some of these sources could for example be included in the vector potential $\vec{A}$ of equation (2).

The influence of rf noise on the beam dynamics in storage rings has been investigated extensively by several authors [114],[115], [116], [117], [118].In the smooth approximation (oscillator model) [7], where one averages $\kappa(s)D(s)$ and $V(s)$ over the circumference L of the storage ring one obtains the following Hamiltonian (see equation (15))

$$\bar{\mathcal{H}}(\bar{\tau}, \bar{p}_\tau) =$$
$$= -\frac{1}{2}\mu \cdot \bar{p}_\tau^2 + \frac{L}{2\pi k} \cdot \frac{e \cdot \bar{V}_0}{E_0} \cdot \cos(\frac{2\pi k}{L} \cdot \bar{\tau}) +$$
$$+ \frac{L}{2\pi k} \cdot \frac{e \cdot \delta \bar{V}_0}{E_0} \cdot \cos(\frac{2\pi k}{L} \cdot \bar{\tau}) - \delta\bar{\tau} \cdot \frac{e \cdot \bar{V}_0}{E_0} \cdot \sin(\frac{2\pi k}{L} \cdot \bar{\tau}) \qquad (146)$$

with

$$\mu = \frac{1}{L} \cdot \int_0^L \kappa(s) \cdot D(s) \cdot ds \tag{147}$$

and

$$\bar{V}_0 = \frac{1}{L} \cdot \int_0^L V(s) \cdot ds. \tag{148}$$

$\delta \bar{V}_0$ and $\delta \bar{\tau}$ denote the amplitude and phase noise (for example Gaussian white noise) of a cavity. The corresponding stochastic differential equations now read

$$\begin{cases} \frac{d}{ds}\, \bar{\tau} = -\mu \cdot \bar{p}_\tau \\ \frac{d}{ds}\, \bar{p}_\tau = \frac{e}{E_0} \cdot (\bar{V}_0 + \delta \bar{V}_0) \cdot \sin(\frac{2\pi k}{L} \cdot \bar{\tau}) + \frac{e \cdot \bar{V}_0}{E_0} \cdot \frac{2\pi k}{L} \cdot \delta \bar{\tau} \cdot \cos(\frac{2\pi k}{L} \cdot \bar{\tau}) \ . \end{cases} \tag{149}$$

Using perturbation methods the corresponding Fokker-Planck equation has been solved in [114], [115], [117]. An alternative to this perturbative approach would be a numerical integration of the exact Fokker-Planck equation or a direct numerical treatment of the stochastic differential equations (149). Numerical methods to solve stochastic differential equations are described in detail in a recently published book [119]. These methods have been applied in [111], [120].

If one treats the cavity as as a strongly localized element, one should investigate the stochastic map, in this case the standard map with explicit stochasticity and an interesting problem, one is then faced with, is:

How is the interplay between deterministic chaos and explicit stochasticity?

Only few results are available for this case [121], [122], and these stochastic nonintegrable Hamiltonian systems remain a challenging and interesting problem not only in accelerator physics.

4. Summary and conclusions

The main aim of this review was to illustrate the problems of nonlinear beam dynamics in storage rings and to introduce some of the concepts and tools to study these systems. Although a lot of facts are known and powerful techniques - especially in the Hamiltonian case - have been developed, designing new accelerators (like the SSC) remains a challenging problem of nonlinear dynamics. The design of such a machine requires a lot of numerical simulations, the knowledge of the basic facts of the qualitative theory of dynamical systems and perturbative investigations. Besides these theoretical issues, future accelerator developments have to rely on the experience with existing machines and on special nonlinear dynamics experiments performed with these machines.

5. Acknowledgement

The author wants to thank the organizers of the workshop, S.Martin and M.Berz , for the opportunity to deliver these lectures and for an interesting week in Gosen. I am also grateful to O. Brüning and A. Pauluhn for many useful discussions, to S. Wipf for carefully reading the manuscript and to M. Böge for his help with TEX. Last, but not least, I would like to thank Antje not only for her tremendous help with the references but also for her encouragement and friendship. She was responsible for my going out to dinner and seeing movies more times than I would have otherwise. Without these distractions I might have finished this review a little bit earlier but with these distractions life was much richer and much more worth living.

References

[1] W.Scandale, G.Turchetti (ed) *Nonlinear problems in future particle accelerators* World Scientific Press (1991)

[2] J.M.Jowett, M.Month, S.Turner (ed) *Nonlinear dynamics aspects of particle accelerators* Springer (1986)

[3] J.D.Jackson *Classical electrodynamics* John Wiley (1975)

[4] G.Ripken "Nonlinear canonical equations of coupled synchro-betatron motion and their solution within the framework of a nonlinear six-dimensional (symplectic) tracking program for ultrarelativistic protons" DESY 85-084 (1985)

[5] H.Mais, G.Ripken "Theory of coupled synchro-betatron oscillations" DESY M-82-05 (1982)

[6] T.Suzuki "Hamiltonian formulation for synchrotron oscillations and Sacherer's integral equation" Part. Accel. 12, 237 (1982)

[7] H.Mais, G.Ripken "Spin-orbit motion in a storage ring in the presence of synchrotron radiation using a dispersion formalism" DESY 86-029 (1986)

[8] C.J.A.Corsten, H.L.Hagedoorn "Simultaneous treatment of betatron and synchrotron motions in circular accelerators" Nucl. Instr. Meth. 212, 37 (1983)

[9] A.Piwinski, A.Wrulich "Excitation of betatron-synchrotron resonances by a dispersion in the cavities" DESY 76-07 (1976)

[10] A.J.Lichtenberg, M.A.Lieberman *Regular and stochastic motion* Springer (1983)

[11] B.V.Chirikov "A universal instability of many-dimensional oscillator systems" Phys. Rep. 52, 263 (1979)

[12] M.Henon "Numerical study of quadratic area-preserving mappings" Quart.Appl.Mathem. Vol. XXVII, 291 (1969)

[13] M.V.Berry "Regular and irregular motion" in *Topics in Nonlinear Dynamics - a Tribute to Sir Edward Bullard* AIP Conf. Proc. 46 (1978)

[14] R.H.G.Helleman "Self-generated chaotic behavior in nonlinear mechanics" in *Fundamental Problems in Statistical Mechanics* Vol. V North Holland (1980)

[15] M.Henon "Numerical exploration of Hamiltonian systems" in *Chaotic Behavior of deterministic Systems* Les Houches 1981 North Holland (1983)

[16] J.Moser *Stable and random motions in dynamical systems* Princeton University Press (1973)

[17] L.E.Reichl *The transition to chaos* Springer (1992)

[18] H.Mais, A.Wrulich, F.Schmidt "Studies of chaotic behaviour in HERA caused by transverse magnetic multipole fields" DESY M-85-08 (1985) and IEEE Trans.Nucl.Sci. 32, 2252 (1985)

[19] F.Schmidt "Untersuchungen zur dynamischen Akzeptanz von Protonenbeschleunigern und ihre Begrenzung durch chaotische Bewegung" DESY HERA 88-02 (1988)

[20] S.N.Rasband *Chaotic dynamics of nonlinear systems* John Wiley (1990)

[21] G.Benettin, L.Galgani, J.M.Strelcyn "Kolmogorov entropy and numerical experiments" Phys. Rev. A14, 2338 (1976)

[22] S.Wiggins *Introduction to applied nonlinear dynamical systems and chaos* Springer (1990)

[23] L.Michelotti private communication

[24] S.Becker, private communication and "Nichtlineare Dynamik in Zirkularbeschleunigern" Diploma thesis TU Berlin (1992)

[25] D.P.Barber, H.Mais, G.Ripken, F.Willeke "Nonlinear theory of synchro-betatron motion" DESY 86-147 (1986)

[26] E.D.Courant, H.S.Snyder "Theory of the alternating-gradient synchrotron" Ann. Phys. 3, 1 (1958)

[27] H.Goldstein *Classical mechanics* (second edition) Addison-Wesley (1980)

[28] E.Forest "A Hamiltonian-free description of single particle dynamics for hopelessly complex periodic systems" J. Math. Phys. $\underline{31}$, 1133 (1990)

[29] E.Forest, K.Hirata "A contemporary guide to beam dynamics" KEK-92-12 (1992)

[30] A.Nayfeh *Perturbation methods* John Wiley (1973)

[31] A.Deprit "Canonical transformations depending on a small parameter" Cel. Mech. $\underline{1}$, 12 (1969)

[32] L.Michelotti "Moser like transformations using the Lie transform" Part.Accel. $\underline{16}$, 233 (1985)

[33] F.G.Gustavson "On constructing formal integrals of a Hamiltonian system near an equilibrium point" The Astron. Journ. $\underline{71}$, 670 (1966)

[34] R.T.Swimm, J.B.Delos "Semiclassical calculations of vibrational energy levels for nonseparable systems using the Birkhoff-Gustavson normal form" J. Chem. Phys. $\underline{71}$, 1706 (1979)

[35] G.Turchetti "Perturbative methods for Hamiltonian maps" in *Methods and Applications of Nonlinear Dynamics* World Scientific Press (1988)

[36] A.Bazzani, P.Mazzanti, G.Servizi, G.Turchetti "Normal forms for Hamiltonian maps and nonlinear effects in a particle accelerator" Il Nuovo Cimento $\underline{102B}$, 51 (1988)

[37] A.Dragt "Lectures on nonlinear orbit dynamics" in *Physics of High Energy Particle Accelerators* AIP Conf. Proc. 87 (1982)

[38] A.Dragt, F.Neri, G.Rangarajan, D.R.Douglas, L.M.Healy, R.D.Ryne "Lie algebraic treatment of linear and nonlinear beam dynamics" Ann. Rev. Nucl. Part. Sci. Vol. 38, 455 (1988)

[39] A.Dragt, J.Finn "Lie series and invariant functions for analytic symplectic maps" J.Math.Phys. $\underline{17}$, 2215 (1976)

[40] J.R.Cary "Lie transform perturbation theory for Hamiltonian systems" Phys. Rep. $\underline{79}$, 129 (1981)

[41] E.Forest "Normal form algorithm on non-linear symplectic maps" SSC-29 (1985)

[42] H.Mais, C.Mari unpublished notes "Lie algebraic methods for nonlinear symplectic maps"

[43] H.Mais, A.Wrulich, F.Schmidt "Particle tracking" DESY 86-024 (1986) and CERN Accelerator School 1985

[44] E.Forest "Canonical integrators as tracking codes (or how to integrate perturbation theory with tracking)" in "Physics of Particle Accelerators" AIP Conf.Proc. 184 (1989)

[45] M.Berz "COSY INFINITY Version 6" these proceedings

[46] L.Schachinger, R.Talman "TEAPOT:a thin element accelerator program for optics and tracking" Part.Accel. $\underline{22}$, 35 (1987)

[47] A.Dragt et al. "MARYLIE 3.0 a program for charged particle beam transport based on Lie algebraic methods" Dept. of Phys. Technical Report, University of Maryland (1987)

[48] K.L.Brown, D.C.Carey, C.Iselin, F.Rothacker "TRANSPORT, a computer program for designing charged particle beam transport systems" SLAC 91 and CERN 80-04

[49] A.Wrulich "RACETRACK:a computer code for the simulation of nonlinear particle motion in accelerators" DESY 84-026 (1984)

[50] E.Forest, R.D.Ruth "Fourth order symplectic integration" Physica D $\underline{43}$, 105 (1990)

[51] F.Feng, Q.Meng-zhao "Hamiltonian algorithms for Hamiltonian systems and a comparative numerical study" Computer Phys. Commun. $\underline{65}$, 173 (1991)

[52] M.Berz "Differential algebraic treatment of beam dynamics to very high orders including applications to space charge" in *Linear Accelerator and Beam Optics Codes* AIP Conf. Proc. 177 (1988)

[53] M.Berz "Arbitrary order description of arbitrary particle optical systems" Nucl.Instr.Methods <u>A298</u>, 426 (1990)

[54] E.Forest, M.Berz, J.Irwin "Normal form methods for complicated periodic systems: a complete solution using differential algebra and Lie operators" Part. Accel. <u>24</u>, 91 (1989)

[55] M.Berz "Differential algebraic description of beam dynamics to very high orders" Part.Accel. <u>24</u>, 109 (1989)

[56] M.Berz "Differential algebraic formulation of normal form theory" these proceedings

[57] R.V.Servranckx "Improved tracking codes: present and future" IEEE Trans. Nucl. Sci. <u>NS-32</u>, 2186 (1985)

[58] P.Wilhelm "Role of rounding errors in beam tracking calculations" Part.Accel. <u>19</u>, 99 (1986)

[59] A.Wrulich "Tracking and special processors" DESY HERA 85-06 published in [2]

[60] J.F.Applegate, M.R.Douglas, Y.Gursel, P.Hunter, C.L.Seitz, G.J.Sussman "A digital orrery" IEEE Trans. on Comp. <u>C34</u>, 822 (1985)

[61] SSC-Conceptual Design, SSC-SR-2020 (1986)

[62] M.Tabor "The onset of chaotic motion in dynamical systems" in *Advances in Chemical Physics* Vol. XLVI (1981)

[63] J.Tennyson "The dynamics of the beam-beam interaction" in *Physics of High Energy Particle Accelerators* AIP Conf. Proc. 87 (1982)

[64] J.Tennyson "The instability threshold for bunched beams in ISABELLE" in *Nonlinear Dynamics and the Beam-Beam Interaction* AIP Conf. Proc. 57 (1979)

[65] F.M.Izraelev "Nearly linear mappings and their applications" Physica <u>D1</u>, 243 (1980)

[66] S.I.Tzenov "Application of master equation technique for the study of nonlinear dynamics of particles in accelerators and storage rings" CERN SL/92-17 (1992)

[67] A.Gerasimov "The applicability of diffusion phenomenology to particle losses in hadron colliders" CERN SL/92-30 (1992)

[68] O.Brüning "An estimate of the diffusion in the HERA-p FODO cell" Int. J. Mod. Phys. A (Proc. Suppl.) <u>2A</u>, 418 (1993)

[69] S.Wiggins *Chaotic transport in dynamical systems* Springer (1992)

[70] G.M.Zaslavsky, R.Z.Sagdeev, D.A.Usikov, A.A.Chernikov *Weak chaos and quasi-regular patterns* Cambridge Univ. Press (1991)

[71] Y.Yan "Applications of differential algebra to single particle dynamics in storage rings" SSCL-500 (1991)

[72] N.N.Nekhoroshev "An exponential estimate of the time of stability of nearly integrable Hamiltonian systems I" Uspekhi Mat. Nauk <u>32</u>(6), 5 (1977), English translation: Russian Math. Surveys <u>32</u>(6), 1 (1977)

[73] R.L.Warnock, R.D.Ruth "Stability of orbits in nonlinear mechanics for finite but very long times" published in [1]

[74] R.L.Warnock, R.D.Ruth "Long-term bounds on nonlinear Hamiltonian motion" SLAC-PUB-5267 (1991)

[75] G.Turchetti "Nekhoroshev stability estimates for symplectic maps and physical applications" in *Number Theory and Physics* Springer (1990)

[76] A.Chao et al "Experimental investigation of nonlinear dynamics in the Fermilab Tevatron" Phys. Rev. Lett. <u>61</u>, 2752 (1988)

[77] X.Altuna et al. "The 1991 dynamic aperture experiment at the CERN SPS" CERN-SL-91-043 (1991)

[78] D.D.Caussyn et al. "Experimental studies of nonlinear beam dynamics" Phys. Rev. <u>A46</u>, 7942 (1992)

[79] A.Chao "Recent efforts on nonlinear dynamics" Int. J. Mod. Phys. A (Proc. Suppl.) 2A, 981 (1993)

[80] W.Horsthemke, R.Lefever *Noise induced transitions* Springer (1984)

[81] C.W.Gardiner *Hanbook of stochastic methods* Springer (1985)

[82] T.T.Soong *Random differential equations in science and engineering* Academic Press (1973)

[83] M.Lax "Classical noise IV: Langevin methods" Rev. Mod. Phys. 38, 541 (1966)

[84] N.G.van Kampen *Stochastic processes in physics and chemistry* North Holland (1981)

[85] J.Honerkamp *Stochastische dynamische Systeme* VCH Verlagsgesellschaft (1990)

[86] H.Risken *The Fokker Planck equation* Springer (1989)

[87] R.L.Stratonovich *Topics in the theory of random noise* Vols. 1,2 Gordon and Breach (1967)

[88] L.Arnold *Stochastische Differentialgleichungen* R.Oldenbourg (1973)

[89] N.G.van Kampen "Ito versus Stratonovich" J. Statist. Phys. 24, 175 (1981)

[90] R.E.Mortensen "Mathematical problems of modeling stochastic nonlinear dynamic systems" J. Statist. Phys. 1, 271 (1969)

[91] J.Jowett "Introductory statistical mechanics for electron storage rings" in *Physics of Particle Accelerators* AIP Conf. Proc. 153 (1987)

[92] C.Bernardini, C.Pellegrini "Linear theory of motion in electron storage rings" Ann. Phys. 46, 174 (1968)

[93] A.A.Kolomensky, A.N.Lebedev *Theory of cyclic accelerators* North Holland (1966)

[94] F.Rohrlich *Classical charged particles* Addison-Wesley (1965)

[95] H.Mais, G.Ripken "Influence of the synchrotron radiation on the spin-orbit motion of a particle in a storage ring" DESY M-82-20 (1982)

[96] A.W.Chao "Evaluation of beam distribution parameters in an electron storage ring" J. Appl. Phys. 50, 595 (1979)

[97] Y.H.Chin "Quantum lifetime" DESY 87-062 (1987)

[98] Yu.P.Virchenko, Yu.N.Grigorev "Equilibrium distribution of charged particles in the phase space of a cyclic accelerator" Ann. Phys. 209, 1 (1991)

[99] A.W.Chao, M.J.Lee "Particle distribution parameters in an electron storage ring" J. Appl. Phys. 47, 4453 (1976)

[100] D.P.Barber, K.Heinemann, H.Mais, G.Ripken "A Fokker-Planck treatment of stochastic particle motion within the framework of a fully coupled 6-dimensional formalism for electron-positron storage rings including classical spin motion in linear approximation" DESY 91-146 (1991)

[101] A.Hofmann, J.Jowett "Theory of the dipole-octupole wiggler" partI: "Phase oscillations" CERN/ISR-TH/81-23 partII: "Coupling of phase and betatron oscillations" CERN/ISR-TH/81-24 (1981)

[102] S.Kheifets "Application of the Green's function method to some nonlinear problems of an electron storage ring, Part IV : Study of a weak-beam interaction with a flat strong beam" Part. Accel. 15, 153 (1984)

[103] F.Ruggiero "Renormalized Fokker-Planck equation for the problem of the beam-beam interaction in electron storage rings" Ann. Phys. 153, 122 (1984)

[104] J.F.Schonfeld "Statistical mechanics of colliding beams" Ann. Phys. 160, 149 (1985)

[105] Y.H.Chin "Renormalized beam-beam interaction theory" KEK-Preprint 87-143a (1988)

[106] A.L.Gerasimov "Phase convection and distribution "tails" in periodically driven Brownian motion" Physica D41, 89 (1990)

[107] H.Mais, C.Mari "Introduction to beam-beam effects" DESY M-91-04 (1991)

[108] A.Piwinski "Dependence of the luminosity on various machine parameters and their optimization at PETRA" DESY 83-028 (1983)

[109] S.Milton "A different approach to beam-beam interaction simulation" in Proc. of 1991 Particle Accelerator Conference, IEEE (1991)

[110] A.Pauluhn, A.Gerasimov, H.Mais "A stochastic map for the one- dimensional beam-beam interaction" Int. J. Mod. Phys. A (Proc. Suppl.) $\underline{2A}$, 1091 (1993)

[111] A.Pauluhn "Stochastic beam dynamics in storage rings" PhD thesis University of Hamburg to be published

[112] A.W.Chao "Evaluation of radiative spin polarization in an electron storage ring" Nucl. Instr. Meth. $\underline{180}$, 29 (1981)

[113] H.Mais, G.Ripken "Theory of spin-orbit motion in electron-positron storage rings - summary of results" DESY 83-062 (1983)

[114] G.Dôme, "Diffusion due to rf noise" in CERN Advanced Accelerator School Oxford, CERN 87-03 (1987)

[115] S.Krinsky, J.M.Wang "Bunch diffusion due to rf-noise" Part. Accel $\underline{12}$, 107 (1982)

[116] H.J.Shih, J.Ellison, B.Newberger, R.Cogburn "Longitudinal beam dynamics with rf noise" SSCL-578 (1992)

[117] A.Pauluhn "Some aspects of rf noise in storage rings" HERA 92-07 (1992)

[118] H.Mais unpublished notes "rf noise in storage rings"

[119] P.E.Kloeden, E.Platen *Numerical solution of stochastic differential equations* Springer (1992)

[120] M.Seesselberg, H.P.Breuer, J.Honerkamp, F.Petruccione, H.Mais "Numerical integration of stochastically driven Hamiltonian systems and their application in particle storage rings" to be published

[121] C.F.F.Karney, A.B.Rechester, R.B.White "Effect of noise on the standard mapping" Physica $\underline{D4}$, 425 (1982)

[122] G.Györgyi, N.Tishby "Destabilization of islands in noisy Hamiltonian systems" Phys. Rev. $\underline{A36}$, 4957 (1987)

Inst. Phys. Conf. Ser. No 131
Paper presented at Int. Workshop Nonlinear Problems in Accelerator Phys. Berlin, 1992

Moment Methods for Nonlinear Maps

Gordon D. Pusch

Center for Transport Theory and Mathematical Physics, Virginia Tech University, Blacksburg, VA USA, 24061-0435

AECL Research, Chalk River Laboratories, Chalk River, Ontario, CANADA K0J 1J0

Abstract. It is shown that DA may be used to push *moments of distributions* through a map, at a computational cost per moment comparable to pushing a single particle. The algorithm is independent of order, and whether or not the map is symplectic. Starting from the known result that moment-vectors transform *linearly* — like a tensor— even under a *nonlinear* map, I suggest that the form of the moment transformation rule indicates that moment-vectors are elements of the *dual* to DA-vector space. I propose several methods of manipulating moments and constructing invariants using DA. I close with speculations on how DA might be used to "close the circle" to solve the *inverse moment problem*, yielding an entirely DA-and-moment-based space-charge code.

1. Introduction

The past decade has seen a renaissance of classical mechanics and nonlinear systems theory, and an important change in its emphasis: rather than studying the *equations of motion*, one now studies the *map* which represents their solution. Accelerator physics has not only participated in, but helped to shape this renaissance, in both the *analysis* and *computation* of maps — most recently by Berz's [1, 2] introduction of *Differential Algebra* (DA), which allows one to compute maps perturbatively to arbitrarily high order in arbitrarily many variables.

Despite advances in the analysis of *accelerators*, there has so far not been a comparable breakthrough in the analysis of *beams* — and unless one takes the narrow view that "accelerators are what accelerator-physicists use to study accelerator physics with," one needs methods of determining what a map does to a beam. One can use "brute force" techniques, such as "pushing" particles from some distribution through a map; however this is not only computationally expensive, but it violates Hamming's dictum that "the goal of computation is not *numbers*, but *insight*."

Now, statisticians have long used *moments* to both parameterize and characterize distribution functions [4]. I suggest that much as DA allows one to sort *maps* into equivalence classes (*e.g.*, maps whose Taylor-expansions agree through *first order*, through *first and second order, etc.*), moments allow one to sort *distributions* into equivalence classes (*e.g.*, distributions with a given *mean*, mean and *variance*, mean, variance, and *skewness, etc.*). I shall present methods of using DA to *transport* moments of beams through maps, and propose methods to *reconstruct* beams from a set of moments. I shall show that transporting moments using DA is almost as easy as transporting particles.

2. Notation and Terminology

2.1. Manifolds and Maps

The goal of this paper is to show how DA may be used to calculate how moments of distribution functions transform under mappings of a manifold, particularly within the context of accelerator physics. Therefore, I shall speak of both the "source" and "target" manifolds as "phase-space," which I shall consider to be mapped continuously and differentiably onto itself *via* some dynamical evolution parameterized by "time." However no result in this paper requires that the map be generated by a Hamiltonian, nor describe a continuous evolution in "time," nor even map the manifold onto itself — I only require that the map be continuous, and sufficiently many times differentiable.

I succumb to a common abuse of notation in specifying a point Q of phase space by its coordinates, $\{\xi^\mu, \mu = 1, \ldots v\}$. On those rare occasions when I need to distinguish between Q and its coordinates, I write $\xi^\mu(Q)$.

I invoke the summation convention for all types of repeated indices, unless otherwise stated, be they coordinate indices, "set" indices, or the "multi-indices" I introduce later. Where no confusion will occur, I shall use bold face type to denote a set of "obviously" missing indices, *e.g.*, $\boldsymbol{\xi}$, and a centered dot to denote contraction: $\boldsymbol{k} \cdot \boldsymbol{\xi} \equiv k_\mu \xi^\mu$.

I write $\mathcal{M}_{t_2,t_1}$ for the "transfer map" which takes a point $\boldsymbol{\xi}_1$ at initial "time" t_1 to a point $\boldsymbol{\xi}_2$ at final "time" t_2, $\mathcal{M}_{t_2,t_1} : \boldsymbol{\xi}_1 \mapsto \boldsymbol{\xi}_2 = \mathcal{M}_{t_2,t_1}\boldsymbol{\xi}_1$; since I shall have no need of more than one map and two times, I shall henceforth simply write "$\mathcal{M}$."

To help distinguish *functions* from their *values,* I shall use upper-case symbols for functions, and lower-case for values, *e.g.*, $F : \mathbf{R}^v \to \mathbf{R}$; $\boldsymbol{\xi} \mapsto f = F(\boldsymbol{\xi})$.

2.2. Multinomials and Multi-indices

I shall make frequent use of multinomial expansions in many variables, for which conventional notations produce a proliferation of clumsy expressions replete with ellipses. In order to simplify such expressions, I shall use two types of "multi-indices" to denote *sets* of indices, with summation conventions for both.

How does one define a monomial in ξ^μ? If the elements of the set of variables ξ^μ *commute,* there are two equivalent ways: I can give a *list of variables* to be multiplied together, or I can give a *list of the number of times* each variable is to be multiplied together. This may best illustrated by an example: $\xi^1\xi^1\xi^1\xi^3\xi^4\xi^4 \equiv (\xi^1)^3(\xi^2)^0(\xi^3)^1(\xi^4)^2$; here the *left-hand side* may be represented by an *unordered list of indices* (unordered since under commutative multiplication any permutation is equivalent, $\{111344\} \equiv \{131144\} \equiv$

$\{411143\} \equiv \dots$), while the *right-hand side* may be represented by an *ordered set of exponents*, (3012). Because each form has its advantages, I shall compromise and use both: "$\{\mu_m\}$" for an *unordered set* of m indices $\{\mu_1, \mu_2, \dots, \mu_m\}$, with any permutation of indices considered to be equivalent to the standard "lexicographic order," and "(m_μ)" for the corresponding *ordered set* of "*exponents*" $(m_1, m_2, \dots, m_v)$, where $v \equiv \max\{\mu\}$ is the number of independent phase-space variables in $\boldsymbol{\xi}$. Along with the multi-indices, three convenient shorthands are the multi-index notation for a *monomial*,

$$\xi^{\{\mu_m\}} := \prod_{k=1}^{m} \xi^{\mu_k} \equiv \prod_{\mu=1}^{\max\{\mu\}} (\xi^\mu)^{m_\mu} =: \xi^{(m_\mu)}; \tag{1}$$

the "multi-index factorials," $(m_\mu)! := m_1! \, m_2! \cdots m_{\max\{\mu\}}!$, $\{\mu_m\}! \equiv (m_\mu)!$, and the multi-index *norm*, $|(m_\mu)| := m_1 + m_2 + \dots + m_{\max\{\mu\}}$, $|\{\mu_m\}| \equiv |(m_\mu)|$. Invoking the summation convention on repeated multi-indices allows a compact representation of power series:

$$f_{\{\mu_m\}} \, \xi^{\{\mu_m\}} := \sum_m \sum_{1 \leq \mu_1 \leq \mu_2 \leq \dots \leq \mu_m \leq v} f_{\mu_1 \mu_2 \dots \mu_m} \, \xi^{\mu_1} \xi^{\mu_2} \cdots \xi^{\mu_m}, \tag{2}$$

$$f_{(m_\mu)} \, \xi^{(m_\mu)} := \sum_m \sum_{\substack{m_1, m_2, \dots, m_v \\ m_1 + m_2 + \dots + m_v = m}} f_{m_1 m_2 \dots m_v} \, \left(\xi^1\right)^{m_1} \left(\xi^2\right)^{m_2} \cdots \left(\xi^v\right)^{m_v}, \tag{3}$$

where the range of the sum over m will be specified as needed. The two sorts of multi-indices are entirely equivalent, being in some sense "dual," and I shall rather cavalierly jump back and forth between them.

2.3. Differential Algebra

One writes $_nD_v$ for the *differential algebra* of nth order in v variables, and $_nN_v := (n+v)!/(n!\,v!)$ for its dimension. My notation for a DA-vector differs from Berz's [2]: I write Df for a DA-valued quantity, f for its *real part*, and df for its *differential part*, where $Df = f + df$.[1] One can expand Df in a particular coordinate basis as

$$Df = \sum_{\substack{(m_\mu) \\ 0 \leq (m_\mu) \leq n}} f_{(m_\mu)} \, d\xi^{(m_\mu)}, \tag{4}$$

where $d\xi^\mu$ is the first-order *unit differential* along the ξ^μ coordinate.

An important class of DA vectors are those resulting from the *DA extension* of a function F, *e.g.*, the extension of $F: \mathbf{R}^j \to \mathbf{R}^k$ to $F: {}_nD_v^j \to {}_nD_v^k$, where $_nD_v^k$ is the k-fold cartesian-product of $_nD_v$ with itself. The coefficients $f_{(m_\mu)}$ of $Df = F(D\boldsymbol{\xi})$ admit a simple interpretation when the argument of F is one of the "privileged" elements

$$D\xi^\mu(Q) := \xi^\mu(Q) + d\xi^\mu, \tag{5}$$

such that $\xi^\mu(Q)$ are the coordinates of some point Q, and $d\xi^\mu$ are again the unit differentials along ξ^μ: they are *partial derivatives* of F,

$$f_{(m_\mu)} = \frac{1}{(m_\mu)!} \left. \frac{\partial^{|(m_\mu)|} F}{\partial \xi^{(m_\mu)}} \right|_{\xi = \xi(Q)}. \tag{6}$$

---

[1] Berz writes $[f]_n$ instead of my Df, because it describes an nth-order *equivalence class*. I choose to use "Df" and "df" instead, because Df also extends the concept of "derivative as tangent-map" to higher-order tangency, while df has the formal properties of an *infinitesimal*. The DA vector Df is also closely related to $j_x^n F$, the *n-jet prolongation* of the function $F(x)$; Omohundro[5] has shown that a jet-space is the natural geometric arena of perturbation theory.

2.4. Dual Differential Algebra

The differential algebra $_nD_v$ forms a vector space. Now every vector space $\mathcal{V}$ may be associated with a *dual vector space* $\mathcal{V}^*$ consisting of the set of all *linear functionals* from vectors in $\mathcal{V}$ to some scalar field (usually $\mathbf{R}$ or $\mathbf{C}$). The rather abstract concept of a "dual vector space" is actually used quite often in physics, albeit seldom by that name: in quantum mechanics, a "bra" is the dual of a "ket;" in solid-state, a "reciprocal lattice vector" is the dual of a lattice vector; in relativity, a "covariant vector" is the dual of a "contravariant vector." In each of the preceeding cases, a vector and a dual-vector ("covector") are combined *via* a bilinear *scalar product* operation, to yield a real or complex scalar. Unless $\mathcal{V}$ is a *metric* vector space,[2] it only makes sense to combine a vector with a covector; one cannot, for example combine two "bras" or two "kets" and get a legitimate scalar quantity. Since a DA is also a vector space, it is therefore natural to inquire what *DA-covectors* are; I will argue in section 3.4 that a *set of moments* of a distribution function through order n form a realization of a DA-covector.

I write D^*f for a *DA covector*, an element of the *dual differential-algebra* $_nD_v^*$. A real DA covector $D^*f \in {_nD_v^*}$ is a *linear map* which takes each $Dg \in {_nD_v}$ into a *real number*: the *scalar product* $D^*f \cdot Dg$ of D^*f and Dg. Therefore there must be some set of coefficients $f^{(m_\mu)}$ such that

$$D^*f \cdot Dg = \sum_{\substack{(m_\mu) \\ 0 \le |(m_\mu)| \le n}} f^{(m_\mu)} g_{(m_\mu)}, \tag{7}$$

where $g_{(m_\mu)}$ are the components of Dg in the basis generated by $d\xi^\mu$. Since $_nD_v^*$ is by definition a vector space of the same dimension as $_nD_v$, there must therefore be a set of linearly independent *dual-basis covectors*, $d^*\xi_{(n_\nu)}$, such that $d^*\xi_{(n_\nu)} \cdot d\xi^{(m_\mu)} = \delta^{(m_\mu)}_{(n_\nu)}$; in terms of the dual basis,

$$D^*f = \sum_{\substack{(m_\mu) \\ 0 \le |(m_\mu)| \le n}} f^{(m_\mu)} d^*\xi_{(m_\mu)}. \tag{8}$$

3. Moments

3.1. Distributions and Moments

For simplicity, I assume that the *Vlasov* ("collisionless") *approximation* is legitimate, *i.e.*, two-body correlations are assumed negligible, so that I only need consider the *single-particle distribution function, $f(\boldsymbol{\xi};t)$*.[3]

The expectation value of a function $G(\boldsymbol{\xi})$ with respect to f is a *linear functional* defined by

$$\langle G \rangle_{f(t)} := \int G(\boldsymbol{\xi}) \, f(\boldsymbol{\xi};t) \, d^v\xi, \tag{9}$$

---

[2] A metric provides a natural map between a vector and its dual, inducing a natural scalar product (the "inner product") between two vectors.

[3] The notation $f(\boldsymbol{\xi};t)$ for the distribution function *appears* to conflict with my "upper-case" convention for functions; this conflict is removed once one realizes that f truly is a *distribution* in the "generalized function" sense [7], and it lives in the *dual* of the vector space of functions over phase space, with $\boldsymbol{\xi}$ serving as a *continuous index*. That the distribution f is a generalized function may easily be seen once one recognizes that f is not itself an observable, but that *integrals* of f over "test functions" *are* observables; this is in essence what one is doing when one makes a "slit and wire-scanner" measurement of emittance.

where $d^v\xi$ is the phase-space volume element; in other words, $\langle\cdot\rangle_f$ is a *linear map* of the *function* G into the *real number* $\langle G\rangle_f$. Consider the vector space $\mathcal{V}$ consisting of the set of all functions over phase space having finite expectation value; let $P^A(\boldsymbol{\xi})$ be a basis for $\mathcal{V}$.[4] The *moments* of $f(\boldsymbol{\xi};t)$ with respect to the basis functions $P^A(\boldsymbol{\xi})$ are defined as the expectation values of the P^A Since the basis P^A spans a linear vector subspace of the functions over phase space, and since $\langle\cdot\rangle_f$ is an element of the dual of the functions over phase space, the set $\langle P^A\rangle_f$ spans the dual of the vector space spanned by P^A.

3.2. Transportation of Moments

In the Vlasov approximation, the distribution functions at any pair of times t_1 and t_2, $f(\boldsymbol{\xi}_1;t_1)$ and $f(\boldsymbol{\xi}_2;t_2)$ (abbreviated f_1 and f_2), can be related by demanding *conservation of particle number* under the *time-evolution map*, $\boldsymbol{\xi}_2 = \mathcal{M}\boldsymbol{\xi}_1$. If the number of particles in a volume element $d^v\xi_1$ is to equal that in its image $d^v\xi_2$ under $\mathcal{M}$, then

$$f(\boldsymbol{\xi}_2;t_2)\,d^v\xi_2 = f(\boldsymbol{\xi}_1;t_1)\,d^v\xi_1; \tag{10}$$

equation (10) will be satisfied if

$$f(\boldsymbol{\xi}_2;t_2) = \left|\frac{\partial\boldsymbol{\xi}_1}{\partial\boldsymbol{\xi}_2}\right| f(\mathcal{M}^{-1}\boldsymbol{\xi}_2;t_1), \tag{11}$$

where $|\partial\boldsymbol{\xi}_1/\partial\boldsymbol{\xi}_2| = |\partial\mathcal{M}^{-1}(\boldsymbol{\xi}_2)|$ is the Jacobian determinant of the inverse map, $\mathcal{M}^{-1}$, at the point $\boldsymbol{\xi}_2$. The solution (11) can be shown to be the complete solution to the Vlasov equation, *independent* of whether or not $\mathcal{M}$ is symplectic, *i.e.*, even for *dissipative systems*; it only requires that $\mathcal{M}$ be invertible. If $\mathcal{M}$ is *volume-preserving*, then $|\partial\mathcal{M}^{-1}(\boldsymbol{\xi}_2)|$ equals unity; symplectic maps are a special case of volume-preserving maps.

In principle, given f_1 at t_1, I could use the solution (11) to directly calculate $\langle P^A\rangle_{f_2}$ at any time t_2; however unless f_1 and $\mathcal{M}$ are very simple, the integral (9) will generally be intractable. Furthermore, (11) requires a precise knowledge of f_1 *as a function*; however by hypothesis, my knowledge of f_1 is incomplete — I only know the *equivalence class* specified by a set of its *initial moments*.

Suppose I instead use (10) and $\mathcal{M}$ to express the final moments in terms of the initial distribution:

$$\langle P^A\rangle_{f_2} := \int P^A(\boldsymbol{\xi}_2)\,f(\boldsymbol{\xi}_2;t_2)\,d^v\xi_1 \equiv \int P^A(\mathcal{M}\boldsymbol{\xi}_1)\,f(\boldsymbol{\xi}_1;t_1)\,d^v\xi_1. \tag{12}$$

Note that equation (12) is *exact*; no expansions or approximations have yet been made.

Now by hypothesis, the set of basis-functions P^A is complete; therefore I can expand the *transformed* set of basis-functions $P^A(\mathcal{M}\boldsymbol{\xi})$ as a *linear combination* of the *initial* set of basis-functions $P^B(\boldsymbol{\xi})$ (recall that A and B are *indices* labeling elements of the set of basis functions),

$$P^A(\mathcal{M}\boldsymbol{\xi}) = \sum_B \mathcal{L}_B^A[\mathcal{M}]\,P^B(\boldsymbol{\xi}), \tag{13}$$

where the coefficients $\mathcal{L}_B^A[\mathcal{M}]$ depend on $\mathcal{M}$, *but not on* f_1. Since $\langle\cdot\rangle_f$ is a *linear* functional, it therefore follows that

$$\langle P^A\rangle_{f_2} = \sum_B \mathcal{L}_B^A[\mathcal{M}]\,\langle P^B\rangle_{f_1}. \tag{14}$$

---

[4] Since the specific basis I shall be most concerned with is the set of *monomials in the phase-space variables*, this means I must assume that f decreases more rapidly than any power of $|\boldsymbol{\xi}|^{-1}$ at large $|\boldsymbol{\xi}|$.

The transformation law (14) states that a complete set of moments transforms *linearly* under $\mathcal{M}$, *i.e.*, like the *components of a vector*, with $\mathcal{L}_B^A[\mathcal{M}]$ as the transformation matrix [8, 19]. It may no doubt seem quite odd that $\mathcal{M}$, even when highly nonlinear, should still act *linearly* on the space of moments; however there is really no contradiction here, because $\mathcal{L}_B^A[\mathcal{M}]$ is a highly nonlinear (but calculable!) function of $\mathcal{M}$. Because this result is so counterintuitive, in the appendix I present a simple one-dimensional example illustrating the distinctive features of moment transport.

Again, note that (14) is true *independent* of whether or not $\mathcal{M}$ is symplectic, *i.e.*, even for *dissipative systems*. Since moment transport does not require $\mathcal{M}$ to be symplectic, moments may therefore prove useful in studies of space-charge-induced emittance-growth or synchrotron-radiation-induced emittance-damping.

3.3. Expansion in Monomials

Let $\mathcal{V}$ be the set of real-analytic functions on phase space. Let me choose the set of *monomials* of phase-space variables, (1), as my basis functions. Expand $\xi_2^\mu = \mathcal{M}\xi_1^\mu$ as a power series in $\boldsymbol{\xi}_1$

$$
\begin{aligned}
\xi_2^\mu &= \sum_{k=0}^{\infty} \sum_{1 \leq \nu_1 \leq \cdots \leq \nu_k \leq v} M_{\nu_1 \cdots \nu_k}^\mu \, \xi^{\nu_1} \cdots \xi^{\nu_k} \\
&= \sum_{n_1 \cdots n_v} M_{n_1 \cdots n_v}^\mu \, (\xi^1)^{n_1} \cdots (\xi^v)^{n_v} \\
&= \sum_{(n_\nu)} M_{(n_\nu)}^\mu \, \xi^{(n_\nu)},
\end{aligned}
\tag{15}
$$

where $M_{(n_\nu)}^\mu$ are the series coefficients (a.k.a. "TRANSPORT-matrix-elements"). Inserting (15) and particle-conservation into the definition of the final moments,

$$
\begin{aligned}
\langle \xi_2^{(m_\mu)} \rangle_{f_2} &:= \int \prod_{\mu=1}^{v} (\xi_2^\mu)^{m_\mu} \, f(\boldsymbol{\xi}_2; t_2) \, d^v\xi_2 \\
&= \int \prod_{\mu=1}^{v} (\mathcal{M}\xi_1^\mu)^{m_\mu} \, f(\boldsymbol{\xi}_1; t_1) \, d^v\xi_1 \\
&= \int \prod_{\mu=1}^{v} \Big(\sum_{(k_\kappa)} M_{(k_\kappa)}^\mu \, \xi^{(k_\kappa)} \Big)^{m_\mu} f(\boldsymbol{\xi}_1; t_1) \, d^v\xi_1.
\end{aligned}
\tag{16}
$$

Equation (16) reduces, after expanding the product and collecting like terms, to

$$
\begin{aligned}
\langle \xi_2^{(m_\mu)} \rangle_{f_2} &= \sum_{(n_\nu)} \mathcal{L}_{(n_\nu)}^{(m_\mu)}[\mathcal{M}] \int \xi_1^{(n_\nu)} f_1 \, d^v\xi_1 \\
&= \sum_{(n_\nu)} \mathcal{L}_{(n_\nu)}^{(m_\mu)}[\mathcal{M}] \, \langle \boldsymbol{\xi}_1^{(n_\nu)} \rangle_{f_1},
\end{aligned}
\tag{17}
$$

where $\mathcal{L}_{(n_\nu)}^{(m_\mu)}[\mathcal{M}]$ is a messy nonlinear combination of the $M_{(n_\nu)}^\mu$, obtained by collecting like terms of the product in equation (16).

Note that for moments with respect to the monomial basis, the Taylor-series is the *natural representation* of $\mathcal{M}$, whereas the Lie-series representation is quite *unnatural*. This strongly suggests that DA methods will be quite useful in moment propagation, as indeed I shall show.

Note also the much less pleasant fact that the sums in (16) and (17) are *unrestricted*: *each final moment* depends on *every initial moment*. Unlike a DA, moments feed "down" as well as "up;" the moment series cannot be truncated at an arbitrary point without loss of accuracy, so questions of consistency and convergence arise. I shall address such questions in detail later; for now, let me just note that when collecting like terms in (16) to obtain the induced-transformation coefficient $\mathcal{L}_{(n_\nu)}^{(m_\mu)}[\mathcal{M}]$, for a given (n_ν), *only those Taylor-series coefficients $M_{(k_\kappa)}^\mu$ such that $|(k_\kappa)| \leq |(n_\nu)|$ contribute*. This observation will be crucial in demonstrating the consistency of the DA version of moment-transport.

3.4. Expansion in DA

Despite the messy nature of $\mathcal{L}_{(n_\nu)}^{(m_\mu)}[\mathcal{M}]$ I won't need an explicit representation of it; because DA multiplication is homomorphic to polynomial multiplication modulo polynomials of order greater than n, I can simply let DA do the bookkeeping!

By what I call the "fundamental theorem of DA," the DA-extension of $\xi_2^\mu = \mathcal{M}\xi_1^\mu$ from $\mathcal{M}: \mathbf{R}^\nu \to \mathbf{R}^\nu$ to $\mathcal{M}: {}_nD_\nu^\nu \to {}_nD_\nu^\nu$ is given by

$$D\xi_2^\mu = \mathcal{M}\,D\xi_1^\mu = \sum_{\substack{(n_\nu) \\ 0 \leq |(n_\nu)| \leq n}} M_{(n_\nu)}^\mu \, d\xi_1^{(n_\nu)}; \tag{18}$$

if $d\xi^\mu$ are the *unit differentials* at $\boldsymbol{\xi}_1$, then the coefficients $M_{(n_\nu)}^\mu$ will be same as those appearing in (15). (Since I am now working in a finite-dimensional DA, henceforth I shall always assume the above restriction on the range of summations.) It therefore follows that a *final* DA monomial $D\xi_2^{(m_\mu)}$ may be expanded in terms of the *initial* DA monomials $D\xi_1^{(n_\nu)}$, $\forall\,(n_\nu)$:

$$\begin{aligned}
D\xi_2^{(m_\mu)} \;\; &:= \;\; \prod_{\mu=1}^{\nu} (D\xi_2^\mu)^{m_\mu} = \prod_{\mu=1}^{\nu} (\mathcal{M}D\xi_1^\mu)^{m_\mu} \\
&= \;\; \prod_{\mu=1}^{\nu} \left(\sum_{(k_\kappa)} M_{(k_\kappa)}^\mu \, d\xi_1^{(k_\kappa)} \right)^{m_\mu} \tag{19} \\
&= \;\; \sum_{(n_\nu)} d\xi_1^{(n_\nu)} \, d^*\xi_{1,(n_\nu)} \cdot \left\{ \prod_{\mu=1}^{\nu} \left(\sum_{(k_\kappa)} M_{(k_\kappa)}^\mu \, d\xi_1^{(k_\kappa)} \right)^{m_\mu} \right\} \tag{20} \\
&= \;\; \sum_{(n_\nu)} \mathcal{L}_{(n_\nu)}^{(m_\mu)}[\mathcal{M}] \, d\xi_1^{(n_\nu)}. \tag{21}
\end{aligned}$$

In equation (20), I have explicitly inserted the completeness identity, $\mathrm{Id}(\ldots) := \sum_{(n_\nu)} d\xi_1^{(n_\nu)} \, d^*\xi_{1,(n_\nu)} \cdot (\ldots) \equiv \{\text{the identity map}\}$, to symbolize collecting like terms of the product in (19) through order n; however this step does not represent any actual labour, *because DA multiplication does the required bookkeeping automatically*.

Now just as noted regarding (16), when collecting like terms in (19) to obtain $\mathcal{L}_{(n_\nu)}^{(m_\mu)}[\mathcal{M}]$, for a given (n_ν) *only those Taylor-series coefficients $M_{(k_\kappa)}^\mu$ such that $|(k_\kappa)| \leq |(n_\nu)|$ contribute;* therefore as long as $|(n_\nu)|$ is less than or equal to n, the order of the DA, *equation (19) will yield precisely the same coefficients appearing in (16).* Therefore, $D\xi_2^{(m_\mu)}$ is given in terms of $D\xi_1^{(m_\mu)}$ by

$$D\xi_2^{(m_\mu)} = \mathcal{M}D\xi_1^{(m_\mu)} = \sum_{(n_\nu)} \mathcal{L}_{(n_\nu)}^{(m_\mu)}[\mathcal{M}] \, d\xi_1^{(n_\nu)}, \tag{22}$$

58   *Nonlinear Problems in Accelerator Physics*

where $\mathcal{L}_{(n_\nu)}^{(m_\mu)}[\mathcal{M}]$ is the *same* messy nonlinear combination of the $M_{(n_\nu)}^\mu$ appearing in (17) for all $|(m_\mu)| \leq n$, $|(n_\nu)| \leq n$. Once I have $\mathcal{L}_{(n_\nu)}^{(m_\mu)}[\mathcal{M}]$, I may use it to determine a final moment $\langle \xi_2^{(m_\mu)} \rangle_{f_2}$ from the set of initial moments $\langle \xi_1^{(m_\mu)} \rangle_{f_1}$, $\forall (m_\mu)$ as follows.

First form a "dual" DA vector D^*f_1 whose components are the moments of the initial distribution:

$$D^*f_1 := \sum_{(m_\mu)} \langle \xi_1^{(m_\mu)} \rangle_{f_1}\, d^*\xi_{1,(m_\mu)}. \tag{23}$$

Now take the scalar product of D^*f_1 with $D\xi_2^{(m_\mu)} \equiv \mathcal{M}D\xi_1^{(m_\mu)}$, (eqn. (22)):

$$\begin{aligned}
D^*f_1 \cdot \mathcal{M}D\xi_1^{(m_\mu)} &= \left[\sum_{(n_\nu)} \langle \xi_1^{(n_\nu)} \rangle_{f_1}\, d^*\xi_{1,(n_\nu)} \right] \cdot \left[\sum_{(\ell_\lambda)} \mathcal{L}_{(\ell_\lambda)}^{(m_\mu)}[\mathcal{M}]\, d\xi_1^{(\ell_\lambda)} \right] \\
&= \sum_{(n_\nu)}\sum_{(\ell_\lambda)} \mathcal{L}_{(\ell_\lambda)}^{(m_\mu)}[\mathcal{M}]\, \langle \xi_1^{(n_\nu)} \rangle_{f_1}\, d^*\xi_{1,(n_\nu)} \cdot d\xi_1^{(\ell_\lambda)} \\
&= \sum_{(n_\nu)}\sum_{(\ell_\lambda)} \mathcal{L}_{(\ell_\lambda)}^{(m_\mu)}[\mathcal{M}]\, \langle \xi_1^{(n_\nu)} \rangle_{f_1}\, \delta_{(n_\nu)}^{(\ell_\lambda)} \\
&= \sum_{(n_\nu)} \mathcal{L}_{(n_\nu)}^{(m_\mu)}[\mathcal{M}]\, \langle \xi_1^{(n_\nu)} \rangle_{f_1} \tag{24}
\end{aligned}$$

— but this is just the sum (17), terminated at order n !!!

Therefore, to the extent that the finite-order result (24) obtained *via* DA accurately approximates the infinite-order result (17), I can compute any final moment $\langle \xi_2^{(m_\mu)} \rangle_{f_2}$ simply by evaluating the scalar product $D^*f_1 \cdot D\xi_2^{(m_\mu)} \equiv D^*f_1 \cdot \mathcal{M}D\xi_1^{(m_\mu)}$.

I do not wish to minimize the fact that there is a significant conceptual difference between the infinite-order sum (17) and the finite-order sum (24). While the monomials $\xi^{(m_\mu)}$ through infinite order form a complete set of functions on $\mathcal{V}$, through finite order they do not; therefore one might rightfully be concerned about the consistency of (24). The key is to realize that the differentials $d\xi^{(m_\mu)}$ *do* form a complete set on $_nD_v$, and obey an algebra homomorphic to that of $\xi^{(m_\mu)}$ so long as the total order of the product does not exceed n. It is this homomorphism which allows me to compute $\mathcal{L}_{(n_\nu)}^{(m_\mu)}[\mathcal{M}]$ exactly through order n. The only remaining questions are those regarding accuracy of the truncated series, and speed of convergence; regarding this, I submit that there is no *fundamental* difference between pushing *moments* through a truncated "moment map" and pushing *particles* through a truncated *Taylor-series* ("Transport") map — as is often done, especially for "single-pass" systems where loss of symplecticity after many iterations is not a problem. In fact, for a perfectly localized initial particle (delta-function distribution!) finding the truncated first-order final moments is *identical* to pushing the particle through the truncated map! Nor do I see any *a priori* reason why calculating, say, the mean-square emittance by pushing *moments* through a truncated map should be any more objectionable than the common practice of pushing a large number of *particles* through a truncated map and averaging — especially since the moment method will not be subject to statistical fluctuations.[5]

---

[5] This is in fact the reason why I first pursued moment methods using DA [6]: I was attempting to minimize the RMS beam-divergence produced by a nonlinear optical element using a monte-carlo-generated initial beam; I found that the solution was sensitive to statistical fluctuations in the beam's halo. By contrast, the moment-transport method produced a good focus using an order of magnitude less computer time.

In summary, to determine a final moment $\langle \xi_2^{(m_\mu)} \rangle_{f_2}$ from the set of initial moments $\langle \xi_1^{(m_\mu)} \rangle_{f_1}$, $\forall(m_\mu)$, first form the "dual" DA vector D^*f_1 whose components are the moments of the initial distribution; Next, evaluate the product $D\xi_2^{(m_\mu)} = \prod_{\mu=1}^{v} (D\xi_2^\mu)^{m_\mu} \equiv \prod_{\mu=1}^{v} (\mathcal{M} D\xi_1^\mu)^{m_\mu}$; its components are the induced-transformation coefficients $\mathcal{L}_{(n_\nu)}^{(m_\mu)}[\mathcal{M}]$ for fixed (m_μ) and all (n_ν). Finally, evaluate the "scalar product"

$$\langle \xi_2^{(m_\mu)} \rangle_{f_2} = D^*f_1 \cdot \mathcal{M} D\xi_1^{(m_\mu)} \equiv D^*f_1 \cdot D\xi_2^{(m_\mu)} = \sum_{(n_\nu)} f_1^{(n_\nu)} \, d^*\xi_{1,(n_\nu)} \cdot D\xi_2^{(m_\mu)}; \qquad (25)$$

to obtain the component $\langle \xi_2^{(m_\mu)} \rangle_{f_2}$ of D^*f_2 obtained by pushing D^*f_1 through $\mathcal{M}$.

I find it highly suggestive that the *monomial-basis* moment-vectors (integrals!) transform like the *duals* of DA vectors (differentials!). This dual relationship might perhaps find a firmer mathematical footing within the larger framework of "almost-finite numbers" into which DA may be imbedded [3]. In the almost-finite numbers, multiplicative inverses of differentials are defined (integrals?), as well as distributions such as the Dirac "delta function." given that distributions are known to live in the dual of function space, it seems highly likely that almost-finite numbers might provide a unified framework for DA-maps and dual-DA moments.

3.5. Computational cost of Moment Transport

The computational cost of pushing a single moment through a map is essentially the cost of forming the monomial $D\xi_2^{(m_\mu)}$; since each DA-multiplication costs $_nN_{2v}$ real-valued multiplications and additions, the additional $_nN_v$ real-valued multiplications and additions required to evaluate the "scalar product" represent a relatively small overhead. Table 1 compares the relative costs of pushing a *second-order moment* through a map, $\propto (_nN_{2v} + {}_nN_v)$, to that of pushing a *single particle* through a map, $\propto v(_nN_v)$.

		v					
		1	2	3	4	5	6
	1	2.50	1.33	0.92	0.70	0.57	0.48
	2	3.00	1.75	1.27	1.00	0.83	0.71
	3	3.50	2.25	1.73	1.43	1.22	1.07
	4	4.00	2.83	2.33	2.02	1.79	1.61
n	5	4.50	3.50	3.08	2.80	2.58	2.40
	6	5.00	4.25	4.00	3.83	3.67	3.52
	7	5.50	5.08	5.10	5.13	5.11	5.06
	8	6.00	6.00	6.40	6.75	7.00	7.16
	9	6.50	7.00	7.92	8.75	9.43	9.95
	10	7.00	8.08	9.67	11.18	12.50	13.63

Table 1. Table of the ratio of the number of multiplications required to push a second-order moment through a map to that of a single particle. The order and number of variables range through $n = 1$–10 and $v = 1$–6, respectively.

3.6. Comparison of the "Scalar Product" and Composition

It is illuminating at this point to compare the "scalar product" to DA composition ("change of basis") under the mapping $\boldsymbol{\xi}_2 = \mathcal{M}\boldsymbol{\xi}_1$. First, examine a single component (*i.e.*, a single monomial) of the DA covector D^*f_2 formed from moments taken in the $\boldsymbol{\xi}_2$-basis; its expansion in terms of D^*f_1 formed from moments taken in the $\boldsymbol{\xi}_1$-basis is given by the "scalar product" formula:

$$
\begin{aligned}
D^*f_2 \cdot d\xi_2^{(n_\nu)} &= D^*f_1 \cdot \left[\mathcal{M} d\xi_1^{(n_\nu)} \right] \\
&= D^*f_1 \cdot \sum_{(m_\mu)} d\xi_1^{(m_\mu)} \left[d^*\xi_{1,(m_\mu)} \cdot \mathcal{M} d\xi_1^{(n_\nu)} \right] \\
&= \sum_{(m_\mu)} \left[D^*f_1 \cdot d\xi_1^{(m_\mu)} \right] \left[d^*\xi_{1,(m_\mu)} \cdot \mathcal{M} d\xi_1^{(n_\nu)} \right] \\
&= \sum_{(m_\mu)} f_1^{(m_\mu)} \mathcal{L}_{(m_\mu)}^{(n_\nu)}[\mathcal{M}].
\end{aligned}
\tag{26}
$$

On the other hand, the DA vector $Dg_2 = g_{2,(n_\nu)} d\xi_2^{(n_\nu)}$ transforms by *DA-composition* under the change of coordinate basis $d\boldsymbol{\xi}_2 = \mathcal{M}d\boldsymbol{\xi}_1$ to yield the components: [6]

$$
\begin{aligned}
d^*\xi_{1,(m_\mu)} \cdot Dg_1 &= d^*\xi_{1,(m_\mu)} \cdot \left[Dg_2 \circ \mathcal{M} d\boldsymbol{\xi}_1 \right] \\
&= d^*\xi_{1,(m_\mu)} \cdot \sum_{(n_\nu)} \mathcal{M} d\xi_1^{(n_\nu)} \left[d^*\xi_{2,(n_\nu)} \cdot Dg_2 \right] \\
&= \sum_{(n_\nu)} \left[d^*\xi_{1,(m_\mu)} \cdot \mathcal{M} d\xi_1^{(n_\nu)} \right] \left[d^*\xi_{2,(n_\nu)} \cdot Dg_2 \right] \\
&= \sum_{(n_\nu)} \mathcal{L}_{(m_\mu)}^{(n_\nu)}[\mathcal{M}] \, g_{2,(n_\nu)}.
\end{aligned}
\tag{27}
$$

Equation (26) and (27) show that *the scalar product is invariant under* $\mathcal{M}$:

$$
\begin{aligned}
D^*f_2 \cdot Dg_2 &= f_2^{(n_\nu)} g_{2,(n_\nu)} = \left[f_1^{(m_\mu)} \mathcal{L}_{(m_\mu)}^{(n_\nu)}[\mathcal{M}] \right] g_{2,(n_\nu)} \\
&= f_1^{(m_\mu)} \left[\mathcal{L}_{(m_\mu)}^{(n_\nu)}[\mathcal{M}] \, g_{2,(n_\nu)} \right] = f_1^{(m_\mu)} g_{1,(m_\mu)} = D^*f_1 \cdot Dg_1.
\end{aligned}
\tag{28}
$$

From the above one may draw the following three conclusions:

- First, (26) shows that DA covectors transform similarly to a "push-forward" under a change of basis, whereas (27) shows that DA vectors transform similarly to a "pull-back" — again consistent with their roles as duals [9, pp. 265 *et. seq.*].

- Second, the components of the scalar product resemble those of composition turned "inside out" — in the former, one sums the product of each component of D^*f with the *corresponding component* of a *particular monomial,* whereas in the latter, one sums the product of each component of Dg with a a *particular component* of the *corresponding monomial. The same transformation tensor* $\mathcal{L}_{(m_\mu)}^{(n_\nu)}[\mathcal{M}]$ *appears in both transformation laws.*

---

[6] The DA-composition operation may be understood as *direct substitution* of a new set of differentials in place of the unit differentials $d\xi^\mu$ appearing in the multinomial expansion (4); the result may be viewed as a *change of variables.* This may be seen in the second line of (27), where $d^*\xi_{2,(n_\nu)}$ picks off the coefficient $g_{2,(n_\nu)}$ of $d\xi_2^{(n_\nu)}$ in Dg_2, which is then multiplied by $\mathcal{M} d\xi_1^{(n_\nu)}$. Formally, composition is a binary operation $\circ : {}_nD_q^r \times {}_nD_p^q \to {}_nD_p^r$; *i.e.*, if $Dx \in {}_nD_q^r$, and $Dy \in {}_nD_p^q$, then $Dx \circ Dy \in {}_nD_p^r$.

- Finally, one notes that it takes *precisely* the same number of operations to compute all of the components of the moment-vector D^*f_2 as it does to compute all of the components of the composition $Dg_1 = Dg_2 \circ \mathcal{M}d\boldsymbol{\xi}_1$; the same operations are performed, merely in a different order.

4. Comments on Polynomial Moments

In contradistinction to DA vectors, an mth-order moment in general depends on *all* other moments — *i.e.*, "feed-down" occurs, in addition to "feed-up;" this raises entirely legitimate concerns about the convergence of the entire moment method, on which I have not yet made any detailed studies. One can obtain some crude ideas about rate of convergence from the simple 1-D example in the appendix. Let σ be an estimate of the "scale" or "physical size" of the distribution (*e.g.*, beam-spot size, or mean-square emittance). Assume the Taylor coefficients behave no worse than the rather extreme example of the appendix, $M_k \sim R^{-k}/k!$, where R is an measure of the "scale" of the map (*e.g.*, a typical magnet aperture). Then one can deduce the following asymptotic behaviors for four commonly assumed distributions:

1.) *Compact distributions* — The moments asymptotically scale *geometrically* with order: $\langle \xi^k \rangle_f \sim \sigma^k$. The final-moment series converges absolutely, and about as fast as the series for $\exp(\sigma/R)$.

2.) *Gaussian-like* distributions — The moments asymptotically scale like $k!! \, \sigma^k$. The moment series again converges absolutely, but rather more slowly.

3.) *Exponential tails* — The moments asymptotically scale like $k! \, \sigma^k$. The moment series is now *conditionally convergent*: slow convergence if the contributions to the moment-map tend to *alternate* with order; divergence if they tend to be *"signed-definite."*

4.) *Power-law tails — Guaranteed divergence* at some finite order k, since by hypothesis $f \sim |\boldsymbol{\xi}|^{-k}$ asymptotically.

In the case of many variables, the scalar product splits up into a sum of orthogonal subspaces of dimension $_kN_{v-1}$ distinguished by the order k; hence the series may be expected to grow much more rapidly with order. However I suspect this growth will be partially offset in realistic accelerator-physics examples by the tendency for both $\langle \xi^{(n_\nu)} \rangle_f$ and $M^\mu_{(n_\nu)}$ to be "sparse" due to symmetry, and for $M^\mu_{(n_\nu)}$ to have an approximately equal number of positive and negative signs due to the symplectic condition; this is certainly an area where more detailed study is needed.

Regarding the above examples, one might legitimately object that *all* physical distributions must be compact. However, what if the beam has an extended "tail" or "halo?" Moments are notoriously sensitive to the tails of distributions; even if the halo is cut off at a finite size, it might still dominate the moments instead of the core, perhaps making the truncated moment-series ill-behaved. I rather expect that no matter what, the accuracy of the truncated series for higher-order moments will be worse than that of lower-order moments; fortunately, it is probably the first- and second-order moments which will be most important in practice.

In the event of convergence problems, at least four alternatives exist which may improve matters. First, one might shift $\mathcal{M}$ to make it *origin-preserving*, if it is not so already; one may then show that $\langle \boldsymbol{\xi}^{(n_\nu)} \rangle_f$ is independent of all lower order moments. Second, one might shift to *central moments,* $\mu^{(n_\nu)} := \langle (\boldsymbol{\xi} - \langle \boldsymbol{\xi} \rangle_f)^{(n_\nu)} \rangle_f$; now all of the final moments depend on all of the intial moments, however one expects the growth-rate of the moment series to be minimized. Third, *cumulants* might be much better behaved than moments, as I shall discuss in the next section. Finally, if the halo is so extended that all else fails, one might still apply *convergence acceleration methods,* such as *Euler's* or *Shank's transformation* [10]. Properly applied, acceleration methods can extract sensible results from even formally divergent series (improperly applied, they can also give garbage, so one must use them with caution).

In summary, I expect that moment methods will provide a good representation of "core-dominated" beams, but may have problems representing "halo-dominated" beams accurately. However I submit that halo-dominated beams are likely to cause problems for nearly *any* beam-representation method; they are not just a problem for moment-methods alone.

5. Cumulants

Cumulants [4] are homogenous polynomials of the moments possessing several advantages: first, all the cumulants are *translationally invariant* (except for the first-order cumulants, which are just the mean) Since they are homogenous polynomials of the moments, they transform identically to the corresponding moments under changes of scale. The second cumulant is just the familiar *sigma matrix*. Finally, the third and higher-order cumulants have the useful property of *vanishing identically* if, and only if, the distribution is a *Gaussian;* the cumulants thus provide a measure of how much a distribution deviates from being a Gaussian.

I shall now show, using the definitions of *cumulant-* and *moment-generating-functions,* that interconverting moments and cumulants is a trivial problem using DA. The moment generating-function (MGF) is defined by

$$\phi(\boldsymbol{s}) = \sum_{(m_\mu)} \frac{1}{(m_\mu)!}\, s_{(m_\mu)} \langle \xi^{(m_\mu)} \rangle_f, \tag{29}$$

where $s_{(m_\mu)} := \prod_{\mu=1}^{v} (s_\mu)^{m_\mu}$, and the s_μ are a set of v order-parameters introduced for bookkeeping purposes; one may think of the s_μ as arising from taking the Fourier- or two-sided Laplace-transform of $f(\boldsymbol{\xi})$, so that in some sense $\boldsymbol{s}$ is "conjugate" to $\boldsymbol{\xi}$.

The *cumulant generating function* (CGF) is the *logarithm* of the moment generating function. One obtains the cumulants from the moments by taking *derivatives* of the CGF with respect to the s_μ at the origin, $\boldsymbol{s} = 0$:

$$\sum_{(m_\mu)} \frac{1}{(m_\mu)!}\, s_{(m_\mu)} \kappa^{(m_\mu)} = \ln \left\{ \sum_{(m_\mu)} \frac{1}{(m_\mu)!}\, s_{(m_\mu)} \langle \xi^{(m_\mu)} \rangle_f \right\}, \tag{30}$$

$$\kappa^{(m_\mu)} = \frac{\partial^{|(m_\mu)|}}{\partial s_{(m_\mu)}} \ln \phi(\boldsymbol{s}) \Big|_{\boldsymbol{s}=0} \tag{31}$$

— but this was exactly what DA was invented for! Likewise, the moments may be obtained from the cumulants by differentiation of $\exp\{\sum_{(m_\mu)} s_{(m_\mu)} \kappa^{(m_\mu)}/(m_\mu)!\}$ at $s = 0$.

Note that since the cumulants can be shown to be translationally independent, the moments may be replaced by *central* moments throughout the above, with absolutely no change in the derivation.

As an example of how cumulants and moments are related, I shall list below the connection between the first few cumulants and central moments for a function of two variables:

$$
\begin{aligned}
\kappa^{(0,1)} &= \mu^{(0,1)} \qquad \kappa^{(0,2)} = \mu^{(0,2)} \qquad \kappa^{(0,3)} = \mu^{(0,3)} \qquad \kappa^{(1,2)} = \mu^{(1,2)} \\
\kappa^{(0,4)} &= \mu^{(0,4)} - 3(\mu^{(0,2)})^2 \\
\kappa^{(1,3)} &= \mu^{(1,3)} - 3\mu^{(1,1)}\mu^{(2,0)} \\
\kappa^{(2,2)} &= \mu^{(2,2)} - \mu^{(0,2)}\mu^{(2,0)} - 2(\mu^{(1,1)})^2 \\
\kappa^{(0,5)} &= \mu^{(0,5)} - 10\mu^{(0,2)}\mu^{(0,3)} \\
\kappa^{(1,4)} &= \mu^{(1,4)} - 4\mu^{(0,3)}\mu^{(1,1)} - 6\mu^{(0,2)}\mu^{(1,2)} \\
\kappa^{(2,3)} &= \mu^{(2,3)} - \mu^{(2,0)}\mu^{(0,3)} - 6\mu^{(1,1)}\mu^{(1,2)} - 3\mu^{(0,2)}\mu^{(2,1)} \\[2ex]
\mu^{(0,4)} &= \kappa^{(1,3)} + 3(\kappa^{(0,2)})^2 \\
\mu^{(1,3)} &= \kappa^{(1,3)} + 3\kappa^{(0,2)}\kappa^{(1,1)} \\
\mu^{(2,2)} &= \kappa^{(2,2)} + \kappa^{(0,2)}\kappa^{(2,0)} + 2(\kappa^{(1,1)})^2 \\
\mu^{(0,5)} &= \kappa^{(0,5)} + 10\kappa^{(0,2)}\kappa^{(0,3)} \\
\mu^{(1,4)} &= \kappa^{(1,4)} + 4\kappa^{(0,3)}\kappa^{(1,1)} + 6\kappa^{(0,2)}\kappa^{(1,2)} \\
\mu^{(2,3)} &= \kappa^{(2,3)} + \kappa^{(2,0)}\kappa^{(0,3)} + 6\kappa^{(1,1)}\kappa^{(1,2)} + 3\kappa^{(0,2)}\kappa^{(2,1)}
\end{aligned}
\tag{32}
$$

One immediately notices two things about the cumulants: first, all the cumulants contribute to a moment via a *positive* sign; contrawise, all the *lower-order moments* contribute to a cumulant via a *negative* sign. These two observations, combined with the fact that the cumulants vanish *identically* for Gaussian distributions, hint that the cumulants of a "near-Gaussian" distributions might grow much more slowly with order than its moments.

6. Moment Brackets

Even when the underlying dynamical system is Hamiltonian, the moments of a distribution do not obey equations of motion having a Hamiltonian form, as can easily be verified by simply counting degrees of freedom. However an underlying Hamiltonian dynamics does induce a kind of "Hamiltonian-*like*" dynamics upon function(al)s of the phase-space variables, called "Lie-Poisson" dynamics [11, 17] *via* the *Lie-Poisson bracket*.

The Lie-Poisson bracket of two functionals $P[f]$ and $Q[f]$ is defined as

$$
[P, Q]_{LP}(f) := \int \left[\frac{\delta P}{\delta f(\boldsymbol{\xi})}, \frac{\delta Q}{\delta f(\boldsymbol{\xi})} \right] f(\boldsymbol{\xi}) \, d^v\xi,
\tag{33}
$$

where $\delta/\delta f(\boldsymbol{\xi})$ is the functional (Fréchet) derivative with respect to f, and $[\,\cdot\,,\,\cdot\,]$ is the usual Poisson-bracket. One can easily show that the LP-bracket inherits the bilinearity, derivation, and "Jacobi" properties of the Poisson-bracket:

$$[AB, CD]_{LP}(f) = AC[B, D]_{LP}(f) + AD[B, C]_{LP}(f)$$
$$+ BC[A, D]_{LP}(f) + BD[A, C]_{LP}(f) \tag{34}$$

$$[A(V), B(W)]_{LP}(f) = \frac{dA}{dU}\frac{dB}{dW}[V, W]_{LP}(f) \tag{35}$$

$$[A, [B, C]_{LP}]_{LP} + [B, [C, A]_{LP}]_{LP} + [C, [A, B]_{LP}]_{LP} \equiv 0. \tag{36}$$

From the above, one can easily show that the LP-bracket of two monomial moments is

$$\left[\langle x^i p_x^j \rangle_f, \langle x^k p_x^l \rangle_f \right]_{LP}(f) = (il - jk)\langle x^{i+k-1} p_x^{j+l-1} \rangle_f; \tag{37}$$

from which one can show that the LP-bracket defines a bilinear map of two DA covectors (moment-vectors!) onto DA covector space having precisely the same form as the Poisson-bracket map of two DA-vectors onto DA-vector space, and may be computed *via* the same algorithm. Therefore, one may easily use DA to integrate the induced Lie-Poisson dynamics for the moments [17] because it has the same "Heisenberg-like form" one is accustomed to seeing in the Lie-algebraic approach to mechanics:

$$\frac{d}{dt}\langle \xi^{(m_\mu)} \rangle_f = \left[\langle \xi^{(m_\mu)} \rangle_f, \langle H \rangle_f \right]_{LP}(f). \tag{38}$$

The "Lie-Poisson Heisenberg equation" (38) represents a possible starting point for a DA version of the "BEDLAM" space-charge dynamics code [12, 13]. Assume that one has somehow obtained a set of self-consistent space-charge potentials to build H with; this requires some method of extracting a smooth estimate of f from D^*f, a problem which I shall postpone until section 8. One may then use DA to expand H about some point $\boldsymbol{\xi}_0$, then perform a componentwise-multiplication of DH by the moments of f about $\boldsymbol{\xi}_0$ to yield $\langle H \rangle_f$:

$$\langle H \rangle_f = \left\langle \sum_{(n_\nu)} \frac{1}{(n_\nu)!} \left(\partial_{(n_\nu)} H \big|_{\xi=\xi_0} (\boldsymbol{\xi} - \boldsymbol{\xi}_0)^{(n_\nu)} \right) \right\rangle_f$$
$$= \sum_{(n_\nu)} \frac{1}{(n_\nu)!} \left(\partial_{(n_\nu)} H \big|_{\xi=\xi_0} \langle (\boldsymbol{\xi} - \boldsymbol{\xi}_0)^{(n_\nu)} \rangle_f = D^*f \cdot DH. \tag{39}$$

Lysenko and Channell [13] choose $\boldsymbol{\xi}_0 = \langle \boldsymbol{\xi} \rangle_f$, *i.e.*, the "centre of mass" frame, so that $\langle H \rangle_f$ is expanded in *central moments;* this choice leads to problems if one wishes to integrate (38) using a *Lie-Poisson integrator* [14]) (the LP-analogue of a symplectic integrator) as discussed by Scoval and Weinstein [15], who instead expand about a "test-particle" within the distribution.

Like the moment map itself, the LP-Heisenberg equation (38) also suffers from "feed down," because $\langle H \rangle_f$ is a sum of moments of all orders; furthermore, equation (37) shows that if one simply truncates D^*f and $\langle H \rangle_f$ at order n, the equation of motion for the moment-vector D^*f would still depend on moments through order $2n - 2$. However as discussed in Lysenko and Channell [13] and Scoval and Weinstein [15], the quotient LP-bracket algebra formed by "modding out" (*i.e.*, setting to zero) all moments of order greater than n forms a closed algebra which still satisfies antisymmetry and the Jacobi identity; one can derive a consistent LP-integrator for this truncated algebra.

7. Moment Invariants

Given that there exist Hamiltonian-like equations of motion for the moments of a Hamiltonian system, one is naturally lead to inquire whether there are any moment *invariants*, *i.e.*, constants of the motion. For example, the *mean-squared emittance* is a *quadratic combination of moments*, and is known to be preserved for any *linear* symplectic map — it is a *kinematic invariant*. Can more general invariants be found? If so, then if one could construct a particle distribution completely characterized by some set of such generalized moment-invariants, it would necessarily remain invariant under transport, generalizing the concept of a "matched" beam [16].

It turns out that for *linear* maps, indeed there are more general kinematic invariants [16, 17, 18, 19]. Unfortunately, even for the "almost-linear symplectic maps" characteristic of a well-designed particle-optical system, no such invariant has yet been found — and perhaps may not even exist.[7] However approximate *dynamic* moment-invariants (invariants of a *particular* Hamiltonian or map) may perhaps exist, as demonstrated in [16].

As first shown in another form by Lysenko [16], and as extended by Rangarajan [18, 19], and by Holm, Lysenko, and Scoval [17], a "pure" $(2n)^{th}$-order kinematic invariant of the set of all symplectic linear maps, $Sp(2N, \mathbf{R})$, can be formed out of quadratic moments:

$$\mathcal{I}_{(2)}^{(2n)} := \langle \xi^{\mu_1} \xi^{\nu_1} J_{\nu_1 \mu_2} \rangle_f \langle \xi^{\mu_2} \xi^{\nu_2} J_{\nu_2 \mu_3} \rangle_f \cdots \langle \xi^{\mu_n} \xi^{\nu_n} J_{\nu_n \mu_1} \rangle_f \tag{40}$$

or symbolically, $\mathcal{I}_{(2)}^{(2n)} = \mathrm{Tr}[\langle \boldsymbol{\xi} \otimes (\mathbf{J} \cdot \boldsymbol{\xi})^{\mathrm{T}} \rangle_f]^n$, where $\mathbf{J}$ is the symplectic tensor, and "superscript-T" denotes transpose. The simplest such invariant, $\mathcal{I}_{(2)}^{(2)}$, generalizes the MS-emittance to fully coupled linear maps; it is a sum of mean-squared and "cross" emittances, which in two dimensions simplifies to

$$\begin{aligned}
\mathcal{I}_{(2)}^{(2)} &= (\langle x^2 \rangle_f \langle p_x^2 \rangle_f - \langle x p_x \rangle_f) + (\langle y^2 \rangle_f \langle p_y^2 \rangle_f - \langle y p_y \rangle_f) \\
&\quad + 2(\langle xy \rangle_f \langle p_x p_y \rangle_f - \langle x p_y \rangle_f \langle y p_x \rangle_f).
\end{aligned} \tag{41}$$

Rangarajan has generalized the "pure" $Sp(2N, \mathbf{R})$ quadratic invariants to pure and "mixed" invariants of arbitrary even order; he has also proved a remarkable theorem that only N of the moment invariants are algebraically independent. In addition, Rangarajan has studied the *dynamic invariants* defined symbolically by $\mathcal{J}_{(2)}^{(2n,m)}[\mathbf{M}] = \mathrm{Tr}[\langle \boldsymbol{\xi} \otimes (\mathbf{J} \cdot \mathbf{M}^m \cdot \boldsymbol{\xi})^{\mathrm{T}} \rangle_f]^n$, which depend upon (and are only invariant under) a *particular* linear map $\mathbf{M}$. Again, only N of the dynamic invariants are algebraically independent.

The linear-map invariants will in general cease to be invariant under nonlinear maps; however for "almost-linear maps," they may provide "slow" variables, changing only on a time-scale much longer than the betatron period of the beamline. Also, Lysenko [16] found that certain combinations of the "pure" invariants could be made insensitive to a *class* of nonlinear maps; for example, Lysenko observed that for the class of "cubic kicks,"

$$x_2 = x_1, \tag{42}$$

$$p_{x,2} = p_{x,1} + \alpha x_1 + \beta x_1^3, \tag{43}$$

---

[7] Neri [20] has argued that *exact* kinematic invariants of the group of analytic symplectic maps cannot exist, because the Killing metric of this group is not invertible.

although $\mathcal{I}_{(2)}^{(2)}$ and $\mathcal{I}_{(2)}^{(4)}$ both depend on β at order $\mathcal{O}(\beta^4)$, this leading behavior can be canceled through order $\mathcal{O}(\beta^8)$ by forming the combination of invariants

$$\mathcal{J}_{(4)} := \mathcal{I}_{(2)}^{(4)} - 10\big(\mathcal{I}_{(2)}^{(2)}\big)^2. \tag{44}$$

Numerical tests showed that (44) is indeed more nearly preserved by "linear plus cubic" forces than any other $\mathcal{I}_{(2)}^{(2n)}$ studied, and is thus a candidate for a "slower" variable. Lysenko's procedure of canceling leading-order nonlinearities strongly resembles the construction of a perturbative invariant by successive canonical transformations of the Hamiltonian ("normalization"), suggesting that one might fruitfully seek a moment-method to normalize (38) analogous to the normal-form methods which have proved so useful in the study of maps [21].

8. The Inverse Problem: Reconstructing f from its Moments

I have presented a method of transporting beam moments from the intial to the final state; one might now ask: "what do these moments *mean?*" As it currently stands, the transported moments are just a collection of numbers, and as quoted earlier, the purpose of computation is *insight,* not numbers. What can one learn from the moments? In particular, is it possible to solve the *inverse* moment problem: to *reconstruct* an f, given its moments? This classic *problem of moments* has a long history in mathematics; for reviews, see [22] and [23].

First of all, note that a *unique f* corresponding to a sequence of moments *may* exist if *all* moments exist and are finite; however this condition, while necessary, is not sufficient [4, probs. 3.12 and 6.21]. Certainly if only a *finite* number of moments are given, one can at best determine an *equivalence-class* of distributions. For many applications it is nevertheless desirable to select a *"plausible"* distribution-function consistent with the given moments; in particular, an *analytic* distribution function will be essential if one wishes to develop a moment-and-DA-based space-charge algorithm.

I am currently investigating two methods of beam reconstruction: "Maximum-Entropy" (MaxEnt), and "Gaussian Beamlets." I shall first briefly discuss a third option which, despite its simplicity, has certain unphysical aspects which I feel are sufficient to merit its rejection.

8.1. The Charlie Ansatz

A simple ansatz allowing a reconstruction of f from its moments is the "Charlie ansatz:"

$$f(\boldsymbol{\xi}; t) = N_0 \exp\Big[-g_2(\boldsymbol{\xi}; t)\Big]\Big[1 + r_1(\boldsymbol{\xi}; t) + r_2(\boldsymbol{\xi}; t) + r_3(\boldsymbol{\xi}; t) + r_4(\boldsymbol{\xi}; t) + \ldots\Big]; \tag{45}$$

here N_0 is a normalization factor, g_2 is a homogenous time-dependent quadratic form, and the r_k are homogenous polynomials of order k in $\boldsymbol{\xi}$, with time-dependent coefficients.

The Charlie ansatz has been successfully employed by R. D. Ryne in his code Charlie [8], which used Lie algebraic methods to self-consistently calculate $\mathcal{M}$ for the assumed f including space-charge effects; it has also been used by van Zeijts and Neri in the code TLie [24], who carry it to very high order using DA methods.

The CHARLIE ansatz has several points in its favour: it is real-analytic; its self-fields can be calculated as a power-series near the origin; and — most important from the moment viewpoint — all its monomial-basis moments exist, can be calculated in closed form, and are *linear* in the coefficients of the r_k, so that in principle once g_2 is fixed (for example, $g_2(\boldsymbol{\xi}; t) = \frac{1}{2}\boldsymbol{\xi}\cdot\boldsymbol{\sigma}^{-1}(t)\cdot\boldsymbol{\xi}$, with $\boldsymbol{\sigma}$ the usual sigma-matrix) one can find the r_k from the moments by solving a set of linear equations.

Unfortunately, the CHARLIE ansatz also has several points against it:

- It is not automatically positive-semidefinite (as a legitimate f must be), regardless of how high an order it is carried to.

- Its form (when truncated to finite order) is not generally preserved under $\mathcal{M}$, so that normalization may be lost (unless reimposed by a constraint).

- Finally, its asymptotic behaviour is inflexible — it is still a power-series times a Gaussian, regardless of how high an order it is carried to; it may therefore have difficulty modeling the sharp-edged ("square") boundaries typical of space-charge dominated beams.

Therefore, despite its simplicity, I believe the CHARLIE ansatz should be rejected (however it provides an excellent initial approximation from which to solve the MAXENT method discussed in the next section).

8.2. MAXENT *Reconstruction*

E. T. Jaynes [25, 26] vigorously advocates dealing with incomplete information *via* the principle of "maximum-entropy inference," or MAXENT. In MAXENT, one maximizes *Shannon's information-theoretic entropy* [28] subject to the constraint of the given data. For discrete unknowns, MAXENT yields the unique "maximally non-committal" probability distribution consistent with the given information, and *only* the given information; "maximally non-committal" meaning that it assumes no parametric or statistical model, and makes no comment about any information not given. The continuum version of MAXENT with moment constraints is [25]:

$$S[f] = -\int_{\text{supp}\{g\}} f(\boldsymbol{\xi}) \ln\left[f(\boldsymbol{\xi})/g(\boldsymbol{\xi})\right] d^v\xi - \sum_{(m_\mu)} \lambda_{(m_\mu)}\left\{\langle \xi^{(m_\mu)}\rangle_f - \langle \xi^{(m_\mu)}\rangle_{f_0}\right\}, \qquad (46)$$

where the $\langle \xi^{(m_\mu)}\rangle_{f_0}$ are the constraining moments, the $\lambda_{(m_\mu)}$ are Lagrange multipliers, the density $g(\boldsymbol{\xi})$ is the *"a priori"* probability density, and $\text{supp}\{g\}$ is the support of g.[8]

[8] The *a priori* density g is an inelegant but inescapable element of nonuniqueness in the continuum problem. MAXENT can only be applied to measurable spaces; the measure g may be viewed as the "degeneracy of phase-space states per unit volume." In the absence of any constraining data, the MAXENT solution for f reduces to g, so it is also the state of "complete ignorance;" this requires g to be normalizable, and $\text{supp}\{g\}$ to have finite volume. Jaynes [27] argues that the *a priori* density is not completely arbitrary; as the state of "complete ignorance" it must be invariant under any symmetries or scale invariances of the problem ((*e.g.*, translations, rotations, change of units) — except perhaps on the boundary of $\text{supp}\{g\}$. If the constraints provide sufficient information, f is usually quite insensitive to g.

The *a priori* distribution g may also be used to incorporate "prior knowledge" about f, such as a known asymptotic behavior or boundary condition, *e.g.*, "the beam is confined to a pipe," or "the beam has a maximum permissible transverse momentum."

Variation with respect to f yields the well-known solution

$$f_{ME}(\boldsymbol{\xi}) = \frac{g(\boldsymbol{\xi})}{Z[\lambda_{(m_\mu)}; g]} \exp\Big[-\sum_{(m_\mu)} \lambda_{(m_\mu)}\, \xi^{(m_\mu)}\Big], \tag{47}$$

while variation with respect to $\lambda_{(m_\mu)}$ yields the moment constraints

$$\langle \xi^{(m_\mu)} \rangle_{f_0} = -\frac{\partial}{\partial \lambda_{(m_\mu)}} \ln Z[\lambda_{(m_\mu)}; g]; \tag{48}$$

the "partition functional" Z is defined by

$$Z[\lambda_{(m_\mu)}; g] := \int_{\mathrm{supp}\{g\}} g(\boldsymbol{\xi}) \exp\Big[-\sum_{(m_\mu)} \lambda_{(m_\mu)}\, \xi^{(m_\mu)}\Big]\, d^v\xi. \tag{49}$$

The MAXENT solution f_{ME} has a number of significant positive features to commend it. Like the CHARLIE ansatz, f_{ME} is *infinitely differentiable*; therefore DA techniques may be used to manipulate it — for example, to calculate the near-origin space-charge potential it produces. Furthermore, by construction it incorporates *all* the given information (the moment constraints and g) *identically*, while being "maximally non-committal" about anything else. Most important, it does not suffer from any of the defects of the CHARLIE ansatz:

- f_{ME} is everywhere positive-semidefinite *identically* — unlike the CHARLIE ansatz, it is impossible for f_{ME} to develop negative "holes," regardless of truncation order. Furthermore, f_{ME} vanishes everywhere that g vanishes, so "prior information" can be incorporated.

- f_{ME} has a form uniquely determined by the given information, rather than imposed *a priori*; furthermore, it is *guaranteed* to be normalized by construction.

- Finally, even though through a given order f_{ME} has exactly the same number of coefficients as the CHARLIE ansatz, its asymptotic behavior can fall off *much* more rapidly — for example, $\exp[-(r/R)^4]$ provides a remarkably good approximation to a "flat-topped spherical distribution," and might provide a good infinitely-differentiable model for a highly space-charge dominated beam.

The MAXENT solution has many other interesting properties. For example, since one can show that the moments and the Lagrange multipliers are *Legendre transforms* of each other [25],

$$\langle \xi^{(m_\mu)} \rangle_f = -\frac{\partial \ln Z}{\partial \lambda_{(m_\mu)}} \tag{50}$$

$$\Updownarrow$$

$$\lambda_{(m_\mu)} = \frac{\partial S}{\partial \langle \xi^{(m_\mu)} \rangle_f} \tag{51}$$

one can derive a number of useful identities, such as

$$-\frac{\partial^2 \ln Z}{\partial \lambda_{(m_\mu)} \partial \lambda_{(n_\nu)}} = \frac{\partial \langle \xi^{(m_\mu)} \rangle_f}{\partial \lambda_{(n_\nu)}} = \frac{\partial \langle \xi^{(n_\nu)} \rangle_f}{\partial \lambda_{(m_\mu)}} = \langle \xi^{(m_\mu)} \xi^{(n_\nu)} \rangle_f - \langle \xi^{(m_\mu)} \rangle_f \langle \xi^{(n_\nu)} \rangle_f \tag{52}$$

$$\frac{\partial^2 S}{\partial \langle \xi^{(m_\mu)} \rangle_f \partial \langle \xi^{(n_\nu)} \rangle_f} = \frac{\partial \lambda_{(n_\nu)}}{\partial \langle \xi^{(m_\mu)} \rangle_f} = \frac{\partial \lambda_{(m_\mu)}}{\partial \langle \xi^{(n_\nu)} \rangle_f} \tag{53}$$

and

$$\frac{\partial^2 \ln Z}{\partial \lambda_{(m_\mu)} \partial \lambda_{(k_\kappa)}} \frac{\partial^2 S}{\partial \langle \xi^{(k_\kappa)} \rangle_f \partial \langle \xi^{(n_\nu)} \rangle_f} \equiv -\delta^{(m_\mu)}_{(n_\nu)} \tag{54}$$

From eqn. (52) one sees that MAXENT provides a way of "analytically continuing" the moments to higher orders [29]:

$$\langle \xi^{(m_\mu)+(n_\nu)} \rangle_f = \langle \xi^{(m_\mu)} \xi^{(n_\nu)} \rangle_f = \langle \xi^{(m_\mu)} \rangle_f \langle \xi^{(n_\nu)} \rangle_f - \frac{\partial^2 \ln Z}{\partial \lambda_{(m_\mu)} \partial \lambda_{(n_\nu)}}. \tag{55}$$

Perhaps (55) might be used to provide an alternative method of truncating the LP-Heisenberg equations.

The primary negative feature of the MAXENT solution f_{ME} and the single remaining barrier to its application is the integral $Z[\lambda_{(m_\mu)}; g]$: it is impossible to evaluate analytically, and appears to be rather hard to accurately compute numerically — especially since in phase space one must perform a six-dimensional "hypercubature." Nor is the nonlinear system of constraint equations (48) particularly easy to solve. Nevertheless, the specific form of Z plus the power of DA might allow special integration methods to be developed; for example, one might perform a high-order Gauss-Hermite quadrature (in DA) after first transforming to the frame in which $\langle \xi^{\mu\nu} \rangle_{f_0}$ is diagonal (the "principle axis" transformation). Alternatively, one might use DA to expand f_{ME} from (47) to very high order, and convert to the CHARLIE ansatz form, eqn. (45); one may then sum the analytic integrals of its coefficients. Finally, though the CHARLIE ansatz has its deficiencies *as a distribution function*, its logarithm (in DA) may be expected to provide an excellent first approximation for $\lambda_{(m_\mu)} D\xi^{(m_\mu)}$, from which an iterative solution for (48) might be obtained.

8.3. Gaussian Beamlet Reconstruction

The method of *Gaussian beamlets* also appears to be a promising method of beam-reconstruction, free of the analytic intractability of MAXENT. This method uses a *regular lattice of variable-amplitude Gaussians*, which Berz, Hartmann, and Wollnik have shown to provide an ideal basis for representing smooth functions of coordinate-space [30, 31]. Furthermore, since the field of a Gaussian can be calculated in closed form, a "Poisson solver" is unnecessary. Gaussian beamlets have been successfully used to describe space-charge density in the codes BEDLAM [13] and TRIBO [31]. Can one apply Gaussian beamlets to the moment-to-beam reconstruction problem? I will argue that the "curse of dimensionality" prevents one from reconstructing a distribution on the *full* six-dimensional phase space; however reconstructions on two- or three-dimensional *projections* of phase space look feasible.

Let us examine some of the properties our hypothetical Gaussian-beamlet-derived distribution function will have: like the MAXENT solution, f_{GB} will be infinitely differentiable, and all its moments are calculable in *closed form*; *unlike* MAXENT, the total moment of f_{GB} will be a *linear* function of the beamlet moments (just an amplitude-weighted sum). One could therefore in principle fit f_{GB} to a set of moments by solving a set of linear equations, much as TRIBO does when fitting f_{GB} to $f(\boldsymbol{\xi}; t)$ at a set of collocation points. However this method has three disadvantages:

1.) the resulting distribution is not guaranteed to be positive-semidefinite;

2.) solving a set of linear equations is a computationally expensive "N-cubed" process (where N is the number of Gaussians);

3.) finally, the number of Gaussians and moments must be equal, which requires one to arbitrarily impose enough constraints on the size or shape of the lattice to make the problem well-determined.

All three of these objection may be overcome by using either *linear-* or *quadratic-programming* [32] instead of matrix-inversion to solve for the beamlet-amplitudes.

Linear programming maximizes the weighted sum J of a set of positive-semidefinite variables $x_i \geq 0$, $J = \sum_i w_i x_i$, subject to a set of linear *equality* and/or *inequality* constraints, $\mathbf{A} \cdot \boldsymbol{x} = \boldsymbol{a}$, $\mathbf{B} \cdot \boldsymbol{x} \leq \boldsymbol{b}$. Linear programming makes no requirement that the number of constraints, $M = \mathrm{Rank}\{\mathbf{A}\} + \mathrm{Rank}\{\mathbf{B}\}$, equal the number of variables, N; if $N > M$, this only means that at least $N - M$ of the variables x_i will vanish.

Quadratic programming maximizes a *quadratic form* $J = \sum_i R_i x_i - \frac{1}{2}\sum_{ij} Q_{ij} x_i x_j$ in the positive-semidefinite variables x_i, with Q_{ij} positive-semidefinite, R_i arbitrary, and the x_i subject to equality and/or inequality constraints as before. A quadratic programming problem can always be reduced to a linear programming problem by introducing auxiliary variables [32].

Suppose one uses linear programming to maximize the weighted sum J of N positive-semidefinite Gaussian beamlet amplitudes g_i,

$$J = \sum_i w_i g_i, \tag{56}$$

or quadratic programming to minimize the *sum of squares* of the g_i,

$$J = \sum_i g_i^{\,2}, \tag{57}$$

in either case subject to M equality-constraints on the moments,

$$\langle \xi^{(m_\mu)} \rangle_{f_{GB}} = \langle \xi^{(m_\mu)} \rangle_{f_0}. \tag{58}$$

Either solution will have the following properties:

1.) All the beamlet amplitudes g_i are *positive-semidefinite; normalization* is guaranteed by imposing the constraint $\sum_i g_i = 1$.

2.) The solution can "almost always" be obtained in $O(N + M)$ time [32] — a substantial savings over the matrix method.

3.) Finally, there is no need to impose arbitrary constraints on the size or shape of the lattice — the linear programming algorithm will automatically select exactly M of the beamlet-amplitudes to be non-zero.

The quadratic-programming method is probably superior, because it gives an "unbiased and democratic" estimate for the g_i, whereas the weights w_i introduce an element of arbitrariness into the linear-programming method.[9]

---

[9] The w_i must not be all equal, because of the normalization constraint $\sum_i g_i = 1$; they are otherwise arbitrary. One interesting choice would be to cause the weights w_i to become more negative the farther one is from the beam centroid; *i.e.*, one "penalizes" the halo, so its extent is in some sense minimized.

The primary disadvantage I foresee for the Gaussian beamlet method is the dreaded *"curse of dimensionality:"* the number of beamlets required to get an adequate resolution along each axis increases far too rapidly with the spatial dimension v. For example, taking equal resolution along all $v = 6$ axes of phase space, one finds that even for as high an order as $n = 6$, the set of $_6N_6 = 924$ moments results in a *very* course grid: $\sqrt[6]{924} \simeq 3.1$. By inspecting table 2, one concludes that for $v = 6$, one may need a *lot* of moments and a very high order $\mathcal{M}$ to get a well-resolved f_{GB}. It would seem that moment-inversion *via* Gaussian beamlets is likely to be limited to reconstructing relatively low dimensional *projections* of f; fortunately, the projections of f most often needed in practice are only two- or three-dimensional, such as a projection on the x–p_x plane, or projection onto three dimensions to compute space-charge distributions.

In addition, one should note that "PIC-codes" also require a very large number of particles to get an accurate model for the space-charge fields, so it is easily conceivable that the n-body force calculations required might actually consume far more computer time than pushing a set of very high order moments through a map; one must therefore try the two methods and compare their results. An encouraging result of Lysenko and Channell suggests that the moment method may be quite competitive: they claim good results compared to a PIC-code when using a mere 35-Gaussian 3-D space-charge model in their 4th-order moment-based space-charge code BEDLAM.

9. Conclusion

In this paper, I have argued that elements of the *dual* of DA-vector space may be interpreted as *sets of moments*. These sets of moments provide a description of *equivalence classes* of particle distribution functions. These DA moment-covectors may be manipulated by DA methods, and transported *via* the "scalar product" of DA-covectors with DA-maps. Since the moment transport algorithm makes no *explicit* use of the symplectic condition, in principle it may be used even when dissipative forces are present; it may therefore prove useful in emittance growth or synchrotron damping calculations. DA also makes conversion between moment and cumulant representations trivial. Finally, I have presented two methods for the inverse problem of reconstructing a "plausible" analytic particle distribution from a set of moments. Such distributions provide a possible approach for self-consistent DA calculation of space-charge dominated beam dynamics.

Acknowledgments

I wish to thank the Workshop Organizers for financial support, and W.G. Davies and S.R. Douglas for their careful reading of this paper. Portions of this paper are based on a dissertation presented by the author to the Graduate School at Virginia Tech University in partial fulfillment of the requirements for the degree of Doctor of Philosophy, and supported under grants NSF DMS-8701050 and DOE DE-FG05-87ER25033.

n	$_nN_2$	$\sqrt[2]{_nN_2}$	$_nN_3$	$\sqrt[3]{_nN_3}$	$_nN_6$	$\sqrt[6]{_nN_6}$
1	3	1.73	4	1.59	7	1.38
2	6	2.45	10	2.15	28	1.74
3	10	3.16	20	2.71	84	2.09
4	15	3.87	35	3.27	210	2.44
5	21	4.58	56	3.83	462	2.78
6	28	5.29	84	4.38	924	3.12
7	36	6.00	120	4.93	1716	3.46
8	45	6.71	165	5.48	3003	3.80
9	55	7.42	220	6.04	5005	4.14
10	66	8.12	286	6.59	8008	4.47
11	78	8.83	364	7.14	12376	4.81
12	91	9.54	455	7.69	18564	5.15
13	105	10.25	560	8.24	27132	5.48
14	120	10.95	680	8.79	38760	5.82
15	136	11.66	816	9.34	54264	6.15
16	153	12.37	969	9.90	74613	6.49
17	171	13.08	1140	10.45	100947	6.82
18	190	13.78	1330	11.00	134596	7.16
19	210	14.49	1540	11.55	177100	7.49
20	231	15.20	1771	12.10	230230	7.83
21	253	15.91	2024	12.65	296010	8.16
22	276	16.61	2300	13.20	376740	8.50
23	300	17.32	2600	13.75	475020	8.83
24	325	18.03	2925	14.30	593775	9.17
25	351	18.73	3275	14.85	736281	9.50
26	378	19.44	3654	15.40	906192	9.84
27	406	20.15	4060	15.95	1107568	10.17
28	435	20.86	4495	16.50	1344904	10.51
29	465	21.56	4960	17.05	1623160	10.84
30	496	22.27	5456	17.60	1947792	11.18

Table 2. Table of the number of moments $_nN_v$ as a function of order, for $n = 1$–30, and $v = 2$, 3, and 6 variables. Also shown is an estimate $\sqrt[v]{_nN_v}$ giving the number of Gaussians per edge of a phase-space hypercube, if one assumes uniform hypercubic packing of the Gaussians.

Appendix: Moment Transport — an Example

Let me choose f_1 to be uniform over the interval $(-L, L)$:

$$f_1(x_1; t_1) := \begin{cases} 0 & x_1 < -L, \\ \dfrac{1}{2L} & -L < x_1 < L, \\ 0 & L < x_1. \end{cases} \tag{59}$$

Let $\mathcal{M}$ be the following map of the real line:

$$x_2 = \mathcal{M} x_1 = \lambda \exp\left(x_1/\lambda\right), \tag{60}$$

where λ is some scale length. Transformation (60) is a rather extreme map: it is neither origin- nor volume-preserving, taking the real line $(-\infty, \infty)$ onto the semiline $(0, \infty)$; in particular, the entire negative semiline $(-\infty, 0)$ maps onto the *finite* interval $(0, \lambda)$. I have not chosen either f_1 or $\mathcal{M}$ because they are *physically reasonable*, but because they are *analytically tractable:* this simple distribution and rather pathological map will yield a moment-series which converges rapidly and absolutely.

The *powers of x* form a complete set of functions on the real line, so let me choose them as my basis; the moments of f_1 are:

$$\langle x_1^n \rangle_{f_1} = \begin{cases} \dfrac{L^n}{(n+1)} & n \in \; \underline{even \; integers}, \\ 0 & otherwise. \end{cases} \tag{61}$$

From (11), f_2 is:

$$f_2(x_2; t_2) := \begin{cases} 0 & x_2 < \lambda \exp\left(-L/\lambda\right), \\ \dfrac{\lambda}{2Lx_2} & \lambda \exp\left(-L/\lambda\right) < x_2 < \lambda \exp\left(L/\lambda\right), \\ 0 & \lambda \exp\left(L/\lambda\right) < x_2. \end{cases} \tag{62}$$

A direct computation of the moments of f_2 gives:

$$\langle x_2^m \rangle_{f_2} = \int_{\lambda\exp(-L/\lambda)}^{\lambda\exp(L/\lambda)} \frac{\lambda}{2Lx_2} x_2^m \, dx_2 = \frac{\lambda^{m+1}}{mL} \sinh\left(\frac{mL}{\lambda}\right). \tag{63}$$

For comparison, let me calculate the final moments *via* equation (17). Under (60) the transformed basis functions become:

$$x_2^m = \left(\lambda \exp\left(x_1/\lambda\right)\right)^m = \lambda^m \exp\left(mx_1/\lambda\right) = \sum_{k=0}^{\infty} \frac{1}{k!} \left(\frac{m}{\lambda}\right)^k x_1^k. \tag{64}$$

Substituting (64) into (12), one obtains:

$$\langle x_2^m \rangle_{f_2} = \lambda^m \sum_{k=0}^{\infty} \frac{1}{k!} \left(\frac{m}{\lambda}\right)^k \langle x_1^k \rangle_{f_1} \tag{65}$$

$$= \lambda^m \sum_{k'=0}^{\infty} \frac{1}{(2k')!} \left(\frac{m}{\lambda}\right)^{2k'} \frac{L^{2k'}}{(2k'+1)} \tag{66}$$

$$= \frac{\lambda^{m+1}}{mL} \sum_{k'=0}^{\infty} \frac{1}{(2k'+1)!} \left(\frac{mL}{\lambda}\right)^{2k'+1} \tag{67}$$

$$= \frac{\lambda^{m+1}}{mL} \sinh\left(\frac{mL}{\lambda}\right). \tag{68}$$

The final result (68) yields (63), as it should. Note that from (65), the final moments are manifestly *linear* in the intial moments, with coefficients depending nonlinearly on the map parameter λ.

Equation (65) also demonstrates a less pleasant aspect of moments: each final moment depends on *every* initial moment. Unlike a differential algebra, the transported moments exhibit "feed-down," and therefore the series cannot be truncated without loss of accuracy. Even the first-order final moment depends on all orders of the initial moments.

Now consider the DA-extension of (60) at $Dx_1 = 0 + dx_1$; by definition,

$$
\begin{aligned}
Dx_2^m &= \lambda^m \exp\left(mDx_1/\lambda\right)\big|_{Dx_1=0+dx_1} && (69)\\
&= \lambda^m \sum_{k=0}^{n} \frac{1}{k!} \left(\frac{m}{\lambda}\right)^k dx_1^k && (70)
\end{aligned}
$$

By (23) and (61), the initial moment DA-covector D^*f_1 is

$$
D^*f_1 = \sum_{k=0}^{n} \langle x_1^k \rangle_{f_1} \, d^*x_{1,k} = \sum_{k'=0}^{\lfloor n/2 \rfloor} \frac{L^{2k'}}{(2k'+1)} \, d^*x_{1,(2k')}. \tag{71}
$$

Taking the scalar product of (71) with the DA-extended basis-vectors (70), I get

$$
D^*f_1 \cdot Dx_2^m = \lambda^m \sum_{k'=0}^{\lfloor n/2 \rfloor} \frac{1}{(2k')!} \left(\frac{m}{\lambda}\right)^{2k'} \frac{L^{2k'}}{(2k'+1)} \tag{72}
$$

which is precisely the truncated version of the expression in equations (66). Therefore, in the limit of large n, $D^*f_1 \cdot Dx_2^m$ rapidly and absolutely converges to the correct value for $\langle x_2^m \rangle_{f_2}$, as advertised.

References

[1] Berz M 1989 *"Differential Algebraic Description of Beam Dynamics to Very High Order,"* Part. Accel. **24** 109

[2] Berz M 1990 *"Arbitrary order description of arbitrary particle optical systems"* Nucl. Instrum. Methods **A298** 426–440

[3] Berz M 1988 *"Analysis auf einer Nichtarchimedischen Erweiterung der Reellen Zahlen"* (Giessen BRD: Universität Giessen Report (in German))

[4] Kendall M and Stuart S 1977 *Advanced Theory of Statistics* (New York: Macmillan)

[5] Omohundro S M 1986 *Geometrical Perturbation Theory in Physics* (Singapore: World Scientific)

[6] Pusch G D 1990 *"Differential Alcgebraic Methods for Obtaining Approximate Numerical Solutions to the Hamilton-Jacobi Equation"* (Blacksburg VA: VPI&SU Ph. D. Dissert.)

[7] Gel'fand I M and Shilov G E 1964 *Generalized Functions* (New York: Academic)

[8] Ryne R D 1987 *"Lie Algebraic Treatment of Space Charge"* (College Park MD: U. M. Ph. D. Dissert.)

[9] Abraham R, Marsden J E and Ratiu T 1988 *Manifolds, Tensor Analysis, and Applications (2ⁿᵈ ed.)* (New York: Springer-Verlag)

[10] Bender C M and Orszag S A 1978 *Advanced Mathematical Methods for Scientists and Engineers* (New York: McGraw-Hill) chap. 8

[11] Marsden J B and Weinstein A 1982 *"The Hamiltonian Structure of the Maxwell-Vlasov Equations"* Physica **4D** 394–406

[12] Channell P J, Healy L M and Lysenko W P 1985 *"The moment Code BEDLAM"* IEEE Trans. Nucl. Sci. **32**(5) 2565

[13] Lysenko W P and Channell P J 1990 *"New BEDLAM"* Proc. Conf. Computer Codes and the Linear Accelerator Community (Los Alamos: Los Alamos National Laboratory Report LA-11857-C) pp 157–169

[14] Channell P J and Scoval J C 1991 *"Integrators for Lie-Poisson dynamical systems"* Physica D **50** 80–88

[15] Scoval J C and Weinstein A 1992 *"Finite Dimensional Lie-Poisson Approximations to Vlasov-Poisson Equations"* (submitted for publication in Comm. Pure Appl. Math.)

[16] Lysenko W P and Overly M S 1988 *"Moment Invariants for Particle Beams"* Proc. Linear Accelerator and Beam Optics Codes, La Jolla (New York: Am. Inst. Phys. Conf. Proc.) **177** 323–335

[17] Holm D D, Lysenko W P and Scoval J C 1990 *"Moment Invariants for the Vlasov Equation"* J. Math. Phys. **31** 1610–1615

[18] Rangarajan G 1990 *"Invariants for Symplectic Maps and Symplectic Completion of Symplectic Jets"* (College Park MD: U. M. Ph. D. Dissert.)

[19] Dragt A J, Neri F and Rangarajan G 1992 *"Kinematic Moment Invariants for Hamiltonian Systems"* Phys. Rev. A **45** 2572–2585

[20] Neri F (Private Communication)

[21] Forest É, Berz M and Irwin J 1989 *"Normal Form Methods for Complicated Periodic Systems,"* Part. Acc. **24** 91

[22] Shohat J A and Tamarkin J D 1943 *The Problem of Moments* (Baltimore: Am. Math. Soc.)

[23] Widder D V 1941 *The Laplace Transform* (Princeton: Princeton University Press) chap. 3

[24] van Zeijts N and Neri F (this proceedings)

[25] Jaynes E T 1963 *"Information Theory and Statistical Mechanics"* in *Statistical Physics* (1962 Brandeis Lectures) ed K W Ford (New York: W A Benjamin) pp 181–218

[26] —— 1983 *Papers on Probability, Statistics and Statistical Physics* (Dordrecht: Reidel)

[27] —— 1968 *"Prior Probabilities"* IEEE Trans. on Systems Science and Cybernetics **SSC-4** 227–241; reprinted in [26]

[28] Shannon C E and Weaver W 1962 *The Mathematical Theory of Communication* (Urbana IL: University of Illinois Press)

[29] Drabold D A and Jones G L 1991 *"Maximum-entropy approach to series extrapolation and analytic continuation"* J. Phys. A: Math. Gen. **24** 4705–4714

[30] Berz M and Wollnik H 1988 *"Simulation of intense particle beams with regularly distributed gaussian subbeams"* Nucl. Instrum. Methods **A267** 25–34

[31] Hartmann B and Wollnik H 1990 *"*Tribo*, a program to determine high-order properties of intense ion beams"* Proc. Conf. Computer Codes and the Linear Accelerator Community (Los Alamos: Los Alamos National Laboratory Report LA-11857-C) pp 431–441

[32] Saaty T L and Bram J 1964 *Nonlinear Mathematics* (New York: Dover)

Differential Algebraic Formulation of Normal Form Theory

Martin Berz

Department of Physics and Astronomy and National Superconducting Cyclotron Laboratory, Michigan State University, East Lansing, Mi 48824

Abstract.
A differential algebraic (DA) formulation of a normal form theory for repetitive systems is presented. Contrary to previous approaches, no Lie algebraic tools are used. The resulting algorithm is very transparent and not restricted to the treatment of symplectic systems.
In the case of symplectic systems, the normal form algorithm provides a nonlinear coordinate transformation in which the motion is confined to circles. The transformation exists if the tunes are not on a resonance; in this case, it can be used to compute tune shifts in a similar way as in the Lie algebraic picture.
In the case of nonsymplectic systems, the motions in the new coordinates are growing or shrinking exponential spirals. In the case all spirals are shrinking, which occurs in electron rings, all amplitude dependent tune shifts vanish and in a formal sense tune resonances do not occur.
The algorithm has been implemented in the code COSY INFINITY. For symplectic systems, which can also be studied with the DA-Lie algorithm also implemented in COSY INFINITY, identical results are obtained at a reduced computational expense.

1. Introduction

The famous Courant Snyder theory [1] completely describes the repetitive behaviour of linear symplectic systems. It provides a unique criterion for stability of the system, it provides an invariant of the system, and allow the calculation of important quantities like the tune.

In the nonlinear case, the situation becomes substantially more involved. The question of stability is very difficult to answer, invariants usually do not exists, and the tune depends on the amplitude of the particle under consideration. Normal form theory [2, 3, 4, 5] comes closest to a nonlinear extension of the Courant Snyder theory in that it answers the questions of the amplitude dependence of the tune. It also produces a set of pseudo-invariants which in special cases are real invariants, and at least allows the description of the motion in coordinates which are more suitable than the original ones.

Normal form ideas were introduced to the field by Dragt and Finn [2] in the Lie algebraic framework [6, 7]. While the original paper [2] contains all of the core ideas, some simplifications were necessary [8] before a first implementation for realistic systems was obtained by Neri and Dragt [4]. Difficulties inherent in the Lie algebraic formulation limited the efforts to relatively low orders, and only the combined DA - Lie approach [5] circumvented this problem, resulting in the first arbitrary order algorithm, and also allowed the use of system parameters.

In the following sections we present an arbitrary order normal form algorithm that does not require any Lie algebraic methods, and in particular does not require ongoing changes between the factored Lie operator representation and the DA Taylor series representation. The resulting algorithm is more direct than the hybrid algorithm, and it allows the treatment of non-symplectic systems.

In the next section we summarize some tools discussed elsewhere in detail. Section 3 shows how to perform a nonlinear change of variable to a rotationally invariant map. Sections 4 and 5 discuss the results of the transformation for symplectic and non-symplectic systems. A summary and an appendix follow.

2. The Linear Transformation of the Map

The goal of the normal form algorithm is to provide a nonlinear change of variables such that the map in the new variables has a significantly simpler structure than before. So we assume we are given the transfer map of a particle optical system

$$\vec{z}_f = \mathcal{M}(\vec{z}_i, \vec{\delta}) \tag{1}$$

where $\vec{z}$ are the $2v$ phase space coordinates and $\vec{\delta}$ are system parameters. While it is in general impossible to obtain the exact map $\mathcal{M}$, the DA methods [9, 10, 11, 12] allow us to compute the partial derivatives $[\mathcal{M}]_n$ of the map to any order n. This can be done in a particularly elegant way using the code COSY INFINITY [13, 14, 15], but also with other DA based codes.

The normal form algorithm consists of a sequence of coordinate transformations $\mathcal{A}$ of the map:

$$\mathcal{A} \circ \mathcal{M} \circ \mathcal{A}^{-1} \tag{2}$$

The first such coordinate transformation is the move to the parameter dependent fixed point $\vec{z}_F$ which satisfies

$$\vec{z}_F = \mathcal{M}(\vec{z}_F, \vec{\delta}) \tag{3}$$

This transformation can be performed to arbitrary order using DA methods. For details we refer to [16]. After the fixed point transformation, the map is origin preserving; this means that for any $\vec{\delta}$, we have

$$\mathcal{M}(\vec{0}, \vec{\delta}) = \vec{0} \tag{4}$$

As explained in [16], we note that the fixed point transformation is possible if and only if 1 is not an eigenvalue of the linear map.

In the next step we perform a linear coordinate transformation that diagonalizes the linear part of the map. For this process, we have to assume that there are $2v$ pairwise distinct eigenvalues. This, together with the fact that no eigenvalue should be unity and that their product is positive are the only requirements we have to demand for the map; under normal conditions, accelerators are always designed such that these conditions are met.

As shown in detail in [16], it is possible to perform a diagonalization such that the linear map assumes the following form:

$$\begin{pmatrix} r_1 e^{+i\mu_1} & & & & & \\ & r_1 e^{-i\mu_1} & & & & \\ & & \cdot & & 0 & \\ & 0 & & \cdot & & \\ & & & & r_v e^{+i\mu_v} & \\ & & & & & r_v e^{-i\mu_v} \end{pmatrix} \tag{5}$$

Here the tunes μ_j are either purely real or purely imaginary. For stable systems, none of the $r_j e^{\pm i\mu_j}$ must exceed unity in modulus.

For symplectic systems, the determinant is unity, which entails that the product of the r_j must be unity. This implies that for symplectic systems, for any $r_j < 1$ there is another with $r_j > 1$. Thus stable symplectic systems have $r_j = 1$ for all j, because otherwise there would be one j for which r_j exceeds unity, and thus at least one of $r_j e^{\pm i\mu_j}$ would have modulus larger than unity. This would also happen if a μ_j were imaginary. So all μ_j are real, and they are even nonzero because we demanded distinct eigenvalues.

To the eigenvector pair $s_j^\pm$ belonging to the eigenvalue $r_j e^{\pm i\mu_j}$, we associate another pair $t_j^\pm$ of variables as follows:

$$\begin{aligned} t_j^+ &= (s_j^+ + s_j^-)/2 \\ t_j^- &= (s_j^+ - s_j^-)/2i. \end{aligned} \tag{6}$$

In case of complex $s_j^\pm$, which corresponds to the stable case, the $t_j^\pm$ are just the real and imaginary parts and thus are real. In the unstable case, t_j^+ is real and t_j^- is imaginary. Obviously the $s_j^\pm$ can be expressed in terms of the $t_j^\pm$ as

$$\begin{aligned} s_j^+ &= t_j^+ + i\, t_j^- \\ s_j^- &= t_j^+ - i\, t_j^- \end{aligned} \tag{7}$$

In the rest of the paper, it is advantageous to perform the manipulations in the $s_j^\pm$, while the results are most easily interpreted in the $t_j^\pm$.

3. The DA Normal Form Algorithm

In this section we will show a map in the $s_j^{\pm}$ can be subjected to nonlinear coordinate transformations that considerably simplify the nonlinear terms. The advertised transformation to the new coordinates is carried out in an iterative manner. The first step consists of the fixed point transformation and the linear diagonalization. All further steps are purely nonlinear and do not affect the linear part anymore. The mth step transforms only the mth order of the map and leaves the lower orders unaffected.

We begin the mth step by splitting the momentary map $\mathcal{M}$ into its linear and nonlinear parts $\mathcal{R}$ and $\mathcal{S}_m$, i.e. $\mathcal{M} = \mathcal{R} + \mathcal{S}_m$. The linear part $\mathcal{R}$ has the form of Eq. 5. Then we perform a transformation using a map that to mth order has the form

$$\mathcal{A}_m = \mathcal{E} + \mathcal{T}_m \tag{8}$$

where $\mathcal{T}_m$ vanishes to order $m - 1$. Because the linear part of $\mathcal{A}_m$ is the unity map, $\mathcal{A}_m$ is invertible. Moreover, inspection of the algorithm to invert transfer maps reveals that up to order m, we have

$$\mathcal{A}_m^{-1} =_m \mathcal{E} - \mathcal{T}_m \tag{9}$$

Of course, the full inversion of $\mathcal{A}_m$ contains higher order terms, which will turn out to be one of the reasons why iteration is needed. It is also worth noting that in principle the higher order parts of $\mathcal{T}_m$ can be chosen freely. It seems to be particularly useful to choose these terms in such a way that they represent the flow of a dynamical system by interpreting $\mathcal{T}$ as the first term in the Lie derivative series [9]. This has the advantages that the computation of the inverse is trivial and that the transformation map comes out to be symplectic if the original map is.

To study the effect of the transformation, we now infer up to order m:

$$\begin{aligned}
\mathcal{A} \circ \mathcal{M} \circ \mathcal{A}^{-1} &=_m (\mathcal{E} + \mathcal{T}_m) \circ (\mathcal{R} + \mathcal{S}_m) \circ (\mathcal{E} - \mathcal{T}_m) \\
&=_m (\mathcal{E} + \mathcal{T}_m) \circ (\mathcal{R} + \mathcal{S}_m - \mathcal{R} \circ \mathcal{T}_m) \\
&=_m \mathcal{R} + \mathcal{S}_m + (\mathcal{T}_m \circ \mathcal{R} - \mathcal{R} \circ \mathcal{T}_m)
\end{aligned} \tag{10}$$

For the first step, we have used $\mathcal{S}_m \circ (\mathcal{E} - \mathcal{T}_m) =_m \mathcal{S}_m$ which holds because $\mathcal{S}_m$ is nonlinear and $\mathcal{T}_m$ is of order m. In the second step we used $\mathcal{T}_m \circ (\mathcal{R} + \mathcal{S}_m - \mathcal{R} \circ \mathcal{T}_m) =_m \mathcal{T}_m \circ \mathcal{R}$ which holds because $\mathcal{T}_m$ is of exact order m and everything in the second term is nonlinear except $\mathcal{R}$.

A closer inspection of the last line reveals that $\mathcal{S}_m$ can be simplified by choosing the commutator $\mathcal{C}_m = \{\mathcal{T}_m, \mathcal{R}\} = (\mathcal{T}_m \circ \mathcal{R} - \mathcal{R} \circ \mathcal{T}_m)$ appropriately. Indeed, if the range of $\mathcal{C}_m$ is the full space, then $\mathcal{S}_m$ can be removed entirely. However, as we shall see, most of the time this is not the case.

Let $(\mathcal{T}_{mj}^{\pm}|k_1^+, k_1^-, ..., k_n^+, k_n^-)$ be the Taylor expansion coefficient of $\mathcal{T}_{mj}$ with respect to $(s_1^+)^{k_1^+}(s_1^-)^{k_1^-} \cdot ... \cdot (s_n^+)^{k_n^+}(s_n^-)^{k_n^-}$ in the j-th component pair of $\mathcal{T}_m$. So $\mathcal{T}_{mj}^{\pm}$ is written as

$$\mathcal{T}_{mj}^{\pm} = \sum (\mathcal{T}_{mj}^{\pm}|k_1^+, k_1^-, ..., k_n^+, k_n^-) \cdot (s_1^+)^{k_1^+}(s_1^-)^{k_1^-} \cdot ... \cdot (s_n^+)^{k_n^+}(s_n^-)^{k_n^-} \tag{11}$$

Similarly we identify the coefficients of $\mathcal{C}$ by $(\mathcal{C}_j^{\pm}|k_1^+, k_1^-, ..., k_n^+, k_n^-)$. Because $\mathcal{R}$ is diagonal, it is easily possible to express the coefficients of $\mathcal{C}$ in terms of the ones of $\mathcal{T}$. One obtains

$$
\begin{aligned}
& (\mathcal{C}_j^{\pm}|k_1, k_1^-, ..., k_n^+, k_n^-) \\
= \ & \left((\prod_{l=1}^n r_l^{(k_l^+ + k_l^-)}) \cdot e^{i\vec{\mu}\cdot(\vec{k}^+ - \vec{k}^-)} - r_j \cdot e^{\pm i\mu_j} \right) \cdot (\mathcal{T}_j^{\pm}|k_1^+, k_1^-, ..., k_n^+, k_n^-) \\
= \ & C_j^{\pm}(\vec{k}^+, \vec{k}^-) \ \cdot \ (\mathcal{T}_j^{\pm}|k_1^+, k_1^-, ..., k_n^+, k_n^-)
\end{aligned}
\tag{12}
$$

Now it is apparent that a term in $\mathcal{S}_j^{\pm}$ can be removed if and only if the factor $C(\vec{k}^+, \vec{k}^-)$ is nonzero; if it is nonzero, then the required term in $\mathcal{T}_j^{\pm}$ is just the negative of the respective term in $\mathcal{S}_j^{\pm}$ divided by $C(\vec{k}^+, \vec{k}^-)$.

So the outcome of the whole normal form transformation depends upon the conditions under which the term $C(\vec{k}^+, \vec{k}^-)$ vanishes. This is obviously the case if and only if the moduli and the arguments of $r_j \cdot e^{\pm i\mu_j}$ and $(\prod_{l=1}^n r_l^{(k_l^+ + k_l^-)}) \cdot e^{i\vec{\mu}\cdot(\vec{k}^+ - \vec{k}^-)}$ are identical. In the next sections we will discuss the conditions of this for various special cases and draw conclusions.

4. Stable Symplectic Maps

As discussed above, in the stable symplectic case all the r_j are equal to one, and the μ_j are purely real. So the moduli of the first and second terms in $C_j^{\pm}(\vec{k}^+, \vec{k}^-)$ are equal if and only if their phases agree modulo 2π. This is obviously the case if

$$
\vec{\mu} \cdot (\vec{k}^+ - \vec{k}^-) = \pm \mu_j \ (\mod 2\pi)
\tag{13}
$$

where the different signs apply for $C_j^+(\vec{k}^+, \vec{k}^-)$ and $C_j^-(\vec{k}^+, \vec{k}^-)$, respectively. This can occur in two possible ways:

1. $k_l^+ = k_l^- \ \forall \, l \neq j$, and $k_j^+ = k_j^- \pm 1$

2. $\vec{\mu} \cdot \vec{n} = 0 \ (\mod 2\pi)$ has nontrivial solutions.

The first case is of mathematical nature and lies at the heart of the normal form algorithm. It yields terms that are responsible for amplitude dependent tune shifts. We will discuss its consequences below. The second case is equivalent to the system lying on a higher order resonance and is of more physical nature. In case the second condition is satisfied, there will be resonance driven terms that cannot be removed and that prevent a direct computation of amplitude tune shifts.

Before proceeding in the discussion, we note that the second condition entails complications even if it is almost, but not exactly, satisfied. In this case, the removal of the respective term produces a small denominator that generates terms that become larger and larger, depending on the proximity to the resonance. In the removal process, this resonance proximity factor is multiplied by the respective expansion coefficient, and so this product obviously is an excellent characteristic of the significance of the resonance.

With higher and higher orders, i.e. larger k^+ and k^-, the number of relevant resonances increases. Since the resonances lie dense in tune space, eventually the growth of terms is almost inevitable and hence produces a map that is much more nonlinear than the underlying one. As we shall see in the next section, this problem is alleviated by damping.

We now discuss the form of the map if no resonances occur. In this case, the transformed map will have the form

$$
\begin{aligned}
\mathcal{M}_j^+ &= s_j^+ \cdot f_j(s_1^+ s_1^-, ..., s_v^+ s_v^-) \\
\mathcal{M}_j^- &= s_j^- \cdot \bar{f}_j(s_1^+ s_1^-, ..., s_v^+ s_v^-)
\end{aligned}
\tag{14}
$$

The variables $s_j^{\pm}$ are not particularly well suited for the discussion of the result, and we express the map in terms of the adjoined variables $t_j^{\pm}$ introduced in 6. Simple arithmetic shows that

$$
s_j^+ \cdot s_j^- = (t_j^+)^2 + (t_j^-)^2
\tag{15}
$$

It is now advantageous to write f_j in terms of amplitude and phase as $f_j = a_j \cdot e^{i\phi_j}$. Performing the transformation to the coordinates $t_j^{\pm}$, we thus obtain

$$
\begin{aligned}
\mathcal{M}_j^{\pm} &= \begin{pmatrix} 1/2 & 1/2 \\ 1/2i & -1/2i \end{pmatrix} \cdot \begin{pmatrix} (t^+ + it^-) \cdot f_j[(t_1^+)^2 + (t_1^-)^2, ..., (t_v^+)^2 + (t_v^-)^2] \\ (t^+ - it^-) \cdot \bar{f}_j[(t_1^+)^2 + (t_1^-)^2, ..., (t_v^+)^2 + (t_v^-)^2] \end{pmatrix} \\
&= a_j \cdot \begin{pmatrix} \cos(\phi_j) & -\sin(\phi_j) \\ \sin(\phi_j) & \cos(\phi_j) \end{pmatrix} \cdot \begin{pmatrix} t^+ \\ t^- \end{pmatrix}
\end{aligned}
\tag{16}
$$

Here $\phi_j = \phi_j[(t_1^+)^2 + (t_1^-)^2, ..., (t_v^+)^2 + (t_v^-)^2]$ depends on a rotationally invariant quantity.

So in these coordinates, the motion is now given by a rotation, the frequency of which depends only on the amplitudes $(t_j^+)^2 + (t_j^-)^2$ and some system parameters and thus does not vary from turn to turn. As we will show now, these frequencies are precisely the tunes of the nonlinear motion.

For any repetitive system, the tune of one particle is the total polar angle advance divided by the number of turns in the limit of turn number going to infinity, if this limit exists. If we now express the motion in the new coordinates, we pick up an initial polar angle for the transformation to the new coordinates; then, every turn produces an equal polar angle ϕ_j which depends on the amplitude and parameters of the particle; at the end, we produce a final polar angle for the transformation back to the old coordinates.

As the number of turns increases, the contribution of the initial and final polar angles due to the transformation becomes more and more insignificant, and in the limit the tune comes out to nothing but ϕ_j. So altogether, we showed that the limit exists and that it can be computed analytically as a by product of the normal form transformation.

5. Stable Non-Symplectic Maps

In the case of stable, non-symplectic maps, all r_j must satisfy $r_j \leq 1$, because otherwise at least one of the $r_j e^{\pm i \mu_j}$ is larger than unity in modulus. Since in the normal form transformation, terms can be removed if and only if the phases or amplitudes for the two contributions in $C(k^+, k^-)$ are different and the amplitudes contribute, more terms can be removed.

Of particular practical interest is the totally damped case in which $r_j < 1$ for all j and all μ_j are real, which describes damped electron rings. In this case an inspection of equation (12) reveals that now every nonlinear term can be removed. Then a similar argument as in the previous section shows that now the motion assumes the form

$$\mathcal{M}_j^\pm = r_j \cdot \left(\begin{array}{cc} \cos(\phi_j) & -\sin(\phi_j) \\ \sin(\phi_j) & \cos(\phi_j) \end{array} \right) \cdot \left(\begin{array}{c} t_j^+ \\ t_j^- \end{array} \right) \tag{17}$$

where now the angle ϕ_j does not depend on the phase space variables anymore but only on the parameters. This means that the normal form transformation of a totally damped system leads to exponential spirals with constant frequency ϕ_j. In particular this entails that totally damped systems do not have any amplitude dependent tune shifts, and that they eventually collapse into the origin. Since in practice the damping is of course usually very small, these effects are usually covered by the short term sensitivity to resonances.

It is quite illuminating to consider the small denominator problem in the case of totally damped systems. Clearly the denominator can never fall below $1 - \max(r_j)$ in magnitude. This puts a limit on the influence of any low order resonance on the dynamics; in fact, even sitting exactly on a low order resonance does not have any serious consequences if the damping is strong enough. In general, the influence of a resonance now depends on two quantities: the distance in tune space and the contraction strength r_j. High order resonances are suppressed particularly strongly because of the contribution of additional powers of r_j.

Because all systems exhibit a residual amount of damping, the arguments here are generally relevant. It is especially noteworthy that residual damping suppresses high order resonances by the above mechanism even for proton machines, which entails that from a theoretical view, ultimately high order resonances become insignificant.

6. Unstable Maps

Clearly the normal form algorithm also works for unstable maps. The number of terms that can be removed will be at least the same as in the symplectic case, and sometimes it is possible to remove all terms. Among the many possible combinations of r_j and μ_j, the most common case in which the μ_j are real is worth studying in more detail. In this case, all terms can be removed unless the logarithms of the r_j and the tunes satisfy the same resonance condition, i.e.

$$\begin{aligned} \vec{n} \cdot (\log(r_1), ..., \log(r_v)) &= 0 \\ \vec{n} \cdot \vec{\mu} &= 0 \ (\bmod 2\pi) \end{aligned} \tag{18}$$

have simultaneous nontrivial solutions. This situation characterizes a new type of resonance, the coupled phase-amplitude resonance.

Phase-amplitude resonances can never occur if all r_j are greater than unity in magnitude. This case corresponds to a totally unbound motion, and the motion in normal form coordinates moves along growing exponential spirals.

Symplectic systems, on the other hand, satisfy $\prod_{l=1}^{n} r_l = 1$. So if there are r_j with both signs of the logarithm, and thus the possibility for amplitude resonances exists. In fact, any symplectic system lies on the fundamental amplitude resonance characterized by $\vec{n} = (1, 1, ..., 1)$. In this light, the stable symplectic case is a degeneracy in which all logarithms vanish and so the system lies on every amplitude resonances. Thus it is susceptible to any phase resonance, and it suffices to study just these.

7. Conclusion

In this paper we have presented a DA normal form algorithm for complex periodic systems. It is applicable as long as the linear transfer map has no multiple eigenvalues, all eigenvalues differ from 1, and their product is positive. All these conditions are basic requirements for linear stability and are usually satisfied by circular accelerators.

The algorithm is very transparent and computationally efficient and does not require Lie algebraic tools. It works to arbitrary order and allows the treatment of system parameters. In the case of symplectic systems, identical results as with the hybrid DA - Lie algorithm discussed in [5] are obtained at a reduced effort. In particular, the algorithms allows the computation of amplitude and parameter tune shifts if the linear tunes are not in resonance. For systems near a resonance, the characteristic small denominator problem occurs.

The algorithm also applies to damped systems. In this case, it can be used to show that formally there are no amplitude dependent tune shifts. In addition, the transformation denominators now also contain a damping dependent term which prevents them from shrinking beyond a certain size, corresponding to the favorable long term behaviour of damped systems. Since any machine has nonzero residual damping, this also explains an old paradox of accelerator physics: in the strict sense, every resonance has to be avoided, but on the other hand, the resonances lie dense in tune space. Under the presence of ever so slight residual damping, resonances of high enough order turn out to be also mathematically irrelevant.

8. Appendix: The DA Normal Form Algorithm in COSY INFINITY Language

Besides the transparency of the normal form algorithm in the DA picture, its strength lies in the possibility to implement it completely and in full detail. As with most DA operations, the resulting programs are not only powerful but also compact and easy to understand. This is particularly true for programs written in the COSY INFINITY language [15, 13, 14].

To stress this point, we present here the COSY INFINITY source of the above normal form algorithm which amounts to about five dozen lines of code. This excludes

the routines for the eigenvalue solver, the closed orbit transformations, and the routines
to compute Twiss parameters as well as low level DA routines. For the sake of comparison we mention that the hybrid DA - Lie program, which is written in precompiled
FORTRAN [17] excluding the same routines is about 20 times longer.

```
PROCEDURE DANF M MN MA IMA EPS ; {Computes Normal Form MN of a map M.
    MA is the transformation map, which is only computed if IMA#0.
    EPS is the tolerance below which resonance denominators are not removed}
    VARIABLE J 1 ; VARIABLE K 1 ; VARIABLE L 2 ; VARIABLE NOM 1 ;
    VARIABLE IER 1 ; VARIABLE XF 100 NV ; VARIABLE T 2*NM1 ;
    VARIABLE F 100 TWOND ; VARIABLE MUU 100 ND ; VARIABLE AA 100 ND ;
    VARIABLE BB 100 ND ; VARIABLE GG 100 ND ; VARIABLE RR 100 ND ;
    VARIABLE MU 2 ; VARIABLE A 2 ; VARIABLE B 2 ; VARIABLE D 2 ;
    VARIABLE PHI 1 TWOND ; VARIABLE R 1 TWOND ;
    VARIABLE M1 2*NM1 NV ; VARIABLE M2 2*NM1 NV ; VARIABLE M3 2*NM1 NV ;
    VARIABLE I 2 ; IMUNIT I ; NOM := NOC ; DSET 1E-14 ;
    FM M  XF MN IER ;  BM MN MN M1 IER ;
    IF IMA#0 ; LOOP J 1 TWOND ; MA(J) := -XF(J) + DD(J) ; ENDLOOP ;
        POLVAL 1 M1 TWOND MA TWOND MA TWOND ; ENDIF ;
        GT MN F MUU AA BB GG RR ;
    LOOP J 1 ND ; K:= 2*J-1 ; L := K + 1 ; MU := CONS(MUU(J))*2*PI ;
        A := CONS(AA(J)) ; B := CONS(BB(J)) ; D := CONS(RR(J)) ;
        IF TYPE(MU)=TYPE(1) ; PHI(K) := MU ; PHI(L) := -MU ;
            R(K) := D ; R(L) := D ;
            M2(K) := (      I*B      *DD(K) +    I*B *DD(L) )/SQRT(2*I*B) ;
            M2(L) := ( (-1-I*A)      *DD(K) + (1-I*A) *DD(L) )/SQRT(2*I*B) ;
            M1(K) := ( ( 1-I*A)/2/I/B*DD(K) -     1/2*DD(L) )*SQRT(2*I*B) ;
            M1(L) := ( ( 1+I*A)/2/I/B*DD(K) +     1/2*DD(L) )*SQRT(2*I*B) ;
        ELSEIF 1=1 ; PHI(K) := 0 ; PHI(L) := 0 ; MU := IMAG(MU) ;
            IF IMAG(D)#0 ; WRITE 6 '$$$ ERROR DANF ' ; ENDIF ; D := REAL(D) ;
            R(K) := D*EXP(-MU) ; R(L) := D*EXP(MU) ;
            M2(K) := DD(K) ; M2(L) := DD(L) ;
            M1(K) := DD(K) ; M1(L) := DD(L) ; ENDIF ;
        ENDLOOP ; ANM MN M2 MN ; CPOLVAL 1 M1 TWOND MN TWOND MN TWOND ;
    IF IMA#0 ; CPOLVAL 0 M1 TWOND MA TWOND MA TWOND ; ENDIF ;
    NOM := NOC ; RS := 0 ;   LOOP J 2 NOM ; LOOP K 1 TWOND ; CO J ;
        CDNFDA MN(K) R PHI K TWOND EPS T ; M3(K) := -T ; CO NOM ;
        IF (K/2)=NINT(K/2) ; RS := RS + T*DD(K-1) ; ENDIF ; ENDLOOP ;
        LOOP K 1 TWOND ; CDFLO M3 DD(K)+0*I T TWOND ; M1(K) := T ; ENDLOOP ;
        LOOP K 1 TWOND ; M3(K) := -M3(K) ; ENDLOOP ;
        LOOP K 1 TWOND ; CDFLO M3 DD(K)+0*I T TWOND ; M2(K) := T ; ENDLOOP ;
        LOOP K TWOND+1 NV ; M1(K) := DD(K) ; M2(K) := DD(K) ; ENDLOOP ;
        CPOLVAL 1 MN TWOND M1 NV M3 TWOND ;
        LOOP K TWOND+1 NV ; M3(K) := M1(K) ; ENDLOOP ;
        CPOLVAL 1 M2 TWOND M3 NV MN TWOND ;
        IF IMA#0 ; LOOP K 1 TWOND ; M3(K) := MA(K) ; ENDLOOP ;
            LOOP K TWOND+1 NV ; M3(K) := DD(K) ; ENDLOOP ;
            CPOLVAL 1 M2 TWOND M3 NV MA TWOND ; ENDIF ;
        ENDLOOP ; LOOP J 1 ND ; K:= 2*J-1 ; L := K + 1 ; IF PHI(K)#0 ;
            M1(K) := (  DD(K) - I*DD(L))/SQRT(2*I) ;
            M1(L) := (  DD(K) + I*DD(L))/SQRT(2*I) ;
            M2(K) := (  DD(K) +   DD(L))*SQRT(I/2) ;
            M2(L) := (I*DD(K) - I*DD(L))*SQRT(I/2) ;
        ELSEIF 1=1 ;
            M1(K) := DD(K) ; M1(L) := DD(L) ; M2(K) := DD(K) ; M2(L) := DD(L) ;
        ENDIF ; ENDLOOP ;  LOOP K TWOND+1 NV ; M1(K) := DD(K) ; ENDLOOP ;
    CPOLVAL 1 MN TWOND M1 NV M3 TWOND ;
```

```
LOOP K TWOND+1 NV ; M3(K) := M1(K) ; ENDLOOP ;
CPOLVAL 1 M2 TWOND M3 NV MN TWOND ;
IF IMA#0 ; CPOLVAL 1 M2 TWOND MA TWOND MA TWOND ; LOOP K 1 TWOND ;
   IF ABS(IMAG(MA(K)))<1E-6 ; MA(K) := REAL(MA(K)) ; ENDIF ; ENDLOOP ;
ENDIF ;  LOOP K 1 TWOND ; IF ABS(IMAG(MN(K)))<1E-6 ;
   MN(K) := REAL(MN(K)) ; ENDIF ; ENDLOOP ; DSET 1E-16 ; ENDPROCEDURE ;
```

References

[1] E.D. Courant and H.S. Snyder. Theory of the alternating gradient synchrotron. *Annals of Physics*, 3:1, 1958.

[2] A. J. Dragt and J. M. Finn. Normal form for mirror machine Hamiltonians. *Journal of Mathematical Physics*, 20(12):2649, 1979.

[3] A. Bazzani, P. Mazzanti, G. Servizi, and G. Turchetti. Normal forms for Hamiltonian maps and nonlinear effects in a particle accelerator. *Il Nuovo Cimento*, 102 B, N.1:51, 1988.

[4] Filippo Neri. Private communication.

[5] E. Forest, M. Berz, and J. Irwin. Normal form methods for complicated periodic systems: A complete solution using Differential algebra and Lie operators. *Particle Accelerators*, 24:91, 1989.

[6] A. J. Dragt and J. M. Finn. *Journal of Mathematical Physics*, 17:2215, 1976.

[7] A. J. Dragt. Lectures on nonlinear orbit dynamics. In *1981 Fermilab Summer School*. AIP Conference Proceedings Vol. 87, 1982.

[8] Etienne Forest. Private communication.

[9] M. Berz. Arbitrary order description of arbitrary particle optical systems. *Nuclear Instruments and Methods*, A298:426, 1990.

[10] M. Berz. Differential algebraic description of beam dynamics to very high orders. *Particle Accelerators*, 24:109, 1989.

[11] M. Berz. Differential algebraic description and analysis of trajectories in vacuum electronic devices including spacecharge effects. *IEEE Transactions on Electron Devices*, 35-11:2002, 1988.

[12] M. Berz. Differential algebraic treatment of beam dynamics to very high orders including applications to spacecharge. *AIP Conference Proceedings*, 177:275, 1988.

[13] M. Berz. Computational aspects of design and simulation: COSY INFINITY. *Nuclear Instruments and Methods*, A298:473, 1990.

[14] M. Berz. COSY INFINITY, an arbitrary order general purpose optics code. *Computer Codes and the Linear Accelerator Community*, Los Alamos LA-11857-C:137, 1990.

[15] M. Berz. COSY INFINITY Version 6 reference manual. Technical Report MSUCL, National Superconducting Cyclotron Laboratory, Michigan State University, East Lansing, MI 48824, 1993.

[16] M. Berz. Direct computation and correction of chromaticities and parameter tune shifts in circular accelerators. In *Proceedings XIII International Particle Accelerator Conference*, Dubna, 1992.

[17] M. Berz. Differential algebra precompiler version 3 reference manual. Technical Report MSUCL-755, Michigan State University, East Lansing, MI 48824, 1990.

Analytical determination of 5th–order transfer matrices of magnetic quadrupole fringing fields

B. Hartmann[1], H. Irnich and H. Wollnik

II. Physikalisches Institut der Justus–Liebig–Universität
W–6300 Giessen, Germany

Abstract. The fringing–field effects on particle trajectories in magnetic quadrupoles are described to 5th order by fringing–field integrals. It is shown that this method improves the description of fringing–field effects noticeably over the so far known use of third–order fringing–field integrals.

1. Introduction

Though the lens action of quadrupoles is mainly due to effects of the main–field region [1], the aberrations are mainly caused by fringing–field effects [2]. These fringing–field effects are accurately described by ray–tracing techniques through known fringing–field distributions [3, 4, 5]. Such calculations are time consuming and their accuracy is limited by the accuracy of the used fringing–field distributions. Alternatively [6, 7, 8, 9] one can describe the particle trajectories by integrals over a still arbitrary fringing–field distribution and determine how this description differs from the more easily calculated trajectories in an idealized field distribution. This idealized field here is assumed to rise abruptly at the so called effective field boundary though this violates Laplace's equation. The resulting differences then can be added as small shifts and bends, which the trajectory under consideration must experience, when crossing this effective field boundary. For quadrupoles [8] such a third–order transfer matrix was determined in an effective fourth–order approximation, so that matrixcoefficients of order n were calculated up to terms of order $(4 - n)$ in Δz, the extension of the fringing field. Here we present the solution of the fifth–order transfer matrix in an effective fifth–order approximation.

2. Fringing–field transfer matrix as a thin–lens approximation

In case of an idealized quadrupole field the overall beam transport can be described by the product of the transfer matrices of two drift lengths (DL) and the magnetic

---

[1] work done in partial fulfillment of the requirements for the doctor's degree, University of Giessen, FRG

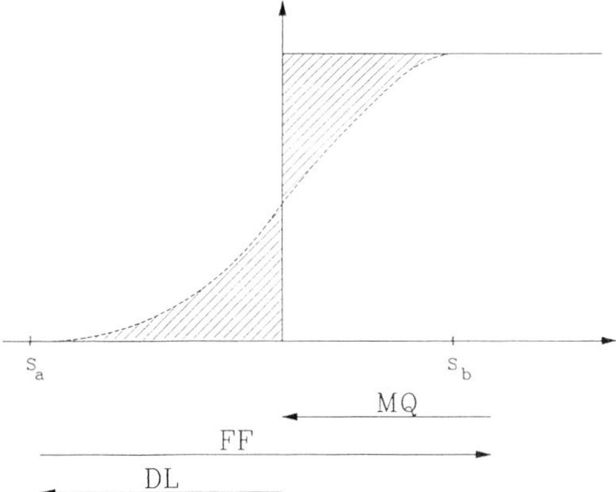

Figure 1. The fringing–field transfer matrix as a product of three transfer matrices.

quadrupole (MQ). Using a real field distribution, this system must be divided into five transfer matrices, with the two additional "fringing–field transfer matrices" (FF) [6, 7, 8] before and behind the quadrupole. These transfer matrices are placed at the effective field boundary and can be obtained by the product of three other transfer matrices [4, 6, 7, 8]. For the entrance fringing field (see fig. 1) these transfer matrices are:

1. a reverse transfer matrix for the field–free region, ranging from z_*, the position of the effective field boundary to z_a, a position where the quadrupole strength and all its derivatives vanish,

2. a forward transfer matrix starting from z_a to z_b, a position in the main field region of the quadrupole, where the field gradient is constant and all its derivatives vanish,

3. a reverse transfer matrix for a region ranging from z_b to z_*, throughout which the quadrupole field gradient is assumed to be constant.

This product matrix has the following properties:

1. many elements of the transfer matrix vanish altogether,

2. several of the elements – though being large – become independent of the specific fringing–field distribution,

3. all elements that depend on the detailed fringing–field distribution are small as compared to the largest elements of that fringing–field transfer matrix.

3. The distribution of the potential in the fringing field of a magnetic quadrupole

Since a magnetic quadrupole [10, 9, 8] has a straight optic axis, it is advantageous to describe the distribution of the flux density $\vec{B}$ in cylindrical coordinates z, r and ϕ. Due to the time–independence and the fact that there are no currents inside the quadrupole element, we infer from Maxwell's equations div $\vec{B} = $ curl $\vec{B} = 0$. Hence there exists a magnetic scalar potential V_Q for which Laplace's equation holds, i.e. $\nabla^2 V_Q = 0$. This V_Q can be expanded in a power series about $r = 0$ as [4, 10]

$$V_Q(r, \phi, z) = \sum_{l=1}^{\infty} q_l(z) \cos(2\phi + \theta) \cdot r^{2l}. \tag{1}$$

Note that the coefficients $q_l(z)$ vary along the z-axis, i.e. the optic axis, without being expanded in a power series in z. In these cylindrical coordinates, Laplace's equation reads

$$\nabla^2 V_Q = \frac{1}{r} \frac{\partial}{\partial r} \left(\frac{r \cdot \partial V_Q}{\partial r} \right) + \frac{1}{r^2} \frac{\partial^2 V_Q}{\partial \phi^2} + \frac{\partial^2 V_Q}{\partial z^2} = 0. \tag{2}$$

Inserting Eq.(1) into Eq.(2) the coefficients $q_l(z)$ can be determined from a known field gradient on the optical axis $g(z)$ and one obtains for the potential [8] up to sixth order:

$$V_Q = -g\,(xy) + \frac{1}{12}g'' \left(x^3 y + xy^3 \right) - \frac{1}{384}g'''' \left(x^5 y + 2x^3 y^3 + xy^5 \right), \tag{3}$$

which shows good agreement to experimental measurements [11]. In a quadrupole of aperture radius G_0 that has a main–field gradient $g(z_b)$ inside the magnet, the z dependence of $g(z)$ can be described [3] by:

$$g(z) = \frac{g(z_b)}{1 + \exp[a_0 + a_1(z/G_0) + a_2(z/G_0)^2 + ...]}, \tag{4}$$

which requires to fit the parameters a_j to a numerically given fringing–field distribution. Here z is assumed to be zero at the effective field boundary, so that the quadrupole strength at this boundary is $g(z_*) = g(z_b)/[1 + \exp(a_0)]$. Differentiating the magnetic potential $V_Q(x, y, z)$ from Eq.(3) the three–dimensional field distribution of the magnetic field can be obtained.

4. Computing the transfer matrix of the fringing field of a magnetic quadrupole by using differential algebra

To calculate a particle trajectory throughout a magnetic quadrupole, we advantageously use particle optical coordinates [1, 4, 12]

$$\begin{aligned}
r_1 &= x, & r_2 &= a = p_x/p_0, \\
r_3 &= y, & r_4 &= b = p_y/p_0, \\
r_5 &= l = v_0(t - t_0), \\
r_6 &= \delta_K = (K/q - K_0/q_0)/(K_0/q_0), \\
r_7 &= \delta_m = (m/q - m_0/q_0)/(m_0/q_0),
\end{aligned} \tag{5}$$

where x and y are the horizontal and vertical distances to the optic axis, respectively, while p_0, v_0, q_0, K_0, m_0 and t_0 denote momentum, velocity, charge, kinetic energy, mass and time of flight of the reference particle, whereas p, v, q, K, m and t stand for the same quantities of the particle under consideration. In these coordinates the equations of motion in a magnetic quadrupole take the form

$$
\begin{aligned}
x' &= a(p_0/\overline{p}) \\
y' &= b(p_0/\overline{p}) \\
l' &= (v_0/v)(p/\overline{p}) - 1 \\
a' &= \frac{q/q_0}{\chi_B}\left(B_z b\frac{T_0}{T} - B_y\frac{\overline{v}}{v_0}\right)(l'+1) \\
b' &= \frac{q/q_0}{\chi_B}\left(-B_z a\frac{T_0}{T} + B_x\frac{\overline{v}}{v_0}\right)(l'+1).
\end{aligned}
\tag{6}
$$

Here $\chi_B = p_0/q_0$ is the magnetic rigidity of a reference particle and we used the abbreviations $\overline{p} = p\sqrt{1 - (p_0/p)^2(a^2+b^2)}$ and $\overline{v} = v\sqrt{1 - (v_0/v)^2(T_0/T)^2(a^2+b^2)}$ as well as

$$
\begin{aligned}
\frac{T}{T_0} &= \left(1 + \frac{2\eta\delta_K + \delta_m}{1+2\eta}\right), \\
\frac{p}{p_0} &= \sqrt{(1+\delta_k)\left(1 + \frac{\eta\delta_K + \delta_m}{1+\eta}\right)}, \\
\frac{v}{v_0} &= \sqrt{(1+\delta_K)\left(1 + \frac{\eta\delta_K + \delta_m}{1+\eta}\right)}\bigg/\left(1 + \frac{2\eta\delta_K + \delta_m}{1+2\eta}\right),
\end{aligned}
$$

where $\eta = K_0/2m_0 c^2$ is used for relativistic corrections with c denoting the speed of light [1].

To obtain the elements of the proper transfer matrix from Eq.(6) one must simply replace all quantities that depend on the particle coordinates (with respect to which we want to differentiate) by DA–vectors. In our case, this includes for instance x, y, l, a, b, and the three components B_x, B_y, B_z of the flux density $\vec{B}$ in Eq.(6), the quantities δ_K, $\overline{p}/p_0$, $\overline{v}/v_0$ and η stay scalars. After replacing the real arithmetic by DA arithmetic a standard integration algorithm like Runge–Kutta can be used to solve the equations of motion [4].

5. Computing the transfer matrix of the fringing field of a magnetic quadrupole by using the technique of fringing–field integrals

An approximate solution of the fringing–field effects of the fifth–order transfer matrix can also be described by fringing–field integrals. The accuracy of the transfer matrix is given in an effective fifth–order approximation. Assuming that Δz, the extension of the fringing field, is of the same magnitude as the diameter of the particle beam, only those integrals for the matrix elements of order n must be taken into account into which Δz enters up to the $(5 - n)$ power. This assumption can be done, because the maximum diameter can be twice the aperture radius and the size of Δz is several G_0.

5.1. The particle trajectory in the real fringing field

The real particle trajectory can be calculated form Eq.(6) by the method of successive approximation assuming that the aperture radius G_0 is small. For example in the x-z plane one obtains to first order with $k = g/\chi_B$ for the geometric elements:

$$
\begin{aligned}
x_f &= x_i\left(\ 1 - \iint k d^2 z + \iint k \iint k d^4 z\right) \\
&\quad + a_i\left(\ z - \iint k z d^2 z\right) \\
a_f &= x_i\left(\ -\int k dz + \int k \iint k d^3 z\right) \\
&\quad + a_i\left(\ 1 - \int k z dz + \int k \iint k z d^3 z\right)
\end{aligned}
\tag{7}
$$

and to third order:

$$
\begin{aligned}
x_f &= x_i^3\left(\ \tfrac{1}{12}k - \tfrac{1}{4}\iint k'' \iint k d^4 z - \tfrac{1}{12}\iint k^2 d^2 z\right) \\
&\quad + x_i^2 a_i\left(\ \tfrac{1}{4}\iint k'' z d^2 z\right) \\
&\quad + x_i a_i^2\left(\ \tfrac{1}{4}\iint k'' z^2 d^2 z - \tfrac{3}{2}\iint k d^2 z\right) \\
&\quad + a_i^3\left(\ \tfrac{1}{2}z\right) \\
a_f &= x_i^3\left(\ \tfrac{1}{12}k' - \tfrac{1}{4}\int k'' \iint k d^3 z - \tfrac{1}{12}\int k^2 dz\right) \\
&\quad + x_i^2 a_i\left(\ \tfrac{1}{4}\int k'' z dz - \tfrac{1}{4}\int k'' \iint k z d^3 z - \right. \\
&\quad\qquad \left. \tfrac{1}{4}\int k \iint k'' z d^3 z - \tfrac{1}{2}\int k'' z \iint k d^3 z\right) \\
&\quad + x_i a_i^2\left(\ \tfrac{1}{4}\int k'' z^2 dz\right) \\
&\quad + a_i^3\left(\ \tfrac{1}{12}\int k'' z^3 dz - \tfrac{1}{2}\int k z dz\right)
\end{aligned}
\tag{8}
$$

and finally to fifth order:

$$
\begin{aligned}
x_f &= x_i^5\left(\ \tfrac{1}{48}\iint k'' k d^2 z - \tfrac{1}{384}k'' + \right. \\
&\quad\qquad \left. \tfrac{5}{384}\iint k'''' \iint k d^4 z + \tfrac{1}{384}\iint k k'' d^2 z\right) \\
&\quad + x_i^4 a_i\left(\ -\tfrac{5}{384}\iint k'''' z d^2 z\right) \\
&\quad + x_i^3 a_i^2\left(\ \tfrac{1}{8}k - \tfrac{5}{192}\iint k'''' z^2 d^2 z\right) \\
&\quad + a_i^5\left(\ \tfrac{3}{8}z\right) \\
a_f &= x_i^5\left(\ \tfrac{1}{48}\int k'' k dz - \tfrac{1}{384}k''' + \right. \\
&\quad\qquad \left. \tfrac{5}{384}\int k'''' \iint k d^3 z + \tfrac{1}{384}\int k k'' dz\right) \\
&\quad + x_i^4 a_i\left(\ \tfrac{1}{24}\int k'' k z dz + \tfrac{1}{16}\int k'' \iint k'' z d^3 z - \tfrac{5}{384}\int k'''' z dz + \right. \\
&\quad\qquad \left. \tfrac{5}{96}\int k'''' z \iint k d^3 z + \tfrac{5}{384}\int k'''' \iint k z d^3 z + \tfrac{5}{384}\int k \iint k'''' z d^3 z\right) \\
&\quad + x_i^3 a_i^2\left(\ -\tfrac{5}{192}\int k'''' z^2 dz\right) \\
&\quad + x_i^2 a_i^3\left(\ \tfrac{1}{8}\int k'' z dz - \tfrac{5}{192}\int k'''' z^3 dz\right).
\end{aligned}
\tag{9}
$$

All integrations started at z_a, far outside of the quadrupole, and ended at z_b, well inside the quadrupole. Thus the quadrupole field strength is z-independent at z_a and at z_b.

5.2. Transferring the fringing–field effects to the effective field boundary

In section 5.1 the equation of motion was solved for the whole fringing–field region. In our fringing–field model, however, one is interested in a reduced description at the effective field boundary. For this reason fringing–field integrals must be defined, which depend on the real field distribution, but are independent of the integration boundaries z_a and z_b. To fifth order 10 standard integrals are needed instead of 4 in ref. [8]:

$$
\begin{aligned}
I_{1a} &= k_0^{-1} \iint k d^2 z - \frac{1}{2} z_b^2 \\
I_{1b} &= k_0^{-2} \iint k^2 d^2 z - \frac{1}{2} z_b^2 \\
I_2 &= k_0^{-1} \int z \int k d^2 z - \frac{1}{3} z_b^3 \\
I_3 &= k_0^{-2} \int \left(\int k dz \right)^2 dz - \frac{1}{3} z_b^3 \\
I_{4a} &= k_0^{-2} \int k^2 dz - z_b \\
I_5 &= k_0^{-2} \int k'^2 dz \\
I_6 &= k_0^{-2} \int k'^2 z dz \\
I_8 &= k_0^{-2} \iint k'^2 d^2 z - I_5 z_b \\
I_{11a} &= k_0^{-2} \int \left(\int k dz \right) \left(\iint k d^2 z \right) dz - \frac{1}{2} I_{1a} z_b^2 - \frac{z_b^4}{8} \\
I_{11b} &= k_0^{-2} \int z \left(\int k dz \right)^2 dz - \frac{z_b^4}{4},
\end{aligned}
\tag{10}
$$

where $k_0 = g(z_b)/\chi_B$ is the normalized field gradient inside the quadrupole. These integrals are of different approximation order in Δz. Note that the integral I_5 is proportional to $(\Delta z)^{-1}$ and produces diverging elements, if the aperture radius becomes smaller and smaller. The advantage of this integrals is that they include the influence of the reverse matrices of the drift length and the ideal magnetic quadrupole, so that automatically the fringing–field transfer matrix is obtained. All the integrals of Eqs.(7,8,9) can be transformed to a linear combination of the standard integrals of Eq. 10, so one gets for the geometric matrix elements:

$$
\begin{aligned}
(X, X) &= 1 - k_0 I_{1a} + k_0^2 (I_{11a} + I_{11b}) \\
(X, A) &= -2 k_0 I_2 \\
(A, X) &= -k_0^2 I_3 \\
(A, A) &= 1 + k_0 I_{1a} + k_0^2 (I_{11a} - I_{11b}) \\
(X, XXX) &= \tfrac{1}{12} k_0 - k_0^2 (\tfrac{1}{4} I_{1a} + \tfrac{1}{3} I_{1b}) \\
(A, XXX) &= -\tfrac{1}{3} k_0^2 I_{4a} \\
(A, XXA) &= -\tfrac{1}{4} k_0 + k_0^2 (\tfrac{1}{4} I_{1a} + I_{1b}) \\
(X, XXXXX) &= k_0^2 (\tfrac{1}{32} - \tfrac{7}{192} I_8) \\
(X, XXXAA) &= -\tfrac{1}{32} k_0 \\
(A, XXXXX) &= -\tfrac{7}{192} k_0^2 I_5 \\
(A, XXXXA) &= k_0^2 (-\tfrac{3}{32} - \tfrac{35}{192} I_6) \\
(A, XXAAA) &= \tfrac{1}{32} k_0
\end{aligned}
\tag{11}
$$

Analogously all other matrix elements can be determined. The transfer matrix of the

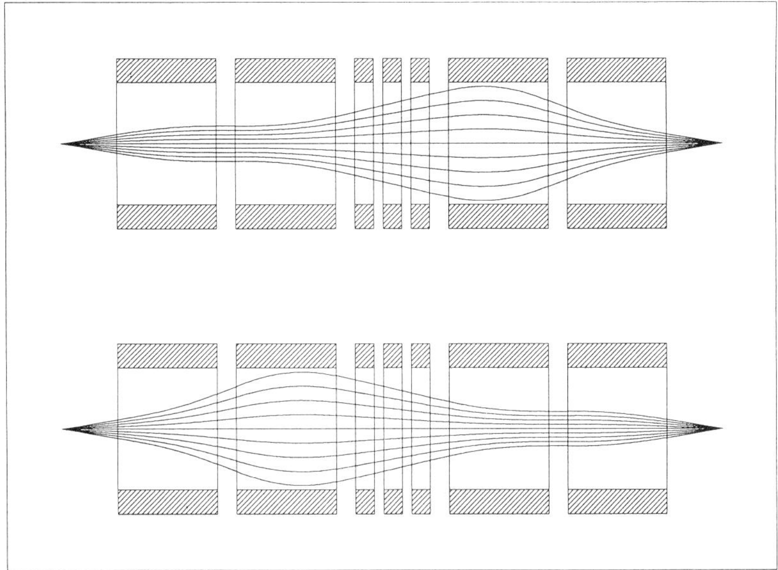

Figure 2. Beam envelopes of the stigmatic focussing system. The system has a total length of 1.768 m, the four quadrupoles have a fieldstrength of 0.8 T/m and a effective length of 0.267 m.

exit fringing field is the reversed matrix of the entrance fringing–field matrix and can be obtained by inverting the entrance matrix and then changing all the signs of a and b [13].

6. Comparison between DA ray tracing and directly calculated fringing–field integrals

To compare the results of ray tracing using DA and the fringing–field integrals method we calculated a quadrupole quadruplet used as a demonstration example in [14]. This system is stigmatic focussing with unity magnification and uses three octupoles to correct for the geometric image aberrations of third order (see fig. 2). For the field distribution of the quadrupole magnets we used Eq.(4) with the coefficients:

$$a_0 = 0.296471 \quad a_1 = 2.266609 \quad a_2 = -0.567746$$
$$a_3 = 0.133578 \quad a_4 = -0.002274 \quad a_5 = 0.000696$$

as reported in ref. [15]. From this fringing–field distribution we calculated the magnitudes of the integrals of Eq.(10) as:

$$I_{1a} = 0.32168\ G_0^2 \quad I_{1b} = 0.23267\ G_0^2 \quad I_2 = -0.08797\ G_0^3$$
$$I_3 = 0.12818\ G_0^3 \quad I_{4a} = -0.43282\ G_0 \quad I_5 = 0.40798\ G_0^{-1}$$
$$I_6 = 0.07112 \quad I_8 = -0.07112 \quad I_{11a} = 0.05174\ G_0^4$$
$$I_{11b} = 0.03565\ G_0^4$$

To analyze the influence of the fringing–field distribution we calculated the entrance fringing–field transfer matrix for single charged particles with mass $m_0 = 4m_p$ and energy $K_0 = 100$ keV. In fig. 3 the variation of some representative first–, third– and fifth–order elements of the fringing–field transfer matrix are shown for aperture radii 0.005 m $\leq G_0 \leq 0.05$ m at a constant field gradient of 1.6 T/m. In table 1 the corresponding matrix elements are listed for an aperture radius G_0 of 0.05 m . The first–order elements

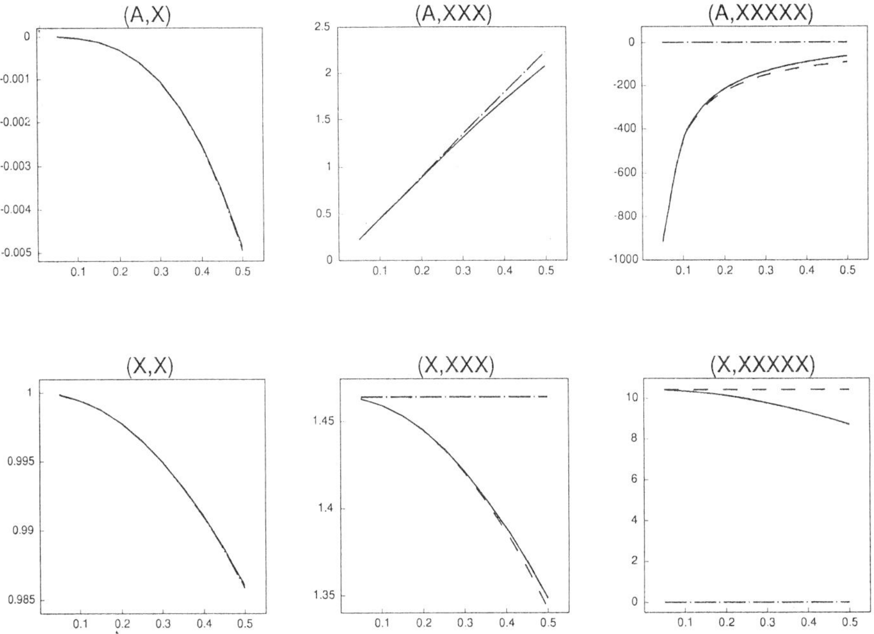

Figure 3. Comparison of representative matrix elements between the results of DA integration (solid), the here presented method (dashed) and the method of ref. [8](dash–dot).

in fig. 3 show no significant difference, though we read from table 1, that the element (X, X) is now better approximated due to the additional integral in Eq.7 as compared to ref [8]. Also the element (X, XXX) is better represented in the new approach. The elements (A, X) and (A, XXX) have the same values in the two approaches, because no additional correction terms occur. For the fifth–order terms only the terms up to zeroth order in Δz are taken into account which may explain that there are still differences to the correct ray tracing result, though these differences reduce considerably for smaller G_0 values. Note here that the element $(A, XXXXX)$ shows a diverging behavior for vanishing aperture, which is well represented in the fringing–field integral description. The difference between the ray tracing value and the approximated value for $G_0 = 0.05$ meter results from a large first–order term in Δz. In this approximating approach the Element $(X, XXXXX)$ is independent of the aperture radius but depends on the detailed field distribution, which is verified for small aperture radii.

Furthermore we investigated the influence of the fringing–field distribution on the focus-size of the beam for this stigmatic focussing example. To correct the image aber-

matrix elements			
	DA integration	our results	results of ref [8]
(X,X)	9.860360E-01	9.860379E-01	9.858693E-01
(A,X)	-4.872697E-03	-4.946890E-03	-4.946890E-03
(X,XXX)	1.348298E+00	1.342328E+00	1.464265E+00
(A,XXX)	2.073417E+00	2.227194E+00	2.227194E+00
(X,XXXXX)	8.701065E+00	1.044888E+01	0.000000E+00
(A,XXXXX)	-6.403380E+01	-9.184756E+01	0.000000E+00

Table 1. Numerical values of some matrix elements of the system as shown in fig.2 with $G_0 = 0.05$ m calculated with the three different methods.

rations of third order we used octupoles and fitted the octupole strengths by using the DA integration technique. These strengths increased by a factor of ≈ 5 by taking the fringing–field effects into account. In fig. 4 the particle density distribution in x–direction in the image plane is shown for the corrected system. To compare the different approximation methods only the transfer matrix up to third order was used thus neglecting all fifth–order terms. As one can see three methods produce almost the same beam shape, but the beam shape computed by ref [8] shows a peak broadening caused by the not fully corrected third–order terms.

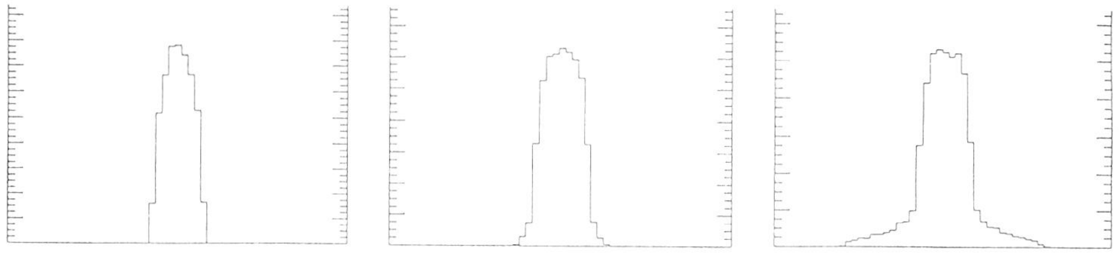

Figure 4. Beam shape in the $x - z$ plane at the focus of the corrected system using DA integration (left), the presented method (center) and the method of ref.[8] (right).

References

[1] H. Wollnik 1987 *Charged Particle Optics* (Orlando, Florida: Academic Press)

[2] H. Wollnik 1991 Nucl. Instrum. Methods **B56/57** 1096

[3] S. Kowalski and H. Enge 1987 Nucl. Instrum. Methods **A258** 407

[4] B. Hartmann, M. Berz, and H. Wollnik 1990 Nucl. Instrum. Methods **297** 343

[5] R. Ryne 1987 PhD thesis, University of Maryland, USA

[6] H. Wollnik and H. Ewald 1965 Nucl. Instrum. Methods **36** 93

[7] H. Wollnik 1965 Nucl. Instrum. Methods **38** 56

[8] H. Matsuda and H. Wollnik 1972 Nucl. Instrum. Methods **103** 117

[9] E. Lee Whiting 1970 Nucl. Instrum. Methods **83** 232

[10] D. Smith 1970 Nucl. Instrum. Methods **79** 144

[11] G. R. Moloney, D. N. Jamieson and G. J. F. Legge 1991 Nucl. Instrum. Methods **B54** 24

[12] M. Berz and H. Wollnik 1987 Nucl. Instrum. Methods **A258** 364

[13] M. Berz 1988 *IEEE Trans. Electron Devices* **ED-35** 2002

[14] M. Berz. 1990 Nucl. Instrum. Methods **A298** 426

[15] K.L. Brown and J. E. Spencer 1981 *IEEE Trans. on Nuclear Sci.* **NS-28,3** 2568

Non-linear beam transport effects in highly charged positive ion beams extracted from ECR ion sources

Timothy A Antaya [1]

National Superconducting Cyclotron Laboratory, Michigan State University, East Lansing MI 48824

Abstract.
"Electron Cyclotron Resonance Ion Sources are now in use at about 40 laboratories world-wide for highly charged ion injection into accelerators for physics research[1]. Using this type of ion source, heavy ions with charge as high as 39+ have been injected directly into cyclotrons at intensities of a few hundred nanoamperes[2], as well as nearly fully-stripped light ions at intensities of a few hundred microamperes for cyclotron or synchrotron injection[3]. When compared to H- and H+ beam transport systems, these are not ion beam intensities normally considered to be space charge dominated, and so a careful measurement of the initial beam properties should be sufficient to specify the beam transport system design and accelerator matching conditions. However, experimentally it is observed that both extremes– highly charged ions at low intensities and low charge ions at high intensities, exhibit sharp non-linear emittance growth in beam transport systems. This is true even in beam transport systems presumably having substantially higher designed acceptance than the initial beam emittance. In extreme cases, the overall beam transmission has actually been observed to decrease sharply with increasing beam intensity[4]. This mis-match has either been ignored or interpreted to imply that the initial beam characteristics were not well known. A correct interpretation of this emittance growth, as being a direct consequence of a non-linear growth in the beam envelope due to space charge forces, will be demonstrated using theory, simulations and experimental measurements. For proper matching of ECR ion source beams to accelerators, an aberration-free maximum beam intensity must be built into the beam transport system. "

---

[1] E-mail: Antaya@MSUNSCL ; Antaya@megapc.nscl.msu.edu

1. First emittance measurements

Highly charged positive ions produced by Electron Cyclotron Resonance (ECR) ion sources are generally transported to accelerators at energies of 5-30 $kV \times Q$, where Q is number of electrons removed. Without making too great a generalization, these low energy beam transport systems (LEBT) share many design assumptions:

1. These beams are dc with particle currents of 0.001-100.0 pμA.

2. These beams are essentially monochromatic, thus $K_Q \ll QV_{ECR}$ where K_Q is the thermal energy of an ion of charge Q.

3. The transverse unnormalized emittances are symmetric, and of magnitude $\epsilon_x = \epsilon_y \leq 200mm \cdot mrad$ $\qquad (\epsilon \equiv A_x/\pi!)$

While generally held to be valid, these assumptions have not been sufficient to properly design and transport ECR beams to accelerators.

Figure 1 shows the RTECR and beam selection system at MSU/NSCL. Essentially, diffusion end loss positive ions are accelerated in a spherical Pierce-type electrode system[5]. All charges are extracted simultaneously at a total current of 0.5-1.0 emA. The total intensity does depend on the details of the RTECR " tune". Some coarse charge selection is obtained in transit of the solenoid lens, particularly for a narrow setting of the 90° magnet object slit. The 90° magnet has de facto unit magnification, as the image slit, shown in the inset in Fig. 1, is normally set to the same opening as the object slit. Beam current can be read at FC# 1, partially resolved in Q/M, or resolved in Q/M at FC# 2. For heavy mass species, like Xenon, forty or more different charges can be extracted simultaneously; hence, the beam current reading at FC# 1 is often meaningless.

Emittance measurements can be made in either transverse coordinate in the Divergence Box or the FC#2 box. We have made 2D 'pepper pot' emittance measurements at each location, but we find a 'wire scanner' measurement at a single M/Q at FC#2, in the dispersion plane of the dipole, to be most useful, as shown in Fig. 1. Such an emittance measurement for Argon 10+ ions is shown in Fig. 2. This measurement shows a central core with a high divergence tail. Similar measurements have been made on other ECR ion sources[6,7] . The 90% emittance is $245mm \cdot mrad$, but there is an equal amount of emittance in the 10% intensity tail. The 90% and 100% intensity emittances for Argon 2+ through 10+ ions are shown in Fig. 3. The emittance increases with Q, while the intensity decreases with Q. In each case, fully 1/2 of the emittance is in the 10% intensity tail. (To make such large emittance measurements, the analysis slits have been fully opened.)

2. Theoretical analysis

Heretofore, ECR ion source beam emittance measurements like those in Fig. 3 have resisted explanation. The first question then is, how large should the emittance from an ECR ion source be?

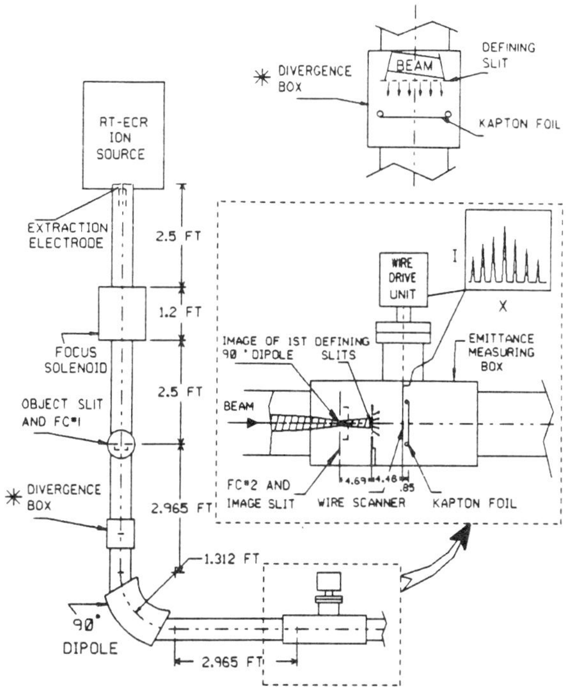

Figure 1. The RTECR charge state analyzed emittance measurement apparatus is shown. Both the focussing solenoid and the 90° dipole magnet have unit magnification. The wire scanner can be moved to permit measurements in the other transverse plane.

The radius of the extraction of ion beams from ECR sources is shown schematically in Fig. 4. Ions are extracted from the approximate maxima of the exit magnetic mirror, in a circular aperture of radius a.

ECR ion sources have 3D minimum B magnetic structures[8], and in the RTECR, this is obtained through the superposition of a 30cm bore $SmCo_5$ hexapole with a warm bore tandem magnetic mirror[9]. Since a=0.8cm for the RTECR, $\vec{A} = A_\phi(r,z)\hat{\phi}$ is valid within the extraction aperture, the hexapole nonwithstanding.

The Hamiltonian for this system is then:

$$H = \frac{1}{2M}\left(P_r^2 + \frac{P_\phi^2}{r^2} + P_z^2 + Q^2 A_\phi(r,z) - \frac{2Q}{r}A_\phi(r,z)\right) + QV \tag{1}$$

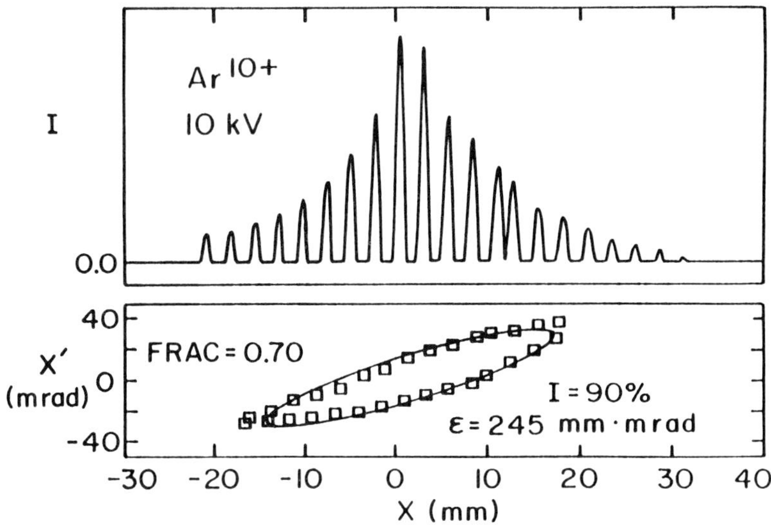

Figure 2. The emittance of Argon 10+ ions as measured by the apparatus shown in Fig. 1. The 90% current level includes only about one half of the total emittance.

The total electrostatic potential (V) should have azimuthal symmetry, and include the extraction fields (ex) and the beam space charge (sp):

$$V(r, z) = V_{ex}(r, z) + V_{sp}(r, z) \tag{2}$$

As indicated, ECR beams are normally extracted continuously, and it is then assumed (but not measured) that such positive dc beam intensities should neutralize in the LEBT. Then we take $V_{sp} = 0$ and we note that

$$\frac{\partial H}{\partial \phi} = \phi \dot{P}_\phi = 0 \tag{3}$$

and hence that the azimuthal canonical momentum is a constant of motion.

$$P_\phi = p_\phi + Qr A_\phi = \text{const} \tag{4}$$

Following this line of analysis further, one can show that

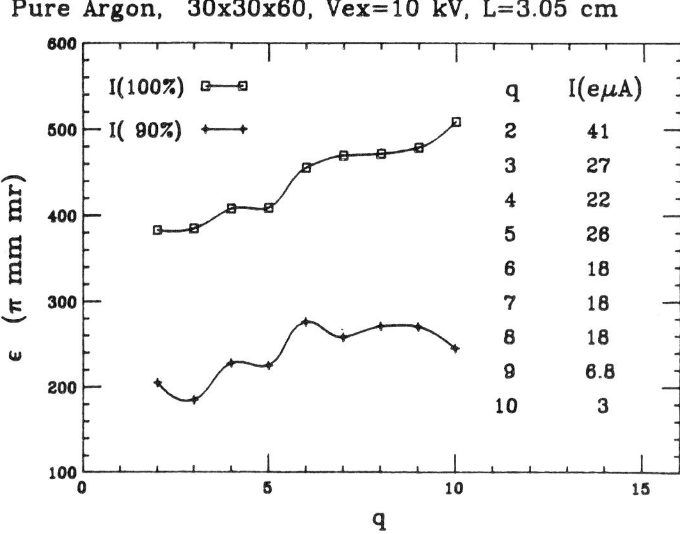

Figure 3. Both 90% and 100% current level emittances are shown for Argon 2+ through 10+ ions extracted from the RTECR ion source.

$$\epsilon_x = \epsilon_y = |P_{\phi_f}/p_{z_f}| \tag{5}$$

where p_{z_f} is the final z mechanical momentum. Furthermore, the maximum emittance is determined by ions starting at radius r=a.

$$\epsilon_{max} = |P_\phi(r = a)/p_{z_f}| = a \left| \frac{Mv_\phi(a,o) + QA_\phi(a,o)}{p_{z_f}} \right| \tag{6}$$

If, as we have said, $K_Q \ll QV_{ex}$, then

$$p_{z_f} \simeq \sqrt{\frac{2QV_{ex}}{M}} \tag{7}$$

Introducing a thermodynamic temperature T_Q as

$$K_Q = \frac{3}{2}T_Q \tag{8}$$

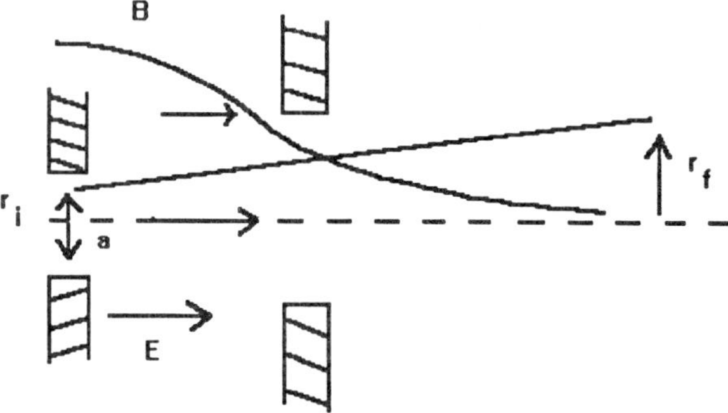

Figure 4. A simplified description of the extraction geometry of an ECR ion
is shown.

then

$$v_\phi = \sqrt{T_Q/M} \tag{9}$$

and we may rewrite the transverse emittance as

$$\epsilon = \epsilon_x = \epsilon_y = a[(\frac{T_Q}{2QV_ex})^{1/2} + A_\phi(a,o)\left(\frac{Q}{2MV_{ex}}\right)^{1/2}] \tag{10}$$

There are two limiting cases:

$$\text{Hot Ions} \qquad \epsilon \approx a\left(\frac{T_\phi}{2QV_{ex}}\right)^{1/2} \tag{11}$$

$$\text{Cold Ions} \qquad \epsilon \simeq a^2 Q B_z(a,o)\left(\frac{Q}{2MV_{ex}}\right)^{1/2} \tag{12}$$

In Equation (12), we have taken $A_\phi(a,o) \simeq aQB_z(a,o)$. Figure 5 shows the calculated effect of temperature on Argon ion beam emittance. In this calculation, other
conditions are the same as the measured argon ion beam emittances of Fig. 3. Only for
$T_i = 0$ does the dependence of emittance on Q agree with Fig. 3 — however, the measured emittances are five times larger. Figure 6 shows the magnetic field and extraction

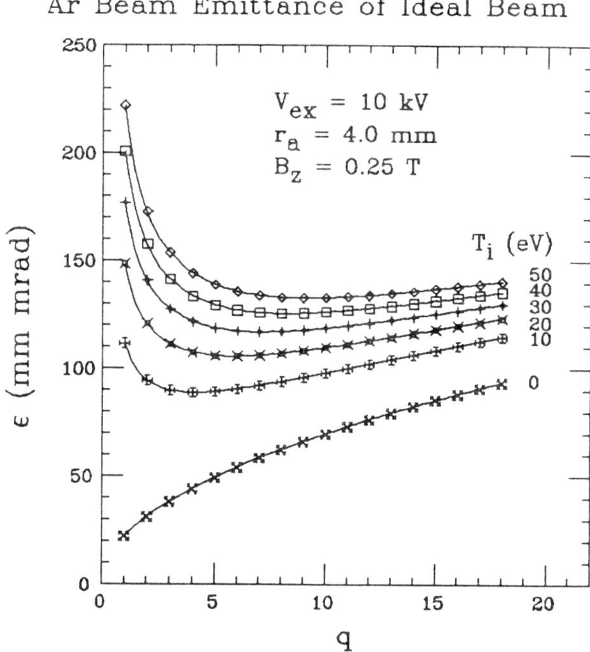

Figure 5. The calculated effects of ion temperature on the ideal argon beam emittance as a function of charge-to-mass ratio. Ions are assumed extracted at 10kV, extraction aperture $r_a = 4mm$ and with $B_z = 0.25T$. (approximately the same conditions as for the RTECR extraction)

voltage dependence of argon ion beam emittance for $T_i = 0$, the limit of Equation (12). At 10 kV, and $B_z(a, o) = 0.25T$, the predicted emittance is $70mm \cdot mrad$ for Ar^{10+}. The experimental numbers (from Fig. 3) are, in comparison $245mm \cdot mrad$ for the core, and $500mm \cdot mrad$ when including the tail. Therefore, if we have hot beams, the theory does not predict the Q dependence; and in the cold beam limit, the predicted emittances are far too small!

3. Extraction simulations

The measured thermal energy of ions extracted from the RTECR is about $6eV \times Q$ [6], and hence the high ion energies alone are not sufficient to explain the large measured emittances. The acceptance of the beam transport system is $200mm \cdot mrad$ in both transverse planes [7], but the measured emittances exceed this acceptance. Therefore, aberrations may be causing an emittance growth —but what is driving the aberrations?

Beam simulations with BEAM3D [8] do explain the measured emittances if we assume that there is zero neutralization of the extracted beam, as we will demonstrate.

Ar Beam Emittance of Ideal Beam

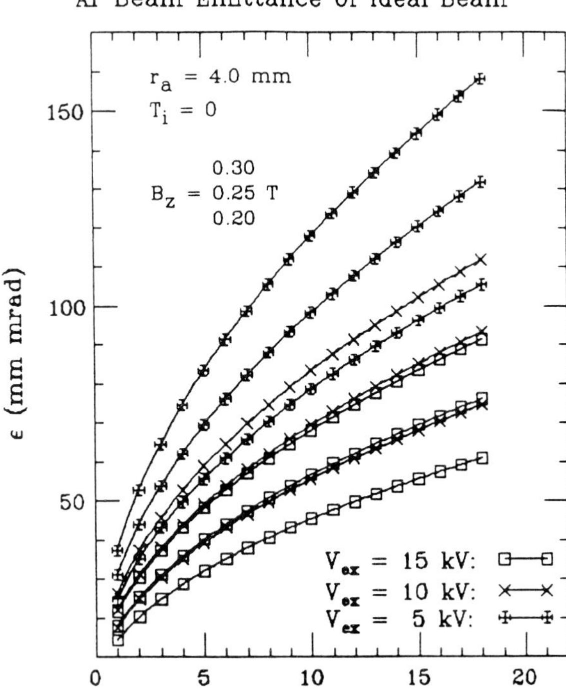

Figure 6. Theoretical emittances of argon ions extracted from the RTECR, in the cold ion limit, for various extraction voltages and magnetic fields.

It is useful first to review how this is possible. If we view the dc beam as an ensemble of ions, then the total radial space charge force is a summation over all particles.

$$E_r = \frac{1}{2\pi\epsilon_o r} \sum \left(\frac{I_i}{v_{zi}}\right) \tag{13}$$

The equation of motion for an ion on the edge of a drifting azimuthally symmetric beam is then

$$P_r = \frac{P_\phi^2}{Mr^3} + \frac{Q}{2\pi\epsilon_o r} \sum \left(\frac{I_i}{v_{zi}}\right) \tag{14}$$

Integration once yields

$$v_r = \left[\frac{P_\phi}{M^2}\left(\frac{1}{r_m^2} - \frac{1}{r^2}\right) + \frac{Q}{\pi M \epsilon_o} \ln(\frac{r}{r_m}) \sum (\frac{I_i}{v_{zi}})\right]^{1/2} \tag{15}$$

Where r_m is the initial beam radius $r = r_m$ at $t = 0$. Observing that $v_\phi = P_\phi / M_r$ and defining $v_\perp \equiv \sqrt{v_\phi^2 + v_r^2}$, then we can write the transverse divergence of the beam as

$$\alpha_\perp = \left[\frac{P_\phi^2}{M^2 v_z^2 r_M^2} + \frac{Q}{\pi M \epsilon_o v z^2} \ln\left(\frac{r}{r_m}\right) \sum \left(\frac{I_i}{v_{zi}}\right) \right]^{1/2} \tag{16}$$

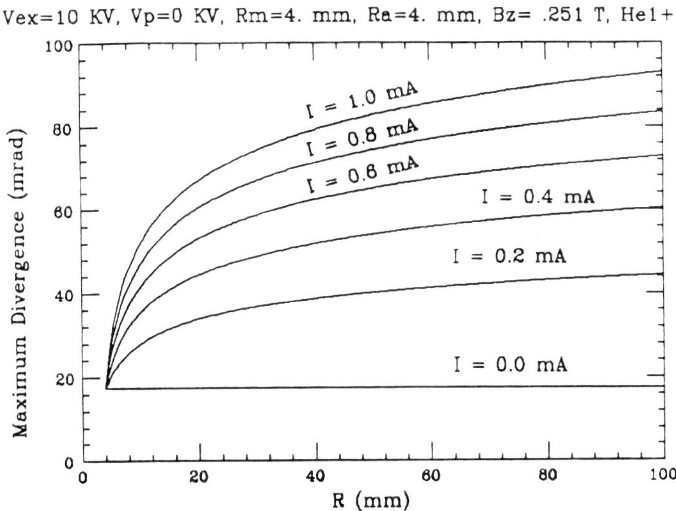

Figure 7. With increasing space charge, the maximum divergence in the radial envelope is seen to grow rapidly as the beam radius grows (with drift distance) .

Equation (16) shows that the beam surface will have a constant maximum transverse divergence if the space charge force is zero, that is $\sum (I_i / v_{zi}) = 0$. This is of course the normal assumption in the LEBT design for ECR sources. One measures (or assumes) an initial maximum divergence at a waist after the ion source, and lets the beam drift, inserting lenses as necessary to limit the maximum beam envelope. However, with space charge, the maximum beam divergence has a strong dependence on the beam radius, as shown in Fig. 7, using Equation (16). (We again use conditions appropriate for beams extracted from the RTECR.) Fig. 7 shows that the beam divergence increases with beam radius, and of course the beam radius increases as the beam drifts in the LEBT system. To determine the beam radius dependence on drift distance, we integrate Equation (16) once with respect to r, and find that

$$\int_{r_m}^{r} \left[\frac{P_\phi^2}{M^2} \left(\frac{1}{r_m^2} - \frac{1}{r^2} \right) + \frac{Q}{\pi M \epsilon_o} \ln\left(\frac{I_i}{v_{zi}}\right) \right]^{-1/2} dr = t \tag{17}$$

There exists no analytical solution for the left side of Equation (17), but numerical integration techniques can be used to estimate the integral. Let $r = r_m + x$ and assume that x is small (i.e., a short drift distance). Then we have

$$\frac{P_\phi}{M^2}\left(\frac{1}{r_m^2} - \frac{1}{r}\right) + \frac{Q}{\pi M \epsilon_o}\ln(\frac{r}{r_m})\sum(\frac{I_i}{v_{zi}}) \approx x\left[\frac{2P_\phi^2}{M^2 r_m^3} + \frac{Q}{\pi M \epsilon_o r_m}\sum(\frac{I_i}{v_{zi}})\right] \quad (18)$$

And by substitution into Equation (17) obtain

$$t = \left[\frac{2P_\phi^2}{M^2 r_m^3} + \frac{Q}{\pi M \epsilon_o r_m}\sum(\frac{I_i}{v_{zi}})\right]\int_o^x dx'/x'^{1/2} \quad (19)$$

Performing the integration and rearranging terms yields

$$r - r_m = \frac{1}{4}\left[\frac{2P_\phi^2}{M^2 r_m^3} + \frac{Q}{\pi M \epsilon_o r_m}\sum(\frac{I_i}{v_{zi}})\right]\frac{z^2}{v_z^2} \quad (20)$$

Thus the beam envelope shows a <u>quadratic dependence</u> on the linear drift distance z.

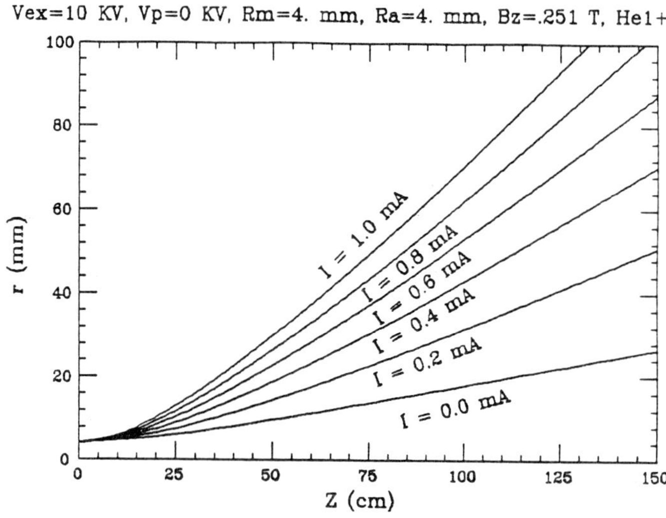

Figure 8. As the maximum divergence increases with drift distance z, the beam envelope radius must increase with z^2. This is a significant effect, even for short drift distances, for high space charge beams.

Fig. 8 shows the growth of a He^+ ion beam envelope for various intensities. In a 1 meter drift, there is a 4-fold increase in beam radius for a 1 emA He^+ beam over a beam with zero net space charge. The consequence of this non-linear radial space charge is that the beam expands faster than would be the case assuming a constant maximum divergence. We conclude that with space charge, both the beam divergence and radius increase with drift distance, as shown schematically in Fig. 9. If this is true, then it is likely that the beam envelopes will fill the lenses and bending magnets of the LEBT system, and large emittance growth due to aberrations during each lens transit will occur.

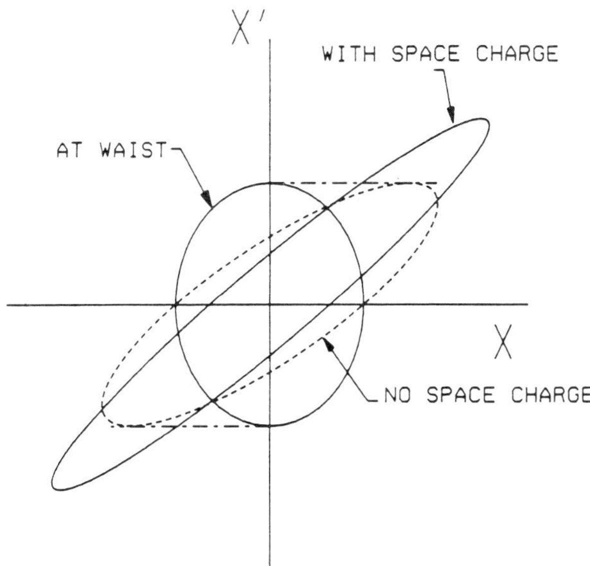

Figure 9. The effect of z dependence of the maximum divergence on the evolution of emittance after a waist is shown. With space charge both X_{max} and X'_{max} increase, however beam transport systems have generally been designed for $X'_{max} = const$. Transmission losses then would increase due to lense aberrations if the space charge is not constant.

4. Helium 1+ emittance measurements

In order to study this phenomena in detail, we developed a new beam code BEAM3D [8]. Beam3D is a 'single pass' ray tracing code with radial space charge. The space charge is carried by an unlimited number of beamlets. Since ECR sources often have many different charge states extracted simultaneously, BEAM3D allows a charge state distribution having unlimited charge states. The integration is written in cartesian coordinates in 3 dimensions, to allow the specification of electric and magnetic fields without requiring symmetries, as well as 3D specification of the beamlet starting conditions. Geometrical structures: slits, caps, etc. that might cut the beam are included, as well as the first element of the LEBT. One can extract $\epsilon_{xx'}, \epsilon_{yy'}, j, n...$ at unlimited arbitrary points along the beam path, in order to compare with measurements. For 5000 rays, BEAM3D requires 4 hours of cpu time on a Vax 8650. A sample BEAM3D calculation is shown in Fig. 10.

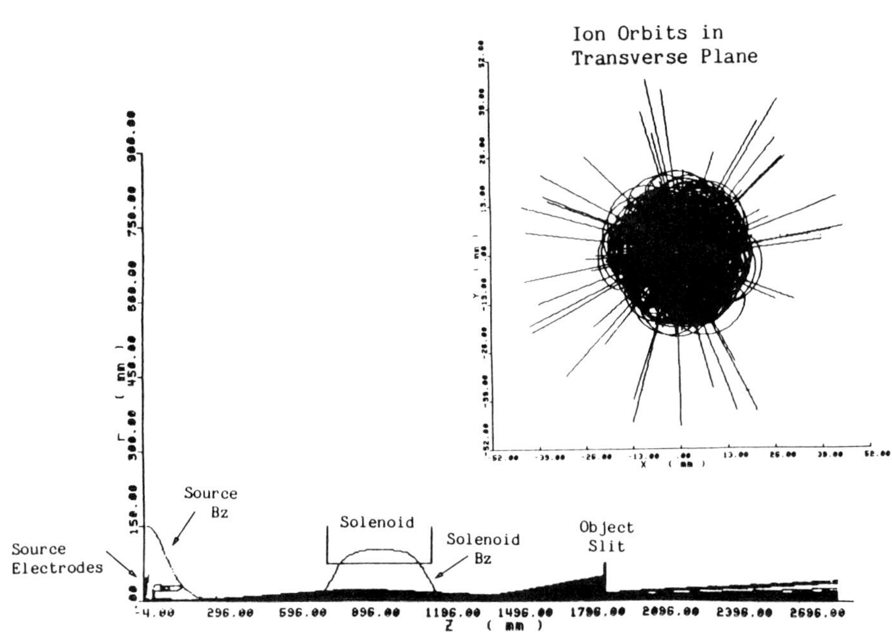

Figure 10. The axial and transverse ion orbits of He^{1+} and He^{2+} calculated by the BEAM3D code.

To study the radial space charge question, we simulated with BEAM3D the beam formation process in the RTECR, through the transit of the focussing solenoid, up to the position of the divergence box (as shown in Fig. 1). To study the extraction from the RTECR, we use He^+ beams. We have found that it is possible to tune this ion source to produce He^+ beams over a wide range of intensities, without producing significant He^{2+} intensities, which would complicate the analysis. Such a He^+ spectrum is shown in Fig. 11. Helium is the only species for which it is possible to produce an approximately mono-charged total extracted current in the RTECR.

Fig. 12 shows the BEAM3D estimate of emittance growth due to space charge for He^+ beams when crossing the focussing solenoid. This emittance growth is the result of spherical aberration due to the space charge growth of the beam envelope <u>before</u> entering the solenoid. As is seen, no emittance growth is observed at $65e\mu A$ intensity, while at 1emA, the emittance growth is nearly 6-fold. Hence theoretically, the radial space charge force could cause a large emittance growth due to aberrations in the transit

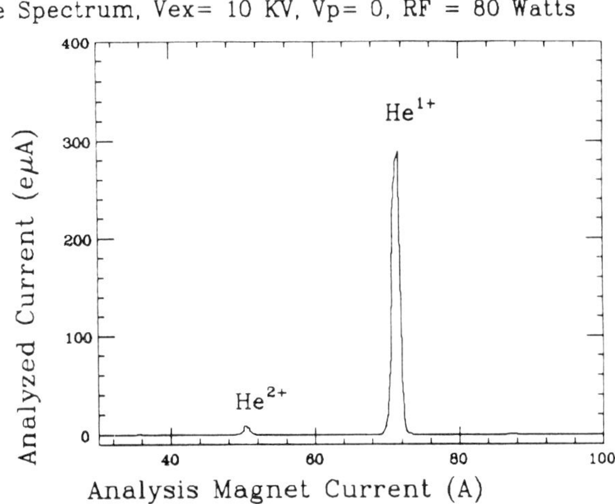

Figure 11. A helium spectrum is shown here. 300 $e\mu A$ helium 1+ was produced by the RTECR source at $V_{ex} = 10kV$, $V_p = 0$, and 80 watts of RF power, while less than 5% helium was produced.

of the focussing solenoid, but does that happen?

Fig. 13 shows actual kapton foil burns for $65e\mu A$ and $550e\mu A$ He^+ beams after transit of the solenoid. To make these burns, the beam is passed through a $10cm \times 10cm$ grid of 0.25mm slits and is imaged on kapton after a drift of 10cm. Beamlets striking kapton cause a sharp-edged discoloration. The $65e\mu A$ He^+ shows a round image of the RTECR extraction aperture, while the $550e\mu A$ beam completely fills the $10cm \times 10cm$ kapton foil, in good agreement with the prediction of Fig. 12. Further, the emittance of the $65e\mu A$ He^+ beam of Fig. 13 can be extracted, as shown in Fig. 14. This measured emittance is $69mm \cdot mrad$, in exact agreement with the starting emittance of Fig. 12. Such an emittance of $69mm \cdot mrad$ for He^+ extracted from the RTECR would result if the ion beam is cold, corresponding to the cold ion limit given in Equation (12). The emittance of the $550e\mu A$ beam cannot be obtained from Fig. 13, since the left boundary of the beam cannot be determined, but is greater than $250mm \cdot mrad$.

5. Summary and conclusions

The main point here is that dc beams extracted from ECR ion sources, like the RTECR, do not neutralize. Then the beam envelope growth will depend quadratically on the drift distance z in the LEBT: $\Delta r \propto \Delta z^2$. All LEBT designs for ECR ion sources assumed $\Delta r \propto \alpha \Delta z$. where α is the starting transverse divergence. Hence all focussing magnets are too far 'downstream'. The beam envelopes are then larger than anticipated,

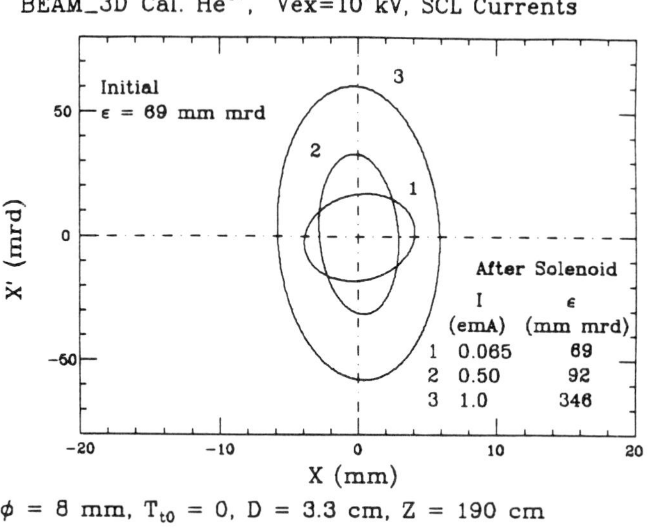

Figure 12. The effective emittance of He^{1+} after crossing the focussing solenoid for various beam intensities. In each case, the extraction is space charge limited, the beam energy is of 10keV, and the emittance after extraction is $69mm \cdot mrad$. The 1.0 emA case shows very large emittance growth due to its large beam profile in the solenoid, thus the aberrations have become very severe.

filling magnet apertures and exciting aberrations, diluting the phase space and LEBT transmission. Proper LEBT requires designing in the desired maximum beam current to be transported.

To summarize our current understanding of ECR extraction and LEBT:

1. Beam transport follows Pierce Theory up to a tune saturated current limit.

2. Assuming a flat plasma emission surface in the simulations best fits experimental measurements.

3. $T_Q \leq 10eV \times Q$.

4. ϵ_x and ϵ_y are dominated by A_ϕ.

5. Extracted beams appear to have low neutralization.

The limitations of this study are as follows. First, we have demonstrated this effect only for He^+ beams extracted from the RTECR. We are now in the process of extending these measurements to highly charged ions, the original goal of this effort. Second, while the beam envelope growth of He^+ is consistent with radial space charge, we have not as yet <u>measured</u> that the degree of neutralization is essentially zero.

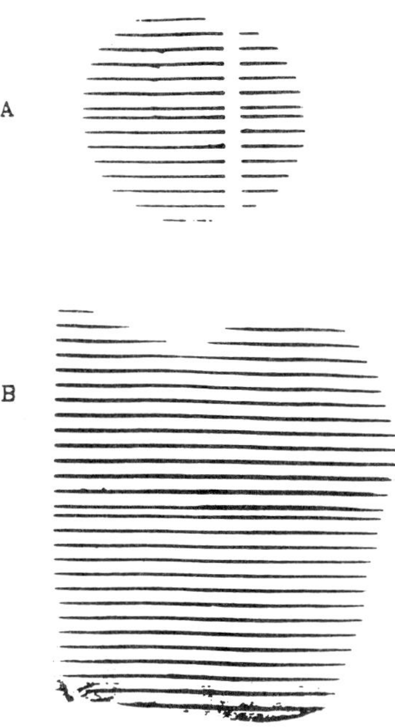

Figure 13. A and B are Kapton foil burns at the RTECR divergence box (see Fig. 1) with He^+ beams of 65 and 550 $e\mu A$ respectively.

References

[1] Meyer F 1990 *Proc. 10th Int. Conf. ECRs, Oak Ridge Tenn.*

[2] MSU/NSCL Beamlist, March 1992

[3] Geller R , private communication

[4] Baron E 1986 *Cyclotron '86 Tokyo, Ionics* 234

[5] Antaya T 1988 *Proc. Int. Conf. ECRs, Grenoble France.* 707

[6] Clark D 1987 *Int. Conf. ECR ion sources, MSU East Lansing, Michigan.* 433

[7] Hagedoorn H 1987 *Int. Conf. ECR Ion Sources, MSU East Lansing, Michigan.* 389

[8] Geller R 1978 *8th Int. Conf. on Cyclotrons, Bloomington Indiana.* 2120

[9] Antaya T 1986 *7th Workshop on ECR Ion Sources, Julich Germany.* 72

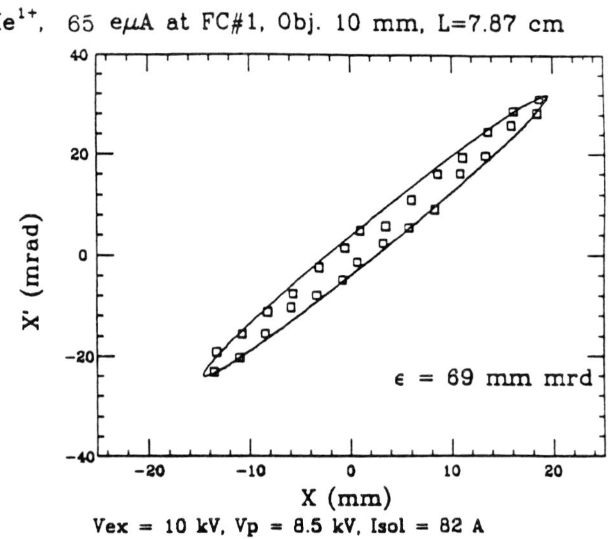

Figure 14. The measured emittance for the 65 $e\mu AHe^+$ beam in Fig. 13 is $\epsilon = 69mm \cdot mrad$, which agrees very well with the BEAM3D calculation, in which the ion thermal energy was taken to be zero.

Inst. Phys. Conf. Ser. No 131
Paper presented at Int. Workshop Nonlinear Problems in Accelerator Phys. Berlin, 1992

Status of MAD (Version 8.5) and Future Plans

F. Christoph Iselin, CERN, SL/AP Group

1. Introduction

The MAD project was started in 1981 at CERN to provide a flexible framework for particle optical computations. From the beginning the following features were considered important:

- Open-ended and modular design of the program to allow easy addition of new features.

- Format-free input language to facilitate data preparation. Later the requirement was added that the language should be understood by a variety of other programs. When moving data to other programs this would avoid extensive translations of the data with the corresponding danger of making errors.

- The internal data structure should describe the machine in a flexible manner and access to this structure should be simple.

- The program should be easy to maintain.

- Canonical variables should be used throughout the computations.

All these features should not hamper the computational efficiency. Section 2 describes the available features of MAD. It is followed by two sections giving a general description of the "standard input language" and extensions to this language. Section 5 introduces the data structures while section 6 summarizes future plans.

2. Features of MAD

At present the MAD program [1] recognizes most physical elements which may constitute an accelerator:

- Marker, a do-nothing element, and drift space,

- Sector or parallel-face bending magnet,

- Long quadrupole, sextupole, or octupole,

- Thin multipole,

- Solenoid without fringing field,

- Orbit corrector, horizontal, vertical, or both,

- Orbit monitor, horizontal, vertical, or both,

- Accelerating cavity,

- Electrostatic separator,

- Rotation about longitudinal or vertical axis.

  New element types can be added with moderate effort.
  The executable commands can be loosely grouped as follows:

- Structure input (definition of physical elements and their order in the machine),

- Structure changes (installation, removal or moving of existing elements),

- Machine geometry calculations (survey),

- Machine imperfections (defined by systematic and/or random errors),

- Closed orbit finding,

- Linear lattice parameters including imperfections,

- Linear lattice matching including imperfections,

- Complex constraints for matching (examples shown below),

- Electron beam parameters (equilibrium emittances and beam sizes),

- Tracking with TRANSPORT or Lie-algebraic methods,

- Orbit correction by the MICADO algorithm,

- Systematic energy loss due to radiation,

- Chromaticity correction by the methods of HARMON,

- Lie algebraic analysis (one-turn maps and normal form analysis),

- Spin kinematics by the SITF program,

- Plotting with various options.

3. Input Language

The "Standard Input Language" which is used in MAD has been defined at a workshop at SLAC in 1984, and described in [2]. This standard attempt to define an accelerator description which may be ported from one beam optics program to another without too much user intervention. It describes in some detail how elements and beam lines are defined. For executable commands however it only describes the format. This standard language has been adopted among others by the programs MAD, TRANSPORT, and DIMAD.

A command (definition or action) has the generic format

```
label:  keyword {attribute=value}
```

where the various parts have the meaning

label	Name given to the definition or command for later reference,
keyword	Selects the item to be defined or the action to be performed,
attribute	Keyword for a command attribute (length or excitation of a magnet, initial conditions for tracking, etc.),
value	Value given to the attribute.

The `attribute=value` group can be repeated as many times as required. The following is a complete example which computes the linear lattice functions for an accelerator cell:

```
TITLE,"Description of a test cell"

!  bending magnets:
B:  RBEND, L=35.09,ANGLE=0.011306116,&
    E1=0.005,E2=-.005,H1=.001,H2=.002
!  drift spaces:
L1: DRIFT,L=1.21
L2: DRIFT,L=0.56
L3: DRIFT,L=0.28
L4: DRIFT,L=1.57
!  quadrupoles:
QF: QUAD,L=1.6,K1=-0.02268553,TYPE=MQ
QD: QUAD,L=1.6,K1= 0.022683642,TYPE=MQ
!  accelerator cavity:
RF: RFCAV,L=0.1,VOLT=100,FREQ=114,LAG=0.4,&
    HARMON=30,TFILL=1,SHUNT=1
!  sextupoles:
SF1:SEXT,L=0.4,K2=-0.13129
SD1:SEXT,L=0.76,K2= 0.26328
!  order of occurrence of elements in the cell:
LIN(SF,SD):LINE=(L4,B,L2,SF,L3,QF,L1,B,L2,SD,L3,QD)
```

```
!  make this sequence active and compute lattice functions:
USE,LIN(SF1,SD1)
TWISS
```

4. Extensions to the Input Language

Over the years a number of extensions to the standard language have been adopted in MAD. They tend to increase the capacity of the language to express complex machines and complicated computations.

4.1. Element Classes

An element class can be thought of as a new keyword defining an element with given default attributes. Any element which has been previously defined can thus be used to define new elements. Let us assume we define the following elements:

```
MQF:  QUADRUPOLE,L=LQM,K1=KQD    ! Defocusing quadrupoles
MQD:  QUADRUPOLE,L=LQM,K1=KQF    ! Focusing quadrupoles
MQT:  QUADRUPOLE,L=LQT,TILT      ! Tilted quadrupoles
```

These definitions specify three classes of quadrupoles with given default lengths; the focusing and defocusing quadrupoles also have default strengths. These classes can be used to define the actual magnets:

```
QD1:  MQD              !  Defocusing quadrupoles
QD2:  MQD
QD3:  MQD
. . .
QF1:  MQF              !  Focusing quadrupoles
QF2:  MQF
QF3:  MQF
. . .
QT1:  MQT,K1=KQT1  !  Skewed quadrupoles
QT2:  MQT,K1=KQT2
. . .
```

Note that the derived magnets *inherit* the properties of their class. During computations, positions in the machine can be referred to by class name or by element names in the same right.

4.2. Output Selection

Additionally, every element carries also a second identifier called TYPE which may also be referred to. Together with the class mechanisms, the TYPE attribute provides a very flexible way to build output tables. Example:

```
    SELECT, OPTICS, {class} {element} {range} {type}
  OPTICS, COLUMNS={name}
```

The above SELECT command sets output flags for the OPTICS command on all elements for which at least one of the following conditions is true:

- The element is a member or belongs to a subclass of one of the classes listed,

- The element's name is listed in element,

- The element occurs in one of the ranges in range,

- The element carries one of the TYPE identifiers in type.

The OPTICS command computes the lattice functions and writes all functions whose name is mentioned as one of the names after the keyword COLUMN. The result is stored as a "dynamic table" as described in the next subsection. This table can later be plotted, tabulated, or otherwise postprocessed. Example:

```
SELECT,OPTICS,MONITOR,KICKER
OPTICS,COLUMN=BETX,MUX,BETY,MUY
```

These commands create a table of the horizontal and vertical β and μ values for all beam position monitors and orbit correction kickers. It could be used in a control program to compute orbit correctors; or it could be tabulated or plotted in various ways.

4.3. Dynamic Tables

Output tables can be created in the form of self-describing table files. These files have column headers, giving the names and formats of the columns, as well as table descriptors which contain global information. Example:

*	S	NAME	BETX	MUX	BETY
$	%f	%16s	%f	%f	%f
@	GAMTR	%f	64.3336		
@	ALFA	%f	.241615E-03		
@	XIY	%f	-.455669		
@	XIX	%f	2.05279		
@	QY	%f	.250049		
@	QX	%f	.249961		
@	CIRCUM	%lf	79.0000000000		
@	DELTA	%f	.000000E+00		
@	COMMENT	%20s	"DATA FOR TEST CELL"		
@	ORIGIN	%28s	"MAD 8.5/4 Apollo - UNIX"		

@ DATE	%08s	"20/03/92"		
@ TIME	%08s	"10.29.18"		
.000000E+00	SEQ	132.276	.000000E+00	23.609
1.57000	[000000]	124.847	.194446E-02	25.182
36.6600	B1	24.8429	.110430	126.37
37.2200	[000001]	24.2715	.114060	129.01
37.6200	SF1	23.8831	.116704	130.92
37.9000	[000002]	23.6210	.118580	132.26
39.5000	QF1	23.6209	.129474	132.26
40.7100	[000003]	24.8112	.137432	126.52
75.8000	B2	124.710	.246071	25.215
76.3600	[000004]	127.328	.246779	24.623
77.1200	SD1	130.934	.247716	23.871
77.4000	[000002]	132.277	.248054	23.609
79.0000	QD1	132.276	.249961	23.609
79.0000	ENDM	132.276	.249961	23.609
79.0000	SEQ	132.276	.249961	23.609

4.4. Arithmetic Expressions

All parameters with a real value can be specified to have the value of an arbitrary arithmetic expression in other parameters. The allowed operations include the basic operations add, subtract, multiply, divide and power, the most common elementary functions like sine, cosine, logarithm, exponential. MAD will evaluate all dependencies created whenever an independent variable used in an expression changes.

Several random functions are also allowed:

RANF()   Random number, uniformly distributed in $[0,1]$,

GAUSS()   Random number, Gaussian distribution with unit standard deviation,

TGAUSS(X)   Random number, Gaussian distribution with unit standard deviation, truncated at X standard deviations,

USER0()   Random number, user-defined distribution without arguments,

USER1(X)   Random number, user-defined distribution with one argument,

USER2(X,Y)   Random number, user-defined distribution with two arguments.

Normally random generators immediately generate a new random value which is then stored as a constant. For machine imperfection definitions, however, new random values are generated as often as needed to assign errors to all specified elements.

The expression mechanism is very powerful. It allows to set up complex dependencies among variables, and to impose almost arbitrary matching conditions.

As a simple example consider the matching of a cell so as to have $\beta_x = \beta_y$:

```
!  define a simple cell:
CEL: LINE=(...)
!  make the cell active and initiate matching
USE,CEL
CELL
    !  impose some constraints
    ...
    !  vary two quadrupoles
    VARY,QF[K1]
    VARY,QD[K1]
    !  impose equality of the two beta values
    VARY,AUX
    CONSTRAINT,#E,BETX=AUX,BETY=AUX
    !  invoke matching
    ...
ENDMATCH
```

The trick is to vary an auxiliary variable AUX, and to request that both β-functions must be equal to this value.

A second example involves a spin matching condition. The positions VB1 and VB2 refer to two vertical bends. The condition imposes that the vertical phase advance between these two positions and the interaction point shall be an odd multiple of $1/2$, and that the transfer matrix element $(2, 1)$ from VB2 to the interaction point IP differs by a given factor from the same element taken from VB1 to IP:

```
!  define the insertion to be matched
INS: LINE=(...,VB1,...,VB2,...,IP)
!  define the factor
FACTOR: CONSTANT=SIN(PI/3)
 !  make insertion the active line and initiate matching
USE,INS
MATCH,LINE=CEL
    !  vary the auxiliary variable AUX and some K1's
    ...
    !  impose conditions on the transfer matrix
    VARY,AUX
    RMATRIX,VB1/IP,RM(3,4)=0,&!  phase advance condition
       RM(2,1)=AUX !  AUX becomes = to RM(2,1)
    RMATRIX,VB2/IP,RM(3,4)=0,&!  phase advance condition
       RM(2,1)=FACTOR*AUX !  RM(2,1) becomes = FACTOR * AUX
    !  invoke matching
    ...
ENDMATCH
```

The two conditions RM(3,4)=0 impose the proper phase advance, while the two conditions involving RM(2,1) and the auxiliary variable AUX impose the required scaling of this matrix element.

4.5. Beam Line Sequences

In a control system environment the sequence of elements in the accelerator is usually kept in a data base. It is convenient to use a more rigid format for the sequence of elements. For this purpose the notion of a beam sequence has been introduced:

```
POSB1=19.115

SEQ: SEQUENCE
   B1:   B,AT=POSB1
   SF1:  SF,AT=37.42
   QF1:  QF,AT=38.70
   B2:   B,AT=58.255,ANGLE=B1[ANGLE]
   SD1:  SD,AT=76.74
   QD1:  QD,AT=78.20
   ENDM: MARKER,AT=79.0
ENDSEQUENCE
```

The sequence must be preceded by *class* definitions. The sequence definition itself combines the process of element definition with their installation at given positions in the ring. The first name on each line is an unique name given to the element, the second name is an element class serving as a template to define its attributes. The two names are followed by the longitudinal position in the machine where the element shall be placed, and optionally by additional element attributes which will override the class attributes. MAD generates the drift spaces automatically.

4.6. Sequence Editor

For machine studies beam sequences must often be modified by adding, removing, or displacing elements. This is made possible by using the sequence editor, which recognizes the following commands:

```
SEQEDIT,SEQUENCE=sequence


! remove single named element
REMOVE,CLASS=element
! remove whole element class
REMOVE,CLASS=class
! remove all elements matching pattern (see below)
REMOVE,PATTERN=pattern


! install at absolute position
INSTALL,ELEMENT=element,AT=expression
! install at position relative to an element
INSTALL,ELEMENT=element,AT=expr,FROM=element
```

```
!   move to absolute position
MOVE,ELEMENT=element,TO=expr
!   move to position relative to an element
MOVE,ELEMENT=element,TO=expr,FROM=element
!   move relative to old position
MOVE,ELEMENT=element,BY=expr


    ENDEDIT
```

4.7. "Wildcard" Strings

Wildcard strings are patterns similar to the patterns used in the UNIX "grep" utility. They can be used in the sequence editor for selectively removing all elements which match a pattern; or when saving matching results to select the definitions to be saved.

5. Data Structures

The internal data structures for MAD are based on the CERN-written ZEBRA package. This package provides:

- Dynamic memory allocation under FORTRAN,

- Data block (bank) format known to the system,

- Automatic garbage collection for deleted blocks,

- Automatic updating of pointers,

- Debugging tools (printing of data structures),

- Reference pointers across trees,

- Saving and retrieving of complete trees on disk,

- Structural pointers in the fashion of linear lists:

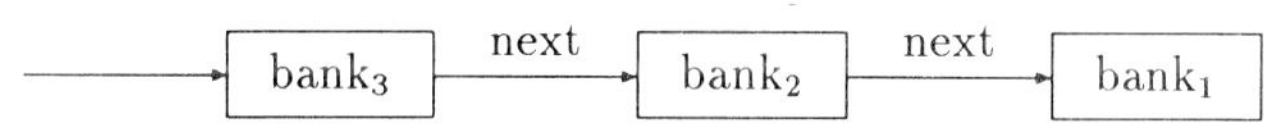

Figure 1. Example of a Linear List

- Structural pointers in the fashion of a tree:

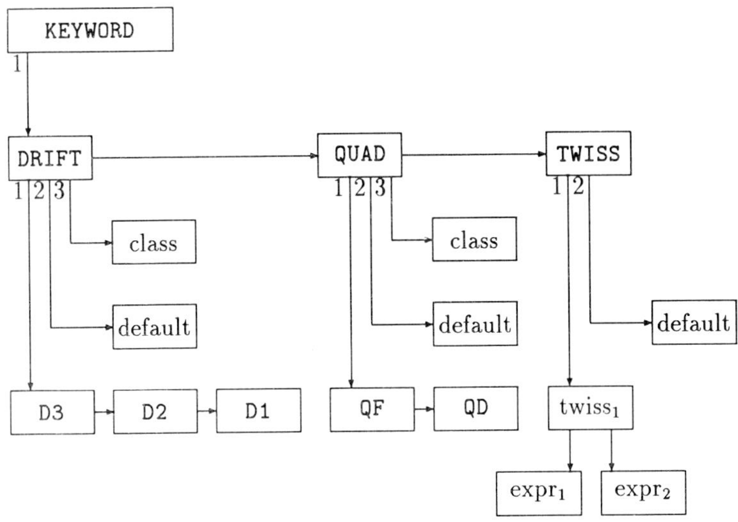

Figure 2. Example of a Tree Structure

Using these features very complex data structures can be built; this makes the implementation of the more exotic features easy.

6. Future Plans

At present we are working on the implementation of Differential Algebra facilities. It is expected that this will allow a simplified approach to the correction of non-linear aberrations. It may also serve to a certain extent help in determining the dynamic aperture, though this seems doubtful at the present time.

Regardless of all its features, the ZEBRA system, being based on FORTRAN, is sometimes painful to use. As it does not support double-precision data with double-word alignment, it is necessary to copy all double-precision data to local storage before use. This can cause loss of computational speed, if care is not taken to organize these data moves properly. Also, bank pointers must always be kept in specific locations, which may also cause considerable difficulty for the programmer. We are therefore also investigating the possibility to rewrite MAD in a modern structured language. The obvious candidate would be C++, but for availability reasons we might have to consider another language, like plain C or FORTRAN-90.

References

[1] Grote H and Iselin F C 1991 *CERN Report* CERN/SL/90/13 (AP), Rev. 2.

[2] Carey D C and Iselin F C 1984 *CERN Report* LEP-TH/84-10; 1984 Summer Study on the Design and Utilization of the Superconducting Super Collider, Snowmass, Colorado, June 23–July 13.

COSY INFINITY Version 6

Martin Berz

Department of Physics and Astronomy and National Superconducting Cyclotron Laboratory, Michigan State University, East Lansing, Mi 48824

Abstract. COSY INFINITY is an arbitrary order map code for the design and analysis of particle optical systems. It is based on differential algebraic methods, which provide an efficient and elegant framework for typical tasks faced in practice. For example, the computation of maps of arbitrary order and their dependence on system parameters, normal form techniques for the analysis and resonance correction of the behavior of repetitive systems, techniques for aberration correction and variety of other features can all be phrased very directly in terms of these methods.

As the user interface, COSY employs a structured object oriented language, which is simultaneously used for the coding of the physics. This approach allows a seamless connection of the work of the user with the tools provided by the code and directly allows the adaptation of the code to specific tasks. Optimization and fitting is supported at the language level, allowing rather direct and general optimization of essentially any quantity. Lattices can be described in terms of COSY's language environment or alternatively in the standard MAD input.

For the sake of portability, the language interface of COSY is standard FORTRAN 77; the compiler and executer of COSY as well as the libraries to which the language links are all written in this language. The language environment supports a variety of interactive and printer or plotter-based graphics drivers, and new drivers are added easily. Currently COSY INFINITY is used by approximately 100 registered users.

1. Introduction

The properties of particle optical systems [1, 2, 3, 4, 5, 6, 7, 8, 9, 10] can be conveniently described by a map relating final phase space coordinates $\vec{z}_f$ to initial coordinates $\vec{z}_i$ and system parameters $\vec{\delta}$ in the following way:

$$\vec{z}_f = \mathcal{M}(\vec{z}_i, \vec{\delta}). \tag{1}$$

Since except for trivial cases it is impossible to express the transfer map $\mathcal{M}$ in closed form, it is customary to represent it in terms of its Taylor expansion, a method used in many codes [11, 12, 13, 14, 15]. With the advent of the differential algebraic methods [16, 17, 18, 19], these Taylor methods were generalized in a rather straightforward way from the orders of two or three used by other programs to any order, and including the dependence on system parameters. COSY INFINITY (see also [20, 21, 22, 23, 15]) is a program that uses these techniques for the computation as well as for the analysis of these maps in a variety of ways.

2. The physics in COSY

The actual computation of maps of particle optical systems is done in several different ways suitable for different situations. The main fields of elements are computed using the Lie derivative [17] $L_f = \vec{f} \cdot \vec{\nabla} + \partial/\partial t$ associated with the differential equation

$$\frac{d}{dt}\vec{x} = \vec{f}(\vec{x}, t), \tag{2}$$

which allows the computation of the higher order time derivatives of any function $g(\vec{x}(t))$ defined on the solution of the differential equation:

$$\begin{array}{rcl} d/dt\, g & = & L_f\, g \\ d^2/dt^2 g & = & L_f^2\, g \\ d^3/dt^3 g & = & L_f^3\, g. \end{array} \tag{3}$$

Hence in a formal sense we have $g(t) = \exp(L_f)\, g(0)$; by passing to DA operations, the convergence of the sequence becomes guaranteed. By choosing g to be the individual phase space coordinates, one can compute their final values from the intial values and hence the transfer map.

The loss of the highest derivative in the derivation operation ∂ in DA entails that the mechanism is only applicable to cases in which $\vec{f}$ does not depend on the independent variable, which in the case of particle optics means that the fields be constant. The treatment of non-constant fields is done by numerical integration with DA operations [24, 16, 17], which is more time consuming. However, for the most important case of such fields, the fringe fields, there is a very efficient perturbative method with high accuracy based on symplectic scaling, which is described in detail in [25].

Besides these general methods, some particle optical elements are treated in a separate way. For example, for multipoles of up to order two, analytical formulas are used in the flavor or TRANSPORT [11], which for these cases gains about a factor of two in speed. The drift and the dipole with circular and tilted edges are treated geometrically, which allows the computation of the aberrations to arbitrary order in an analytical way and sometimes a tenfold speed gain.

Besides the computation of the maps, COSY allows the analysis of maps in a variety of ways. For repetitive systems, normal form methods play an important role in the analytical computation and correction of tune shifts, pseudo invariants and resonance strengths [26]. They are based on an iterative order by order simplification of the map by nonlinear changes of variables

$$\mathcal{M}_n = \mathcal{A}_n^{-1} \circ \mathcal{M}_{n-1} \circ \mathcal{A}_n. \tag{4}$$

An example of the correction of resonances with COSY can be found in a companion paper [27].

For the long term analysis of systems, it is important to be able to perform symplectic tracking, which is achieved by means of one of a variety of generating function representations of the map

$$
\begin{aligned}
F_1(q_i, q_f) \text{ satisfying } (\vec{p}_i, \vec{p}_f) &= (+\vec{\nabla}_{q_i} F_1, -\vec{\nabla}_{q_f} F_1) \\
F_2(q_i, p_f) \text{ satisfying } (\vec{p}_i, \vec{q}_f) &= (+\vec{\nabla}_{q_i} F_2, +\vec{\nabla}_{P_f} F_2) \\
F_3(p_i, q_f) \text{ satisfying } (\vec{q}_i, \vec{p}_f) &= (-\vec{\nabla}_{p_i} F_3, -\vec{\nabla}_{q_f} F_3) \\
F_4(p_i, p_f) \text{ satisfying } (\vec{q}_i, \vec{q}_f) &= (-\vec{\nabla}_{p_i} F_4, +\vec{\nabla}_{p_f} F_4),
\end{aligned}
\tag{5}
$$

which in the DA framework can be computed in a very direct way [17, 18].

For the case of single pass systems, there are tools to output the aberrations of systems as well as to compute resolutions of spectrographs in a variety of operating modes. Particularly useful is the method of reconstructive correction, which computationally eliminates all aberrations via measurements in two planes [28, 29]

3. The COSY language

The COSY language is a structured language with the flavor of PASCAL. Different from PASCAL, it allows the use of arbitrary data types and operations thereon that can be specified. In particular, this allows a direct use of the DA (differential algebra) data type, which is essential for an efficient use of DA in practice. It is also helpful for the manipulation of pictures, which are stored as picture objects. Type checking is performed at execution time, which allows the use of the same procedures with DA or real arguments, a feature which makes the computation of maps depending on parameters all but trivial. The commands of the COSY language are as follows:

```
BEGIN ;            END ;            { Begins and ends program            }
INCLUDE ;          SAVE ;           { Includes and saves compiled code   }
VARIABLE ;                          { Declares a local variable          }
PROCEDURE ;        ENDPROCEDURE ;   { Declares a local procedure         }
FUNCTION ;         ENDFUNCTION ;    { Declares a local function          }
< assignments > ;                   { Sets value of variable             }
< procedure calls > ;               { Calls previously defined  procedure }
IF ;               ENDIF  ;         { Executed once if argument is true  }
WHILE ;            ENDWHILE ;       { Executed while argument is true    }
LOOP ;             ENDLOOP ;        { Stepping argument                  }
FIT ;              ENDFIT ;         { varying arguments to fit conditions }
```

Except for the last one, the flow control statements are rather standard. The FIT block is executed over and over again as long as the sum of squares of the objective variables listed in the ENDFIT statement can be reduced by modifying the values of the free variables listed in the FIT statement. The ENDFIT statement also contains the number of the optimizer to be used as well as the tolerance and the maximum number of iterations allowed.

Module Name	Contents	Language	Lines
FOXY.FOR	Compiler/Executor	FORTRAN 77	5,000
DAFOX.FOR	Low-level libraries	FORTRAN 77	12,000
FOXFIT.FOR	Various optimizers	FORTRAN 77	5,000
FOXGRAF.FOR	Graphics interfaces	FORTRAN 77	2,000
MADCOSY.FOR	MAD processor	FORTRAN 77	5,000
COSY.FOX	Physics library	COSY	2,000
DEMO.FOX	Demos of key tools	COSY	1,000

Table 1. The modules comprising COSY INFINITY

The nesting of procedures and functions as well as the visibility of modules and variables follows standard practice of modular languages. The flow control structures IF, WHILE, LOOP and FIT can be nested. In practice, the direct availability of various optimization methods at the language level via the FIT command proves very helpful. Another feature of the COSY language is that it is rather directly possible to interface to FORTRAN subroutines, which allows easy connection to other codes.

The physics in COSY is written almost entirely in COSY's own input language and contained in the macro file COSY.FOX. Using the COSY commands SAVE and INCLUDE, this macro file can be compiled ahead of time and included as needed in user code. The use of the features in COSY is via the call of procedures and functions.

For example, particle optical elements are just procedures, which when invoked update the current transfer map of the system, the coordinates of rays to be tracked, and a variety of other quantities. In this framework it is rather directly possible to build lattices in a convenient way. For example, if we want to build a simple FODO consisting of a dipole (DI), a quadrupole (MQ) as well as some drifts (DL), which is then used six times with slightly different values of quad strengths due to random errors, the input would look like this:

```
PROCEDURE CELL PHI Q ;
    DL 5 ; DI 10 PHI .1 ; DL 5 ; MQ .5 Q .1 ; DL 5 ;
    ENDPROCEDURE ;
    .

    .

    .

UM ;                             { Sets the transfer map to unity }
LOOP I 1 6 ; CELL 60*(1+.1*RANF()) .2*(1+.01*RANF()) ; ENDLOOP ;
TS MUX MUY MUZ ;    { Computes the tune dependence on emittance }
```

Altogether, the COSY INFINITY package consists of six different modules plus a demo file showing the use of the most important features, with a total of about 30,000 lines of code, a breakdown of which is shown in Table 1.

4. Elements in COSY

The library of macros in COSY.FOX contains a large list of pre-defined beamline elements, each of which can be calculated to arbitrary order including the dependence on

parameters. Besides the ones listed here, it is possible to develop new ones in a variety of
ways, for example using the procedure GE, which allows the computation of the transfer
map from a table of data for multipole strengths. The following elements can be used
directly in COSY:

- Bending magnets including inhomogeneities and edge angles

- Electrostatic deflectors including inhomogeneities

- Magnetic and electric multipoles, any order and superimposed

- Wien filters, with or without net deflection, including inhomogeneities

- Wigglers with various field models

- Thin cavities including nonlinearities

- General element GE, described by a table of multipole strengths

- Magnetic and electric round lenses including various solenoids

- Fringe fields for all magnetic and electric elements

- Glass lenses, spherical and aspherical, mirrors, prisms

- Misalignments (shifts, tilts of reference orbit, rotation around reference orbit)

5. Analysis with COSY

The macro package of analysis tools contained in COSY.FOX contains a large variety
of tools that are useful for various problems. Using the open approach of COSY, it is
also rather easily possible to develop new analysis tools. The currently available features
include

- Transfer maps and aberrations of arbitrary order including system parameters

- Various generating function and Lie factorization representations

- Closed orbit and linear lattice parameters including parameter dependence

- Tracking of rays through the system and layout of the system

- Analytical calculation of amplitude and parameter dependent tune shifts

- Analytical calculation and correction of resonance strengths of repetitive systems

- Repetitive Tracking, if desired with one of several symplectification methods

- Computation of resolutions of spectrographs

- Reconstructive correction of aberrations

6. Distribution of COSY

COSY INFINITY has been made available to users since 1989. Currently, version 6 of the code is being distributed. The code can be obtained from the author by electronic mail after signing a simple registration form including a nonproliferation agreement. Currently, the code is being used by about 100 registered users in about 50 laboratories. Figure 1 (which incidentally was created using COSY's graphics features and output directly using the LaTeX graphics driver) shows the history of the number of users as well as the number of procedures since its first release.

Due to the rigorous use of plain FORTRAN 77, COSY can easily be adjusted to any FORTRAN environment, and in most cases the necessary changes concern only file handling as well as very few low-level library routines like the ones returning CPU time. For most popular environments, a dedicated version can be generated from the master version by running a small program supplied with COSY which selectively comments and un-comments certain commands. The list of environments that are currently contained in the master version in this way are

- VAX (VMS and UNIX)

- SUN (Unix)

- HP (Unix)

- CRAY

- IBM 6000 (Unix)

- IBM PC (Lahey)

- IBM Mainframe

COSY graphics at its lowest level is based on seven elementary routines which can be interfaced rather easily with any environment. The routines are BEGIN, END, MOVE, DRAW, CHARACTER, LINEWIDTH, and COLOR, and so it is rather easy to write drivers for a variety of environments. With the normal shipment of COSY, the following drivers are supplied automatically:

- GKS-based X-Windows

- GKS-based VMS-Windows

- IBM PC VGA (Lahey)

- Direct Tektronix

- Direct Postscript

- LaTeX

- GKS-based HP7475

- GKS-based HP Paintjet

- Low Resolution ASCII

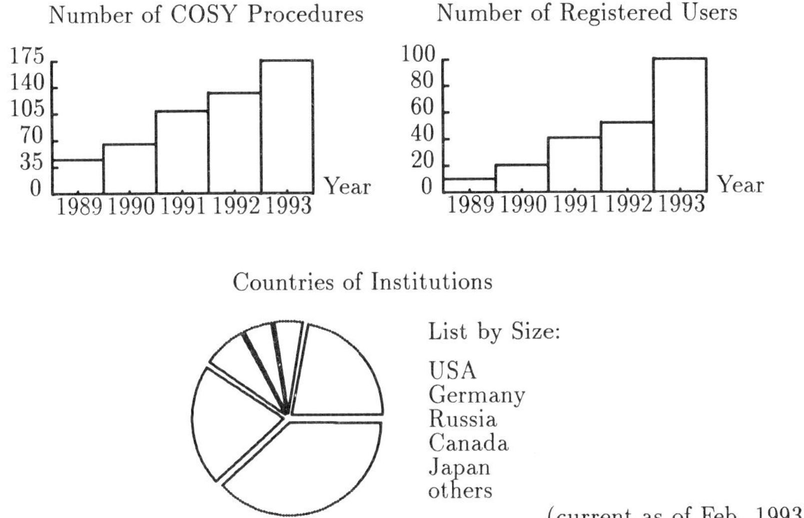

Figure 1. Statistics about Users of COSY (Picture generated in LaTeX with COSY's graphics environment)

7. Acknowledgements

For increasing help as well as stimulating and valuable discussions, I would like to thank my colleagues Ralf Degenhardt, Kyoko Fuchi, Georg Hoffstätter, Weishi Wan, and Meng Zhao. For financial support, I would like to thank the National Science Foundation as well as the Alfred P. Sloan Foundation.

References

[1] H. Wollnik. *Charged Particle Optics*. Academic Press, Orlando, Florida, 1987.

[2] D. C. Carey. *The Optics of Charged Particle Beams*. Harwood, 1987.

[3] P. W. Hawkes and E. Kasper. *Principles of Electron Optics*. Academic Press, London, 1989.

[4] Ed. A. Septier. *Focusing of Charged Particles*. Academic Press, New York, 1967.

[5] A. P. Banford. *The Transport of Charged Particle Beams*. Spon, London, 1966.

[6] H. Buchdahl. *An Introduction to Hamiltonian Optics*. Cambridge University Press, Cambridge, MA, 1970.

[7] R. K. Luneburg. *Mathematical Theory of Optics*. University of California, Berkeley Press, Berkeley, CA, 1964.

[8] A. B. El-Kareh and J. C. J. El-Kareh. *Electron Beams, Lenses and Optics, 2 Vols.* Academic Press, New York, 1970.

[9] J. Grosser. *Einführung in die Teilchenoptik*. Teubner, Stuttgart, 1983.

[10] P. Dahl. *Introduction to Electron and Ion Optics*. Academic Press, New York, 1973.

[11] K. L. Brown. The ion optical program TRANSPORT. Technical Report 91, SLAC, 1979.

[12] H. Wollnik, B. Hartmann, and M. Berz. Principles behind GIOS and COSY. *AIP Conference Proceedings*, 177:74, 1988.

[13] T. Matsuo and H. Matsuda. Computer program TRIO for third order calculations of ion trajectories. *Mass Spectrometry*, 24, 1976.

[14] A. J. Dragt, L. M. Healy, F. Neri, and R. Ryne. MARYLIE 3.0 - a program for nonlinear analysis of accelerators and beamlines. *IEEE Transactions on Nuclear Science*, NS-3,5:2311, 1985.

[15] M. Berz. Computational aspects of design and simulation: COSY INFINITY. *Nuclear Instruments and Methods*, A298:473, 1990.

[16] M. Berz. Differential algebraic description of beam dynamics to very high orders. *Particle Accelerators*, 24:109, 1989.

[17] M. Berz. Arbitrary order description of arbitrary particle optical systems. *Nuclear Instruments and Methods*, A298:426, 1990.

[18] M. Berz. Differential algebraic treatment of beam dynamics to very high orders including applications to spacecharge. *AIP Conference Proceedings*, 177:275, 1988.

[19] M. Berz. Differential algebraic description and analysis of trajectories in vacuum electronic devices including spacecharge effects. *IEEE Transactions on Electron Devices*, 35-11:2002, 1988.

[20] M. Berz. COSY INFINITY Version 6 reference manual. Technical Report MSUCL-869, National Superconducting Cyclotron Laboratory, Michigan State University, East Lansing, MI 48824, 1993.

[21] M. Berz. COSY INFINITY Version 5 reference manual. Technical Report MSUCL-811, National Superconducting Cyclotron Laboratory, Michigan State University, East Lansing, MI 48824, 1991.

[22] M. Berz. New features in COSY INFINITY. In *Third Computational Accelerator Physics Conference*. AIP Conference Proceedings, 1993.

[23] M. Berz. COSY INFINITY. In *Proceedings 1991 Particle Accelerator Conference*, San Francisco, CA, 1991.

[24] M. Berz. The method of power series tracking for the mathematical description of beam dynamics. *Nuclear Instruments and Methods*, A258:431, 1987.

[25] G. Hoffstätter and M. Berz. Efficient computation of fringe fields using symplectic scaling. In *Third Computational Accelerator Physics Conference*. AIP Conference Proceeings, 1993.

[26] M. Berz. Differential algebraic formulation of normal form theory. *in: M. Berz, S. Martin and K. Ziegler (Eds.), Proc. Nonlinear Effects in Accelerators*, 1993.

[27] R. Servranckx. Optics programs at Triumf. *in: M. Berz, S. Martin and K. Ziegler (Eds.), Proc. Nonlinear Effects in Accelerators*, 1993.

[28] M. Berz, K. Joh, J. A. Nolen, B. M. Sherrill, and A. F. Zeller. On-line correction of residual aberrations in spectrographs. In *Proceedings 1991 Particle Accelerator Conference*, San Francisco, CA, 1991.

[29] M. Berz, K. Joh, J. A. Nolen, B. M. Sherrill, and A. F. Zeller. Reconstructive correction of aberrations in nuclear particle spectrographs. *Physical Review C*, 47,2:537, 1993.

Inst. Phys. Conf. Ser. No 131
Paper presented at Int. Workshop Nonlinear Problems in Accelerator Phys. Berlin, 1992

The arbitrary order design code Tlie 1.0

Johannes van Zeijts†[1] and Filippo Neri‡[2]

† Physics Department, University of Maryland, College Park, MD-20742, USA

‡ AT-Division, MS-H829, Los Alamos National Laboratory, Los Alamos, NM-87545, USA

Abstract. We describe the arbitrary order charged particle transfer map code TLIE. This code is a general $6D$ relativistic design code with a MAD compatible input language and among others implements user defined functions and subroutines and nested fitting and optimization. First we describe the mathematics and physics in the code. Aside from generating maps for all the standard accelerator elements we describe an efficient method for generating nonlinear transfer maps for realistic magnet models. We have implemented the method to arbitrary order in our accelerator design code for cylindrical current sheet magnets. We also have implemented a self-consistent space-charge approach as in CHARLIE. Subsequently we give a description of the input language and finally, we give several examples from productions runs, such as cases with stacked multipoles with overlapping fringe fields.

1. Introduction

In some applications higher order correction are needed than are provided by a third order code, like MARYLIE. Tlie has been derived as an arbitrary order version of MARYLIE. It is, like MARYLIE a map code in which maps are derived that represent a mapping from the initial to the final conditions for a particular transport line, made up of magnetic elements. The maps in Tlie are represented internally as Taylor series, but they can be translated into a factorized Lie algebraic reprensentation when it is more convenient (which it is almost always.) The main strength of the program is the ability to generate maps for complex sets of magnets with overlapping fringe fields.

[1] E-mail: johannes@quark.umd.edu
[2] E-mail: filippo@filippo-mac.atdiv.lanl.gov

2. Hamiltonian dynamics

Let $z = (X, P_x, Y, P_y, \tau, P_\tau)$ be the six dimensional vector describing the location of a particle in phase space. The effect of a linear Hamiltonian system on this particle can be expressed as the action of a matrix M that takes the particle from its initial state z^{in} to the final state z^{fin}:

$$z_i^{fin} = M_{ij}\, z_j^{in}. \tag{1}$$

It can be shown that M satisfies the symplectic condition

$$\tilde{M} J M = J \tag{2}$$

where $\tilde{M}$ is the transpose of M and J is an antisymmetric matrix defined as follows:

$$J = \begin{pmatrix} 0 & 1 & 0 & 0 & 0 & 0 \\ -1 & 0 & 0 & 0 & 0 & 0 \\ 0 & 0 & 0 & 1 & 0 & 0 \\ 0 & 0 & -1 & 0 & 0 & 0 \\ 0 & 0 & 0 & 0 & 0 & 1 \\ 0 & 0 & 0 & 0 & -1 & 0 \end{pmatrix}. \tag{3}$$

Matrices M satisfying Eq. (2) are called symplectic matrices. The set of all such matrices can be shown to form the symplectic group Sp(6,R)[2]. We describe the linear motion by such a matrix.

2.1. Orbit hamiltonian

We consider the relativistic motion of a particle with charge q and rest mass m in an electromagnetic field described by vector potential $\mathbf{A}$ and scalar potential Φ.

The electromagnetic fields $\mathbf{E}$ and $\mathbf{B}$ are given by the standard relations

$$\begin{aligned} \mathbf{B} &= \nabla \times \mathbf{A} \\ \mathbf{E} &= -\nabla\Phi - \frac{\partial \mathbf{A}}{\partial t} \end{aligned} \tag{4}$$

The orbit Hamiltonian is given by

$$H_{orbit} = \sqrt{m^2 c^4 + c^2 \left[(p_x - qA_x)^2 + (p_y - qA_y)^2 + (p_z - qA_z)^2 \right]} + q\Phi . \tag{5}$$

2.2. Hamiltonian in cartesian scaling variables

In magnetic optics the longitudinal position z of the reference particle is taken as the independent variable instead of the time. Correspondingly, the time becomes a dependent variable, and a momentum p_t conjugate to the time is introduced. The time along the design orbit is given by the relation

$$t(z) = \frac{z}{v_z^0}, \tag{6}$$

with

$$v_z^0 = -\frac{c^2 p_0}{p_t^0} . \tag{7}$$

Here $p_t^0 = -H|_{\text{design orbit}}$ is the value of p_t on the design orbit and where p_0 is the design momentum.

We introduce the deviation variables $\mathbf{z} = (X, P_x, Y, P_y, \tau, P_\tau)$ defined by

$$
\begin{aligned}
X &= x \,, \\
Y &= y \,, \\
\tau &= c(t - \tfrac{z}{v_z^0}) \,, \\
P_x &= \frac{p_x}{p_0} \,, \\
P_y &= \frac{p_y}{p_0} \,, \\
P_\tau &= \frac{p_t - p_t^0}{p_0 c} \,.
\end{aligned}
\tag{8}
$$

In these new phase space coordinates the hamiltonian becomes

$$
H = -\frac{P_\tau}{\beta} - \frac{A_z}{G} - \sqrt{1 - \frac{2P_\tau}{\beta} + P_\tau^2 - (P_x - \frac{A_x}{G})^2 - (P_y - \frac{A_y}{G})^2} \,.
\tag{9}
$$

G is the magnetic rigidity '$B\rho$' of particles on the design orbit. It is given by the relation

$$
G = \frac{p_0}{q} \,,
\tag{10}
$$

and has units of Tesla meters. Also, β and γ are the standard relativistic factors for the design orbit. They are related to p_0 and p_t^0 by the equations

$$
\begin{aligned}
p_0 &= \beta\gamma mc \,, \\
p_t^0 &= -\gamma mc^2 \,.
\end{aligned}
\tag{11}
$$

The Hamiltonian H is dimensionless. The dimension of X, Y, and τ is meters, whereas P_x, P_y, and P_τ are dimensionless. We will often use the expansion

$$
H = \sum_i H_i = H_1 + H_2 + H_3 + \cdots \,,
\tag{12}
$$

where H_i is a homogeneous polynomial in $\mathbf{z}$ of order i.

3. Transfer maps

The Hamiltonian defined in the previous section can be used to numerically integrate particles by using Hamiltons equations. However, it is more usefull to generate a 'time t' map of the flow generated by the Hamiltonian. Maps are generated for every element in the lattice, subsequently concatenated into a one-turn map, and used in the analysis of the system.

3.1. Map representation

The orbit dynamics is described as a nonlinear map on the 6-dim. phase space $\mathbf{z} = \{X, P_x, Y, P_y, \tau, P_\tau\}$.

The orbit dynamics is represented alternatively as polynomials in the initial phase-space coordinates $\mathbf{z}$

$$z_a^f = T_a(\mathbf{z}) = \sum_b R_{ab} z_b + \sum_{b,c} T_{abc} z_b z_c + \sum_{b,c,d} U_{abcd} z_b z_c z_d + \cdots \tag{13}$$

or as a Lie algebraic polynomial

$$e^{:F(\mathbf{z}):} = e^{:f_1:} e^{:f_2:} e^{:f_3:} \cdots e^{:f_n:} \tag{14}$$

where the f_i are homogeneous polynomials in $\mathbf{z}$ of order i. Here we used the Dragt-Finn factorization of the Lie algebraic map. The $:f:$ notation means that this is an operator to be applied on phase space coordinates.

We do not use the $e^{:f_2:}$ representation but instead use a symplectic $6*6$ matrix M, that is related to f_2 as follows

$$e^{:f_2:}(\mathbf{z}) = M \cdot \mathbf{z} \tag{15}$$

3.2. Map generation for constant hamiltonians

In most models the field is treated as constant inside the element with hard-edge fringe fields. In that case the formal solution of Hamiltons equation is

$$e^{-\theta:H:} \tag{16}$$

3.2.1. Exponentiation. Transfer map generation for the case where H does not depend on the longitudinal variable is most readily done by direct exponentiation of the formal solution

$$T(\mathbf{z}) = e^{-\lambda:H:}(\mathbf{z}) = \mathbf{z} - \lambda[H, \mathbf{z}] + \frac{1}{2!}\lambda^2[H, [H, \mathbf{z}]] + \cdots , \tag{17}$$

where $[,]$ is the Poisson bracket operator. This algorithm will eventually converge due to the $n!$ term in the denominator of the expansion of the exponential operator.

3.3. Map generation for realistic hamiltonians

The method presented in the previous section can also be used in the case that the Hamiltonian does depend on z. The equations for the Lie polynomials have to be integrated order by order at every integration step.

Here we present a method which allows for a more efficient implementation and is easily extended to arbitrary order. We denote the independent variable along the machine by λ. Given the initial coordinates $\mathbf{z}$ at λ, the final coordinates $\mathbf{z}_f$ at λ_f can be written as 6 functions $T_k(\mathbf{z}, \lambda, \lambda_f)$, so that

$$\begin{aligned} X^f &= T_x(\mathbf{z}, \lambda, \lambda_f) , \\ P_x{}^f &= T_{P_x}(\mathbf{z}, \lambda, \lambda_f) , \\ &\cdots \quad . \end{aligned} \tag{18}$$

Or, as a vector equation,

$$\mathbf{z}_f = T(\mathbf{z}, \lambda, \lambda_f) \ . \tag{19}$$

Since $\mathbf{z}$ has 6 components, so does T. Eventually we will consider the expansion of T in Taylor series, but for the moment we will regard T as a general 6 dimensional vector of functions. We can derive a partial differential equation for T as a function of $\mathbf{z}$ and λ by observing that a change in λ will not affect $\mathbf{z}_f$ if $\mathbf{z}$ is also changed so that it stays on the trajectory from $\mathbf{z}$ to $\mathbf{z}_f$. In other words, the total derivative of T with respect to the independent variable λ, keeping λ_f fixed is zero:

$$\frac{dT}{d\lambda} = 0 = \frac{\partial T}{\partial \lambda} + \frac{\partial T}{\partial \mathbf{z}} \frac{\partial \mathbf{z}}{\partial \lambda} \ . \tag{20}$$

Using the equation of motion for $\mathbf{z}$:

$$\frac{\partial \mathbf{z}}{\partial \lambda} = [\mathbf{z}, H] = J \frac{\partial H}{\partial \mathbf{z}} \ , \tag{21}$$

the equation for T can be written as

$$\dot{T} = -\frac{\partial T}{\partial \mathbf{z}} J \frac{\partial H}{\partial \mathbf{z}} = [H, T] \ , \tag{22}$$

where we have used a dot to denote a derivative with respect to λ. Eq. (22) is a linear partial differential equation for T. Note that the independent variable λ in Eq. (19) is the initial value of λ, so that Eq. (22) gives the evolution of T backward, starting from the identity at $\lambda = \lambda_f$.

Alternatively, one can define the function vector $I(\mathbf{z}, \lambda_i, \lambda)$, which is the inverse of T in the sense that it gives the initial coordinates $\mathbf{z}_i$ at λ_i in term of the final coordinates $\mathbf{z}$ at λ.

$$\mathbf{z}_i = I(\mathbf{z}, \lambda_i, \lambda). \tag{23}$$

The functions I satisfy an equation analog to (22):

$$\dot{I} = [H, I], \tag{24}$$

where a dot is a derivative with respect to the final λ, and the Poisson bracket is taken with respect to the final coordinates. Eq. (24) can be integrated forward from the initial λ to the final λ but the resulting functions give the initial coordinates as a function of the final coordinates, and have to be inverted if one wants to find the evolution of the system.

Taking T to be the truncated Taylor series expansion of the final coordinates in terms of the initial coordinates, we turn eq.(22) into an ordinary differential equation and find T by integrating backwards starting with the identity map.

$$T_{\lambda_1 \to \lambda_2}(\mathbf{z}) = \int_{\lambda_2}^{\lambda_1} [H(\mathbf{z}, \lambda), T(\mathbf{z})] \, d\lambda \ . \tag{25}$$

For 3'th order orbit motion this is a set of $6 * (1 + 6 + 21 + 56) = 504$ coupled differential equations. For 5'th order we have $6 * (1 + 6 + 21 + 56 + 126 + 252) = 2772$ equations. In the practical implementation, we actually reduce the number of equations by only integrating the nonzero coefficients of T, which are determined by a rough initial integration. Subsequently we use the algorithm shown in the next section to transform the Taylor series into a Lie algebraic map

$$e^{:F:}(\mathbf{z}) = T(\mathbf{z}) \tag{26}$$

4. Magnet models

We do not discuss zero-length approximations, 'Kick' maps, here. All the models used here are finite length.

4.1. Non constant potential models

4.1.1. Solenoid ([10]). It can be shown that if $B_z(0,0,z)$ (the on-axis longitudinal component of $\mathbf{B}$) is known, then Maxwell's equations plus the requirement of axial symmetry are sufficient to determine all components of $\mathbf{B}$ everywhere. In particular, $\mathbf{B}$ can be obtained from a vector potential $\mathbf{A}$ given by the expressions

$$
\begin{aligned}
A_x &= -yU(z,\rho^2) \\
A_y &= xU(z,\rho^2) \\
A_z &= 0
\end{aligned}
\tag{27}
$$

Here U is defined by the relation

$$
U = \sum_{n=0}^{\infty} \frac{(-\rho^2)^n b_{2n}}{2^{2n+1} n!(n+1)!}
\tag{28}
$$

where b_{2n} denotes

$$
b_{2n} = \frac{\partial^{2n} B_z(0,0,z)}{\partial z^{2n}}
\tag{29}
$$

and $\rho^2 = x^2 + y^2$.

The choice of the on-axis longitudinal field can be a soft-edge bump function.

4.2. Realistic potential models ([5]).

The vector potential off axis, for a given multipole symmetry, is determined from the appropriate magnetic field gradients and their longitudinal derivatives on axis.

Subroutines to compute the required gradients are available for Halbach REC quadrupoles, and for general multipoles, with the current distribution on a cylindrical surface specified by a shape function. This function can be supplied by the user, or selected from internal options.

4.2.1. Representation of the vector potential. In the following we will write expressions only for the normal multipoles (for $m \neq 0$). Skew multipoles correspond to $cos(m\theta)$ terms in Eq.30. Given the Fourier expansion of the scalar potential

$$
V(r,\theta,z) = \sum_{m=1}^{\infty} U_m(r,z) sin(m\theta),
\tag{30}
$$

a vector potential giving the same field is

$$
\begin{aligned}
A_z &= -\sum_{m=1}^{\infty} \frac{cos(m\theta)}{m} r \frac{\partial}{\partial r} U_m(r,z) \\
A_r &= \sum_{m=1}^{\infty} \frac{cos(m\theta)}{m} r \frac{\partial}{\partial z} U_m(r,z).
\end{aligned}
\tag{31}
$$

Here we have chosen a gauge where $A_\theta = 0$. The scalar potential off-axis may be written

$$U_m(r, z) = r^m \sum_{l=1}^{\infty} \frac{(-1)^l (m-1)!}{l!(l+m)!} \left(\frac{r}{2}\right)^{2l} \left(\frac{\partial}{\partial z}\right)^{2l} g_m(z), \tag{32}$$

where

$$g_m(z) = \lim_{r \to 0} \frac{m U_m(r, z)}{r^m} \tag{33}$$

represents the profile of the m'th multipole.

This is a general solution to Maxwell equations order by order, for arbitrary $g_m(z)$. The problem is thus reduced to computing the generalized field gradients on axis for realistic magnet models. We report here on the implementation of a family of magnet models representable by current sheets on a cylindrical surface.

4.2.2. Current sheet magnets. Cylindrical current sheets can be used not only to represent radially thin windings on a cylinder, but also to calculate arbitrary fields in the source-free volume bounded by a cylindrical surface. In the latter case, fictitious surface currents that produce the same interior fields as complicated outside sources replace them for rapid field calculation. (For some simple volume field sources, it can be better to work directly with analytical expressions for the field from the source). Continuous currents on the surface of a cylinder of radius a can be represented by a stream function $\Psi(\phi, z)$ of the surface coordinates as follows [4]:

$$j_\phi = -NI \frac{\partial \Psi(\phi, z)}{\partial z}, \qquad j_z = \frac{NI}{a} \frac{\partial \Psi(\phi, z)}{\partial \phi}. \tag{34}$$

The quantity NI is introduced to make Ψ dimensionless. The above prescription produces currents that are automatically divergence-free. The stream function can be used to find a set of discrete turns that approximate a continuous distribution, since current streamlines are contours of constant Ψ; to make spiral windings, successive closed turns are cut and transitions between them are inserted. The stream function can be written as a sum of Fourier components as follows:

$$\Psi(\phi, z) = \sum_{m=1}^{\infty} w_m(z) \sin(m\phi). \tag{35}$$

The $w_m(z)$ are called shape functions. The $m = 0$ case is excluded here because it requires special treatment. If only a single m value is present in Eq.35 and the boundaries are rotationally symmetric, the resultant field has m-pole θ dependence everywhere. For open boundaries, the scalar potential produced by the currents of Eq.34 is

$$V(r, \theta, z) = \frac{\mu_0 N I a}{\pi} \sum_{m=1}^{\infty} \sin(m\theta) \int_{-\infty}^{+\infty} w_m(x) G_m(r, z, x) dx. \tag{36}$$

The Green's function $G_m(r, z, x)$ has the form

$$G_m(r, z, x) = \frac{\partial}{\partial a} \left[\frac{1}{2\sqrt{ar}} Q_{m-\frac{1}{2}} \left(\frac{a^2 + r^2 + (z-x)^2}{2ar} \right) \right], \tag{37}$$

where $Q_{m-\frac{1}{2}}$ is a Legendre function of the second kind of half-integral order. (For the closed-boundary case of a surrounding cylinder of infinitely permeable material, hybrid series solutions containing both ordinary and hyperbolic Bessel functions have been obtained). Expressions for the magnetic field and vector potential are easily derived from Eq.36. For calculation of fields, etc., the $Q_{m-\frac{1}{2}}$ are not directly used; instead, the $\frac{1}{2\sqrt{ar}}$ factor cancels a factor in the hypergeometric series expression for $Q_{m-\frac{1}{2}}$; the resultant expression is regular as $r \to 0$. The generalized on-axis gradient for a single multipole m was defined in Eq.33. This limit of Eq.36 is

$$g_m(z) = \frac{\mu_0 N I a (2m-1)!!}{(m-1)! 2^{m+1}} \int_{-\infty}^{+\infty} w_m(x) K_m(z,x)\,dx \qquad (38)$$

where

$$K_m = \frac{\partial}{\partial a}\left\{ \frac{a^m}{[(z-x)^2 + a^2]^{m+1/2}} \right\}. \qquad (39)$$

If $w_m(x)$ is a piecewise continuous polynomial, exact analytic integration of Eq.38 can be done. An exact quadrature subroutine for polynomials of up to the 4th degree in Eq.38 is used to calculate g_m and its z derivatives to arbitrary order. This routine is called by routines for shape functions of several types, including flattops with square or rounded edges, a symmetric quartic, etc. Lambertson coils are represented by two shape functions, the first for the fundamental with a given m value for the angular dependence, and second for the first allowable harmonic, with an angular dependence of $3m$. The two shape functions are represented by a series of parabolic arcs precomputed in an initialization call. Finally, a user may provide a shape function in the form of a set of points; these points are then interpolated by parabolic arcs. The above routines have been implemented along with a routine for Halbach REC quadrupoles.

5. Examples

5.1. Optimization and fitting example

We give an example of a fitting call inside an optimization function. The BeamKick() routine is a Tlie language subroutine (not shown here) which returns the RMS kick of the lie map.

```
# Fit-opti example
use 24 MeV protons

real beamkickvector[12]

throw   : drift, L = 3.04
q1              : quad, L = 1.0, K= -6.38e-2
q2              : quad, L = 1.0, K= -6.38e-2
q3              : quad, L = 1.0, K= -6.38e-2
q4              : quad, L = 1.0, K= -5.5e-2
q5              : quad, L = 1.0, K= -5.5e-2
q6              : quad, L = 1.0, K= 8.55e-2
q7              : quad, L = 1.0, K= 8.55e-2
soft  : line = (throw, q1, q2 , q3, q4, q5, q6, q7)

func Kick3()
{
fit soft.F[Px,Px], soft.F[Py,Py]
change q1[K], q7[K]
BeamKick(F,beamkickvector)
type "q1[K] = ", q1[K]
type "q2[K] = ", q2[K]
type "q3[K] = ", q3[K]
type "q4[K] = ", q4[K]
```

```
type "q5[K] = ", q5[K]
type "q6[K] = ", q6[K]
type "q7[K] = ", q7[K]
return(beamkickvector[3])
}

optimize error = 1.e-6, Kick3()
change q2[K], q3[K], q4[K], q5[K], q6[K]
```

5.2. Multipole example

Here we exercise the multipolé capabilities.

```
order = 7
use 24 MeV protons

bigap: drift, L = 3.584

mqa: multipole, Halbach, L = 2.5, K = -0.116968075285486,
                m = 2, radius = 0.22

moa1: multipole, Halbach, L = 1.0,  K = -0.165455578374042,
                m = 4, radius = 0.22

moa3: multipole, Halbach, L = 1.0, position = 1.5, K = 0.163635917292597,
                m = 4, radius = 0.22

mda1: multipole, Halbach, L = 0.5, K = 0,
                m = 6, radius = 0.22

mda2: multipole, Halbach, L = 0.5,position = 1.0, K = 0,
                m = 6, radius = 0.22

mda3: multipole, Halbach, L = 0.5, position = 2.0, K = 0,
                m = 6, radius = 0.22

mqb: multipole, Halbach, L = 1.0, position = 3.0, K = 0.205615366067756,
                m = 2, radius = 0.22

mob: multipole, Halbach, L = 1.0, position = 3.0, K = -4.422979283494387E-02,
                m = 4, radius = 0.22

mdb: multipole, Halbach, L = 1.0 ,position = 3.0, K = 0,
                m = 6, radius = 0.22

qcoild: multipolelist = (mqa, mda1,mda2,mda3,mdb,moa1, moa3, mqb, mob, zmin =
-1,
                zmax = 5,error = 1.0e-10)

softel: line = (bigap, qcoild)

type " matrix entries: ",F[Px,Px],F[Py,Py]
fit error = 1.0e-14, F[Px,Px],F[Py,Py] change mqa[K],mqb[K]

type mqa[K],mqb[K]
type " matrix entries: ",F[Px,Px],F[Py,Py]

type " F4 entries: ",F[x^4],F[x^2 y^2],F[y^4]

fit error = 1.0e-14,F[x^4],F[x^2 y^2],F[y^4] change moa1[K],moa3[K],mob[K]
type moa1[K],moa3[K],mob[K]
type " F4 entries: ",F[x^4],F[x^2 y^2],F[y^4]

type " F6 entries: ",F[x^6],F[x^4 y^2],F[x^2 y^4],F[y^6]
Fu = softel.F

fit error = 1.0e-14, F[x^6],F[x^4 y^2],F[x^2 y^4],F[y^6]
change mda1[K],mda2[K],mda3[K],mdb[K]

type mda1[K],mda2[K],mda3[K],mdb[K]
type " F6 entries: ",F[x^6],F[x^4 y^2],F[x^2 y^4],F[y^6]
```

```
type F
```

We show the quadrupole and octupole profile for this case in the figure below.

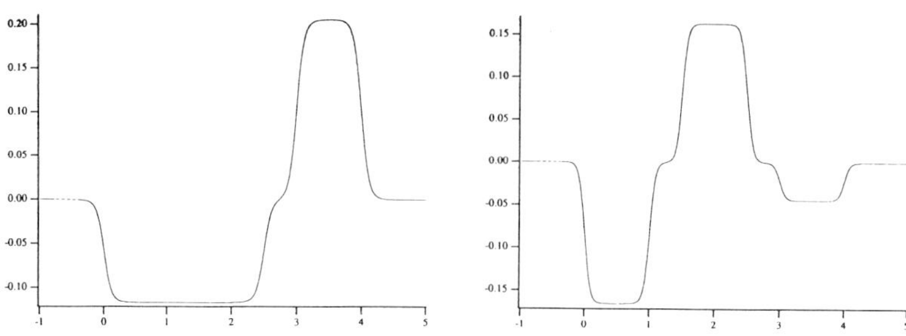

Figure 1. Quadrupole profile   **Figure 2.** Octupole profile

5.3. Subroutines

Here we give an example of a Tlie subroutine which is interpreted by the program and accessible as any other build in function. In this case we implement the sigma matrix. In the real example we also implemented the dependence on energy of the sigma matrix. The routine is too large to show here in its entirety.

```
# beam, map -> Sigma

real sigma[6]

proc beamsigma(liemap $1, vector $2) {
    cf = $1[1,1]
    sf = $1[1,2]
    cp = $1[2,1];
    sp = $1[2,2];
    xs = $2[1]*$2[1];
    xa = -$2[1]*$2[2]*$2[3]/sqrt(1.0e0 + $2[2]*$2[2]);
    xps = $2[3]*$2[3];
    sigma[1] = cf*cf*xs + 2.0*cf*sf*xa + sf*sf*xps;
    sigma[2] = cf*cp*xs + (cf*sp+sf*cp)*xa + sf*sp*xps;
    sigma[3] = cp*cp*xs + 2.0*cp*sp*xa + sp*sp*xps;
}
```

6. Acknowledgements

This program was originally written as an arbitrary order companion to MARYLIE 3.0 [1] and most of the physics and several good ideas are taken from that program. We would like to thank Peter Walstrom for providing routines to evaluate the on-axis gradient and its derivatives for several magnet types. This work was supported in part by US DOE Grant No. DE-AS05-80ER-10666. JvZ would like to thank the AT division at Los Alamos National Laboratory and in particular Thomas Mottershead for its hospitality during several periods when this program was written.

References

[1] A.J. Dragt, *MaryLie 3.0, A Program for Charged Particle Beam Transport Based on Lie Algebraic Methods.*

[2] For a review of the properties of symplectic matrices, see A. J. Dragt, in *Physics of High Energy Particle Accelerators*, edited by R. A. Carrigan, F. R. Huson, and M. Month, AIP Conference Proceedings No. 87 (American Institute of Physics, New York, 1982), p. 147; see also A. J. Dragt *et. al.*, Annu. Rev. Nucl. Part. Sci. **38**, 455 (1988).

[3] A.J. Dragt and E. Forest, J. Math. Phys 24:2734 (1983).

[4] J.B.J. van Zeijts, *Tlie, An Accelerator Design Code to Arbitrary Order, including Misalignments*, University of Maryland, unpublished.

[5] P. Walstrom, Filippo Neri and Tom Mottershead *High Order Optics of Multipole Magnets*, LINAC Meeting Albuquerque 1990.

[6] P. Walstrom, "Magnetic Fields and Inductances of Cylindrical Current Sheet Magnets", to be published as a Los Alamos National Laboratory Unclassified Report (1990).

[7] In practice, a dipole term could be present either intentionally or due to misalignment. If the dipole term is present intentionally, the element in question is actually a combined-function dipole. Normal entry and exit combined-function dipoles are treated in reference [11]. The case of misaligned elements can be treated by combining the maps for aligned element with maps of the form $e^{:f_1:}$.

[8] The effect of a skew quadrupole term can be obtained from the transfer map described in this paper simply by pre- and postmultiplying with the transfer maps for suitable axial rotations, and making associated changes in the sextupole and octupole strengths. Thus, there is no loss of generality in assuming the absence of a skew quadrupole term.

[9] D.D. Šiljak, *Nonlinear Systems*, pp. 384-393, John Wiley (1969).

[10] A.J. Dragt, *Numerical third-order transfer map for solenoid* N.I.M. A298 (1990) 441-459.

[11] A.J. Dragt, F. Neri, J.B.J. van Zeijts and J.Diamond, *Numerical third-order transfer map for Combined Function Dipole* UMD preprint 90-074 (1990).

[12] H. Wollnik, *Optics of Charged Particles*, Academic Press (1987).

[13] A.J. Dragt and R. Ryne, Proceedings of the 1987 IEEE Particle Accelerator Conference, Vol. 2, p. 1063 (1987).

[14] R. Ryne and A.J. Dragt, Particle Accelerators, 35, p. 129 (1991).

[15] P.J. Channell, Acc. Theory Note, Los Alamos National Laboratory, AT-6:ATN-86-6, (1986).

[16] I. Gjaja, *Exact Evaluation of Arbitrary Symplectic Maps* submitted to Phys. Rev. A, (1992).

[17] R.P. Brent, *Algorithms for Minimization without Derivatives*, Prentice-Hall (1973).

[18] B.W. Kernighan and R. Pike, *The Unix Programming Environment*, Prentice-Hall (1984).

The Chalk River Differential Algebra Code "DACYC" and the Role of Differential and Lie Algebras in Understanding the Orbit Dynamics of Cyclotrons

W G Davies, S R Douglas, G D Pusch, and G E Lee-Whiting

AECL Research, Chalk River Laboratories,
Chalk River Ontario, Canada K0J 1J0

Abstract. A new orbit dynamics code, DACYC, is being developed for the TASCC superconducting cyclotron that makes use of *differential algebra* and *Lie algebra* to calculate and analyze partial, one and/or multi-turn maps to very high order. Accurate, three-dimensional, analytic models of the magnetic and RF fields are used, which satisfy Maxwell's equations exactly. The maps can be analyzed by *normal-form* methods or to produce linear or high-order phase space plots.

1. Introduction

A new orbit dynamics code, DACYC, is being developed for the TASCC [1,2] superconducting cyclotron at Chalk River, a cutaway diagram of which is shown in Fig. 1. Our purpose is to develop a code for the high-precision study of the ion-optics of the cyclotron, using "state-of-the-art" numerical and analytical tools. The design goals of DACYC are:

(1) Precision calculation of central trajectories;

(2) Precision calculation of high-order transfer-maps;

(3) Analysis of resonances;

(4) Accurate modeling of RF fields;

(5) Accurate modeling of magnetic fields.

The natural symbiosis of differential algebraic [3] map computation and Lie-algebraic [4] map analysis provides very powerful tools with which to accomplish these goals.

Although cyclotrons are only quasi-periodic devices under operational conditions, the use of maps is still a very efficient way of exhibiting and analyzing important orbit-dynamical features, especially resonances. The Lie-algebraic formulation of *Hamiltonian*

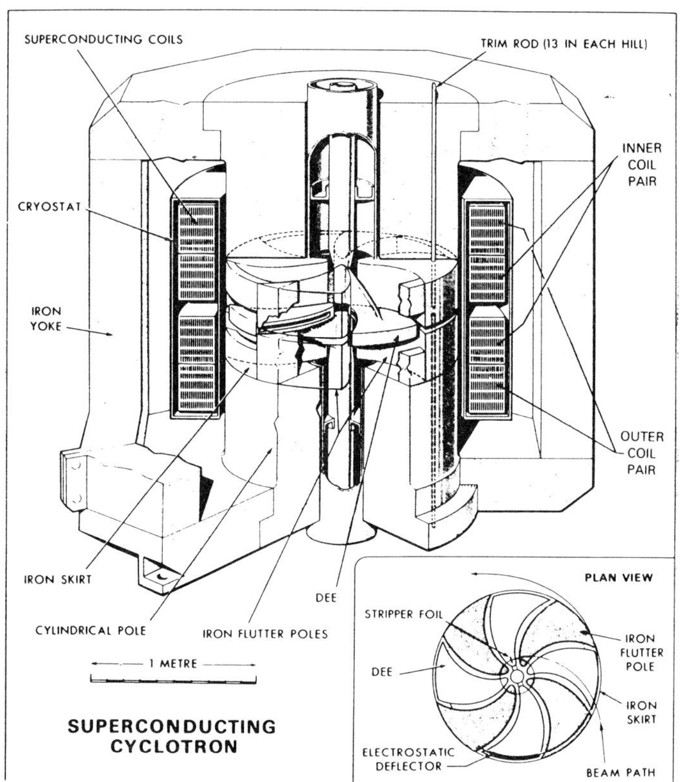

Figure 1. Cutaway view of the superconducting cyclotron showing the major magnet and RF components.

mechanics provides powerful methods for representing and analyzing maps, especially the transformation to *normal form*, but requires that the coordinates be truly canonical. Thus, we integrate Hamilton's equations directly.

The high-order Taylor-series map (the usual aberration expansion) is very efficiently computed from Hamilton's equations with differential algebra, but the Lie map is not. Fortunately, *DA* provides, through its ability to compute high-order derivatives with high accuracy, efficient means for converting the Taylor-series map into a Lie-map when desired. However, for *DA* techniques to be used, the equations of motion and the electromagnetic potentials must be in analytic form.

2. Equations of Motion

The relativistic Hamiltonian in cylindrical coordinates, the "natural" coordinate system for cyclotrons, is [5]

$$H = -erA_\theta - r\left\{ \left(\frac{p_0 + \varphi}{c}\right)^2 - m^2c^2 - [P_r - eA_r]^2 - [P_z - eA_z]^2 \right\}^{1/2} \tag{1}$$

where H is the momentum in the azimuthal direction, θ is the independent variable, r is the radius of curvature, p_0 is the total energy, φ is the electromagnetic scalar potential, and P_r, P_z and A_r, A_z are the radial and vertical components of the momentum and the electromagnetic vector potential, respectively; A_θ is the azimuthal component of the vector potential and c is the velocity of light. Since we will work to some predetermined order of truncation, we make a DA expansion of the Hamiltonian:

$$H \Rightarrow (H, H_r, H_{P_r}, H_Z, H_{P_Z}, \cdots) \tag{2}$$

where the DA vector can be written as

$$(H, H_r, H_{P_r}, H_Z, H_{P_Z}, \cdots) = \sum_{s=0}^{N} \frac{1}{s!} \frac{\partial^{|s|} H}{\partial q^s} \, dq^s \tag{3}$$

with $q^s = r^i P_r^j Z^k P_Z^l T^m P_T^n$, $|s| = i + j + k + l + m + n$, and $s! = i!j!k!l!m!n!$. In this expansion, the derivatives are taken with respect to the canonical coordinates, $\mathcal{Z} = (r, P_r, Z, P_Z, T, P_T)$, of the fundamental cylindrical coordinate system. The differential quantities $dq = (x, p_x, z, p_z, t, p_t)$ are the familiar canonical coordinates of the expanded Hamiltonian. Hamilton's equations are:

$$\left. \begin{aligned} T' &= \frac{\partial H}{\partial P_T} & P_T' &= -\frac{\partial H}{\partial T} \\[4pt] r' &= \frac{\partial H}{\partial P_r} & P_r' &= -\frac{\partial H}{\partial r} \\[4pt] Z' &= \frac{\partial H}{\partial P_z} & P_z' &= -\frac{\partial H}{\partial Z} \end{aligned} \right\} \tag{4}$$

where the "$'$" denotes differentiation with respect to θ. The DA expansion of Hamilton's equations has the form

$$r' = \frac{\partial}{\partial P_r}(H, H_r, H_{P_r}, H_Z, H_{P_Z}, H_T, H_{P_T} \cdots)$$

$$= (H_{P_r}, H_{P_r r}, H_{P_r P_r}, H_{P_r Z}, H_{P_r P_Z}, H_{P_r T}, H_{P_r P_T} \cdots) \tag{5}$$

for the r' equation and similarly for the other coordinates. The resulting DA "vectors" of the expanded form of the equations are integrated numerically by a Bulirsch-Stoer integrator [6]. The result is the reference trajectory, along with the usual Taylor series map, or aberration expansion,

$$\zeta_i^{FIN} = R_{ij}\zeta_j^{IN} + T_{ijk}\zeta_j^{IN}\zeta_k^{IN} + U_{ijkm}\zeta_j^{IN}\zeta_k^{IN}\zeta_m^{IN} + \cdots \tag{6}$$

where $\zeta = (x, p_x, z, p_z, t, p_t)$ are the canonical coordinates relative to the reference trajectory; repeated indices are summed. The superscripts, IN and FIN, refer to the initial and final coordinates connected by the map. The order of the map (6) is determined at run-time. Eq. (6) plus the reference trajectory are sufficient if we are interested only in simple phase-space plots and/or the behavior of the reference trajectory. For the study of the non-linear behavior, in particular *resonances*, we transform to the Lie representation [4].

3. Magnetic Field Representation

For *DA* to be used, the vector potential, $\mathbf{A}$, must be in analytic form. We have constructed a 3-dimensional analytic model of the magnetic field which satisfies Maxwell's equations exactly. The model is divided into four parts:

1. The superconducting coils ($1 - 3$ Tesla);
2. The saturated iron flutter-poles (~ 2 Tesla);
3. The partially saturated yoke ($\sim 10\%$ main field);
4. The saturated iron trim-rods ("dimples" of up to ~ 0.7 Tesla).

3.1. Field from the coils

Because we have cylindrical symmetry, the vector potential of a pair of circular loops of radius a carrying a current I located at $r = a$, $z = \pm b$, and valid for $0 \leq r \leq \infty, \mid z \mid < b$, can be derived from the integral expression:

$$A_\theta = \mu_0 a I \int_0^\infty e^{-bs} \, J_1(rs) \, J_1(as) \, \cosh(zs) \, ds. \tag{7}$$

Instead of a "brute-force" integration over the volume of the coils, that is,

$$A_\theta = \mu_0 \mathcal{J} \int_{-\alpha}^{\alpha} d\alpha (a + \alpha) \int_{-\beta}^{\beta} d\beta \int_0^\infty e^{-(b+\beta)s} \, J_1(rs) \, J_1[(a + \alpha)s] \, \cosh(zs) \, ds, \tag{8}$$

we partition the coils and make a moment expansion about the center of each cell:

$$A_\theta = \mu_0 \mathcal{J} \sum_{\substack{n,m \\ even}} \frac{1}{n!m!} \, T_{n,m} \int_0^\infty e^{-bs} \, J_1(rs) \, \cosh(zs) \, f^{(n)}(as) \, s^{n+m-1} \, ds. \tag{9}$$

Here $\mathcal{J}$ is the current density, $T_{nm} = \iint \alpha^n \beta^m \, d\alpha \, d\beta$; $f(as) \equiv (as)J(as)$, $f^{(n)}(as)$ is obtained by recursion from J_0 and J_1, and $\int \ldots ds$ is obtained by recursion from two elliptic integrals, which are evaluated in *DA*. This moment expansion reduces the computing time by a factor of 65 over a "brute-force" application of (8).

3.2. Field from the flutter poles

The vector potential for the saturated iron flutter poles is derived from first principles from the equivalence between uniformly magnetized iron and a solenoid. Here, both A_θ and A_r components are needed. The pole-boundary is approximated by a polygon, with a gap of $2b$ in the midplane; a view of the polygonal model of the lower pole is shown in Fig. 2. By Ampère's law,

$$\mathbf{A}(\mathbf{r}) = \frac{\mu_0 M_0}{4\pi} \left\{ \int_{-\infty}^{-b} dz' + \int_b^\infty dz' \right\} \oint_{\Gamma'} \frac{1}{\mid \mathbf{r} - \mathbf{s}' - z'\hat{e}_{z'} \mid} \, ds' \tag{10}$$

where s', ds' is a point and a line segment, respectively, on the edge of the polygon Γ' and M_0 is the saturation magnetization of the iron. The potential is expressed in terms of *logarithms* and *arctangents* of algebraic functions, both of which have known

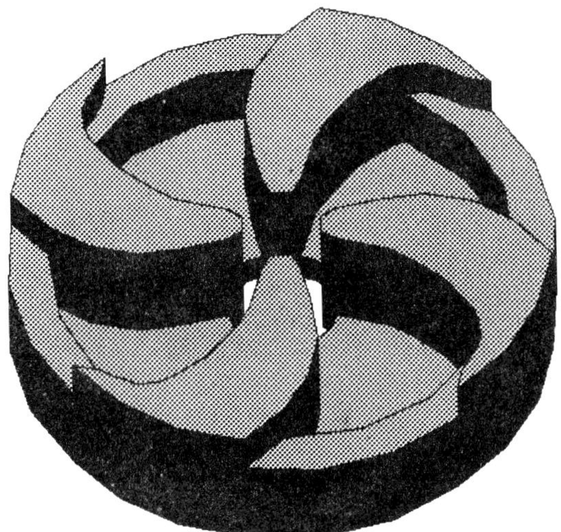

Figure 2. Polygonal model of the lower pole of the cyclotron. The hill and valley regions are easily seen as are the iron "skirts," which reduce the fall-off of the field at large radius.

DA representations. Each hill, valley and "skirt" (see Fig. 2) region is represented by such a model.

The edges of the hill, valley, and skirt are "softened" by linear dipole-densities, to help compensate for non-uniform magnetization effects. The vector potential for the linear dipole-density is

$$\mathbf{A}\left(\mathbf{r}\right) = \frac{\mu_0}{4\pi} \int \mathbf{m} \times \nabla' \frac{1}{\mid \mathbf{r} - \mathbf{r}' \mid} \, dl' \tag{11}$$

where $\mathbf{m}$ is the dipole strength per unit length. The fields of the edge-dipoles involve only algebraic functions of quantities already computed, and so provide a number of physically-motivated fitting parameters at almost no extra cost to help fit the details of the field.

3.3. Yoke field

The yoke terminates the poles at a finite height, and completes the magnetic circuit. The true octagonal yoke is approximated by a structure with axial symmetry — a hollow cylinder with the same mean radius and cross-sectional area as the real yoke capped by a pair of circular disks. The magnetic field in the iron is assumed to be radial and non-uniform in the end-rings and vertical and uniform in the yoke walls. The field in the yoke is estimated using a magnetic circuit model, where the $B - H$ curve is idealized as being piecewise-linear up to saturation and constant above saturation.

The vector potential and mid-plane induction are calculated from formulas based on

$$\mathbf{A}\left(\mathbf{r}\right) = \frac{\mu_0}{4\pi} \int_V \mathbf{M}\left(\mathbf{r}'\right) \times \nabla' \frac{1}{\mid \mathbf{r} - \mathbf{r}' \mid} \, d^3\mathbf{r}'. \tag{12}$$

The r' and z' integrals are done analytically, while the θ' integral is done numerically by means of the trapezoidal rule. Nevertheless, Maxwell's equations are satisfied identically, because of the form of the integrand.

A least-squares fit of items 1, 2, and 3 of this model to the measured midplane field is shown in Fig. 3. The quality of the fit is quite impressive; the *RMS* error in Fig. 3 is already only 0.237%, and we expect this to be reduced even further by additional refinements to our model, such as the "z-shims" at outer radius.

3.4. Trim rods

The TASCC cyclotron [1] uses 52 pairs of saturated iron "trim-rods" instead of trim-coils to fine-tune the magnetic field (see Fig. 1). Retracting a trim-rod is assumed to be equivalent to superimposing a slug of iron with magnetization $-\mathbf{M}$. Outside the sphere enclosing the slugs for a pair of retracted rods (upper and lower poles), the potential of a pair of trim-rods is well represented by a low-order multipole expansion. Inside this sphere, the vector potential is represented by a current-sheet, or uniform magnetization approximation, which is treated in a similar manner as the yoke, (12); the r' and z' integrals are done analytically, and the θ' integral is done numerically. Nevertheless, Maxwell's equations are again satisfied identically, by virtue of the form of the integrand. Performance can be increased by lumping clusters of distant rods together as in "fast multipole methods." Computational efficiency of the trim-rod model is of great importance, due to the large number of trim rods.

In summary, the magnetic field model outlined above provides a full 3-dimensional description of the field that satisfies Maxwell's equations exactly. This includes not only the acceleration region between the poles, but the fringe-field region out to the vicinity of the yoke wall. It was originally thought that it would be necessary to include a residual-field correction [2], but the model field is now of sufficient quality that this is unlikely.

4. RF Field Representation

Due to peculiarities of the RF cavity design of the TASCC cyclotron, there are distributed vertical components of the electric field, especially in push-pull, or "π-mode." Thus the usual kick approximation cannot be used. The success of the magnet pole model, section 3.2, motivated us to search for a similar model for the RF field, *i.e.*, a model that is closely tied to the physics of the cavity. The geometry of the RF structure in the TASCC cyclotron, as in many cyclotrons, tends to concentrate the RF currents on the "dees" and "hills" in the vicinity of the accelerating gap. Thus one looks for an RF model that takes advantage of this fact, which would result in a much simpler and computationally more efficient model.

If we assume a simple harmonic time dependence and the Lorentz gauge, a solution of Maxwell's equations is given by

$$\mathbf{A}(\mathbf{r}, T) = \mathbf{A}_0(\mathbf{r})\sin(\omega T); \qquad \varphi(\mathbf{r}, T) = \varphi_0(\mathbf{r})\cos(\omega T) \tag{13}$$

where

$$\mathbf{A}_0(\mathbf{r}) = \frac{\mu_0}{4\pi} \int_\Sigma \frac{\cos(k\,|\,\mathbf{r} - \mathbf{r}'\,|)}{|\,\mathbf{r} - \mathbf{r}'\,|}\, \mathbf{K}_0(\mathbf{r}')\, d^2 S'$$

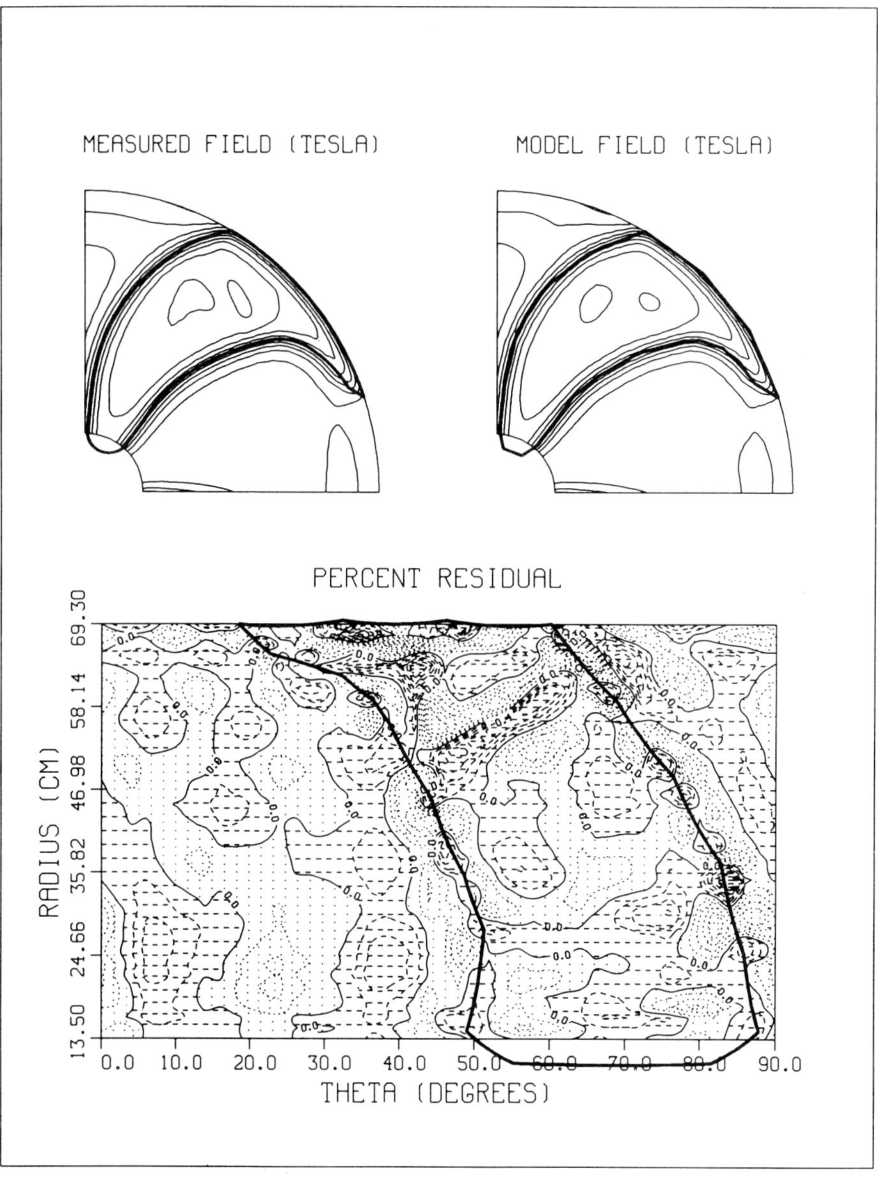

Figure 3. a) Measured mid-plane field (Tesla). The heavy line outlines the physical pole boundary. b) Model mid-plane field (Tesla). The heavy line outlines the polygonal approximation to the pole. Contours are in 0.1 Tesla steps for both a and b. c) Percent deviation of the fitted magnetic field from the measured mid-plane magnetic field; dotted (dashed) lines denote positive (negative) contours, respectively. Contours are spaced 0.2% apart. The heavy solid line outlines the polygonal "hill" boundary.

$$\varphi_0(\mathbf{r}) = \frac{1}{4\pi\epsilon_0} \int_\Sigma \frac{\cos(k\,|\,\mathbf{r} - \mathbf{r}'\,|)}{|\,\mathbf{r} - \mathbf{r}'\,|}\, \sigma_0(\mathbf{r}')\, d^2 S' \tag{14}$$

and $\mathbf{K}_0(\mathbf{r})$, $\sigma_0(\mathbf{r})$ are surface currents and charge-densities respectively; μ_0 and ϵ_0 have their standard meanings. In an isochronous cyclotron the time, T, is a function of the azimuthal angle θ and the relative time, t. We seek polygonal [7] (triangular or quadrilateral) interpolative current elements such that the integrals in (14) can be done analytically. We determine the surface-current and charge-density constants, $\mathbf{K}_0$ and σ_0, from a combination of experimental measurements, electro-static calculations and the continuity equation. The form of the Green's function in (14), which is half advanced plus half retarded, guarantees that there will be zero time-averaged power flux *i.e.* we produce a standing-wave solution, independent of $\mathbf{K}_0$.

5. Lie-Algebraic Analysis

The Taylor series map (6), written as $\zeta^{FIN} = \mathcal{M}(\theta)\,\zeta^{IN}$, can be transformed [8] into an equivalent Lie-map

$$\mathcal{M} \Rightarrow \mathcal{M}_f; \qquad \mathcal{M}_f = e^{:f:}; \qquad :f{:}\,g = [f,g] \tag{15}$$

where

$$e^{:f:} \stackrel{\text{def}}{=} \sum_{n=0}^{\infty} \frac{:f:^n}{n!}; \qquad :f{:}^0 = \mathcal{I} = \{identity\ map\}, \tag{16}$$

and $[\cdot,\cdot]$ is the Poisson bracket [8] with respect to the canonical coordinates (x, p_x, z, p_z, t, p_t). The Lie-map $\mathcal{M}_f$ can be factored [8] into

$$\mathcal{M}_f = \mathcal{M}_2 \mathcal{M}_3 \mathcal{M}_4 \ldots = e^{:f_2:}\, e^{:f_3:}\, e^{:f_4:} \ldots \tag{17}$$

where $\mathcal{M}_f$ and $\mathcal{M}_n$ are *symplectic maps* and f_n are homogeneous polynomials of degree n; there is no $\mathcal{M}_1$ factor in the map because DA automatically expands about the reference trajectory. The maps $\mathcal{M}_2$ are elements of the symplectic group $Sp(6,\mathbf{R})$, while the $\mathcal{M}_n$ are elements of the infinite-dimensional group of symplectic maps. The Lie-polynomials f_n provide a much more compact representation of the map than (6), and preserve the symplectic nature of the map, $\mathcal{M}$, exactly, which (6), when truncated, does not do. With the help of DA, (17) can be transformed into *normal form* [4], which provides a very powerful method for identifying and analysing non-linear behaviour and resonances in particle accelerators.

5.1. Normal Form

The normal form is the transformation $\mathcal{N} = \mathcal{A}\,\mathcal{M}\,\mathcal{A}^{-1}$ such that $\mathcal{N}$ is in its *"simplest"* form; $\mathcal{M}$ must be expanded about a *"center manifold"* or *"closed orbit"* i.e. no H_1 terms in H, which implies no $\mathcal{M}_1$ terms in $\mathcal{M}$.[1] The symplectic map $\mathcal{A}$ is an nth order

---

[1] If an $\mathcal{M}_1$ factor is present, it may always be removed as the first step of the normalization procedure *via* the "fixed-point transformation" [8]. If ζ_* is a point on the center manifold such that $\zeta_* = \mathcal{M}\zeta_*$ (a "fixed point"), the canonical change of variables $\bar{\zeta} = \mathcal{A}_1\zeta$ such that $\bar{\zeta}_* = \mathcal{A}_1\zeta_* = 0$ yields the map $\bar{\zeta}_* = \mathcal{A}_1 \mathcal{M}\mathcal{A}_1^{-1}\bar{\zeta}_* = \overline{\mathcal{M}}\bar{\zeta}_* = 0$; *i.e.*, the center manifold becomes the new origin, which is a fixed point of the transformed map $\overline{\mathcal{M}}$.

canonical transformation that *isolates* the *tune shifts* and *resonances* to n^{th} order. Using the factored map, we can normalize $\mathcal{M}$ order by order to obtain

$$\mathcal{N} = \mathcal{A}_n \cdots \mathcal{A}_4 \big\{ \mathcal{A}_3 \{ \mathcal{A}_2 \mathcal{M} \mathcal{A}_2^{-1} \} \mathcal{A}_3^{-1} \big\} \mathcal{A}_4^{-1} \cdots \mathcal{A}_n^{-1} \tag{18}$$

We proceed as follows: From the eigenvectors of R, we get a canonical transformation that *rescales and block diagonalizes R*;

$$\mathcal{R} = \mathcal{A}_2 R \mathcal{A}_2^{-1} \quad \Longrightarrow \quad \mathcal{R} = \begin{pmatrix} \mathcal{R}_1 & & 0 \\ & \mathcal{R}_2 & \\ 0 & & \mathcal{R}_3 \end{pmatrix} \tag{19}$$

where

$$\mathcal{R}_i = \begin{pmatrix} \cos(\mu_i) & \sin(\mu_i) \\ -\sin(\mu_i) & \cos(\mu_i) \end{pmatrix}, \quad \begin{pmatrix} 1 & L_i \\ 0 & 1 \end{pmatrix}, \quad \text{or} \quad \begin{pmatrix} 1 & 0 \\ a_i & 1 \end{pmatrix}.$$

Next, we normalize $\mathcal{M}_3$

$$\mathcal{N} = \mathcal{A}_3 \mathcal{A}_2 \, R e^{:f_3:} \cdots \mathcal{A}_2^{-1} \mathcal{A}_3^{-1} = e^{:G_3:} \, \mathcal{R} e^{:g_3:} \cdots e^{-:G_3:}.$$

$$= \mathcal{R} \, \exp\big\{ : -(\mathcal{I} - \mathcal{R}^{-1}) \, G_3 + g_3 : \big\} \cdots . \tag{20}$$

A further simplification results if we expand $(\mathcal{I} - \mathcal{R}^{-1})$, and g_3 in a complex or "resonance" basis as follows:

$$h_\pm = x \pm i p_x, \qquad v_\pm = z \pm i p_z. \tag{21}$$

Hence,

$$(\mathcal{I} - \mathcal{R}^{-1}) \, | \, \mathbf{n}, \mathbf{m} > \; = \Big[1 - \sum_{\mathbf{n}+\mathbf{m}=3} e^{i(\mathbf{n}-\mathbf{m}) \cdot \mu} \Big] \, | \, \mathbf{n}, \mathbf{m} >, \tag{22}$$

and

$$g_3 = \sum_{\mathbf{n}+\mathbf{m}=3} A_{\mathbf{n}\mathbf{m}} \, | \, \mathbf{n}, \mathbf{m} > \quad \text{where} \quad | \, \mathbf{n}, \mathbf{m} > \; = h_+^{n_1} h_-^{m_1} v_+^{n_2} v_-^{m_2} \cdots \tag{23}$$

and $\mathbf{n}, \mathbf{m}, \mu$ are vectors with components $(n_1, n_2, \cdots)$ *etc.* Substituting (22) and (23) into (20) and solving for G_3, we obtain

$$G_3 = \sum_{\mathbf{n}+\mathbf{m}=3} \frac{A_{\mathbf{n}\mathbf{m}}}{\{ 1 - \exp[\, i(\mathbf{n} - \mathbf{m}) \cdot \mu] \}}. \tag{24}$$

All terms in g_3 can be incorporated into G_3 except when $\mathbf{n} - \mathbf{m} = 0$, or $(\mathbf{n} - \mathbf{m}) \cdot \mu = 2\pi \ell$; $\ell = $ integer; that is $(n_1 - m_1)\mu_x + (n_2 - m_2)\mu_z \neq 2\pi \ell$; $\ell = 0, \pm 1, \pm 2 \cdots$. The integer ℓ characterizes the resonance. In a similar fashion, we could proceed to find G_4 *etc.* A further substitution of $J_x = h_+ h_-$, $J_z = v_+ v_-$, the *action invariants*, leads to

$$\mathcal{N} = \exp : \big\{ (\mu_x + \mu_x' p_t + \mu_x'' p_t^2 + \cdots) J_x + (\mu_z + \cdots) J_z + a_x J_x^2 + a_z J_z^2 + \cdots \big\} : \tag{25}$$

where we have assumed no acceleration and no explicit resonances; μ are the *phase advances*, μ' *etc.* are the *chromaticities* and all terms proportional to J^n are non-linear *tune shifts*. That we can incorporate all the other non-linearities into $\mathcal{A}$ and "get rid of them" makes the analysis of the tune-shifts and resonances much easier! The Lie-algebra library of Forest [9] is used for this analysis. If explicit resonances occur with $(\mathbf{n} - \mathbf{m}) \cdot \mu = 2\pi \ell$; $\ell = $ integer, these must be added to (25). Because $[J_i^n, J_j^m] = 0$, that is all powers of the actions commute, the transformation from factored to single-exponent form is straightforward in (25). However, if there are resonances in the system, the *Campbell-Baker-Hausdorff* formula [8] must be used; alternatively, a much easier solution is to convert the equivalent Taylor-series map to single exponent form with *DA*. Finally, the *normal form* has the remarkable property that

$$\mathcal{M}^n = \mathcal{A}^{-1} \mathcal{N}^n \mathcal{A} \tag{26}$$

which also results in an enormous saving of work when applicable.

6. Examples and Discussion

Like all isochronous cyclotrons,[2] the TASCC cyclotron is nearly resonant in the bending plane ($\nu_r \simeq 1$). Hence it is very sensitive to a first harmonic, $b_1 \cos(\theta)$, term in the Hamiltonian which drives decentering of the orbits. In addition, the beam passes through many resonances in the vertical plane because the vertical tune, ν_z, varies as a function of r in the range ($0.2 \lesssim \nu_z \lesssim 1$) for low-field beams and ($0 \lesssim \nu_z \lesssim 1$) for high-field beams. The principal resonances are:

a) $\nu_z = 1/5$ driving term — very weak

b) $\nu_z = 1/4$ driving term — *octupole* — strong

c) $\nu_z = 1/3$ driving term — *sextupole* — moderate

d) $\nu_z = 1/2$ driving term — *octupole* — strong

e) $\nu_z = 1$ driving term — $b_1(r)\cos(\theta)$ — weak

f) $\nu_r = 2\nu_z$ (Walkinshaw resonance)   driving term — *average sextupole at outer radii* — very strong.

In our cyclotron we sometimes have $\nu_z \simeq 1/2$, $\nu_r \simeq 1$, and $2\nu_z \simeq \nu_r$ simultaneously!

As an example, let us investigate briefly the "topology" of the maps in the TASCC cyclotron, and in particular that of the Walkinshaw resonance near extraction. As was mentioned in sect. 4, the RF cavity generates distributed vertical fields that break mid-plane symmetry. Thus, we would expect the map to be dense, *i.e.* there will be no non-zero components in the linear or non-linear parts of the map. This is confirmed by calculations with a simplified model of the magnetic and RF fields, valid only near the inner region of the cyclotron. Calculations of the first 5 turns in the cyclotron (in order to include one complete betatron oscillation in the vertical plane) were performed to second order ($\mathcal{M}_3$ or T) for $^{35}C\ell$ at 30 MeV/u, with a preliminary version of the *DA* program. The effect of the vertical RF fields on the reference trajectory is shown in Fig. 4. These results are quite sensitive to the details of the RF field, and especially to the RF phase when the beam passes through the carbon stripping foil at injection (see [1] and references therein). From the "first order" map R, we find that the radial

---

[2] An "isochronous" cyclotron is not isochronous in the sense that the period of *every* orbit is independent of energy, but refers only to the static equilibrium orbits. In practice, only a finite set of static equilibrium orbits at specific radii are made isochronous. Hence the one-turn maps of such machines are only approximately isochronous, although they will be close.

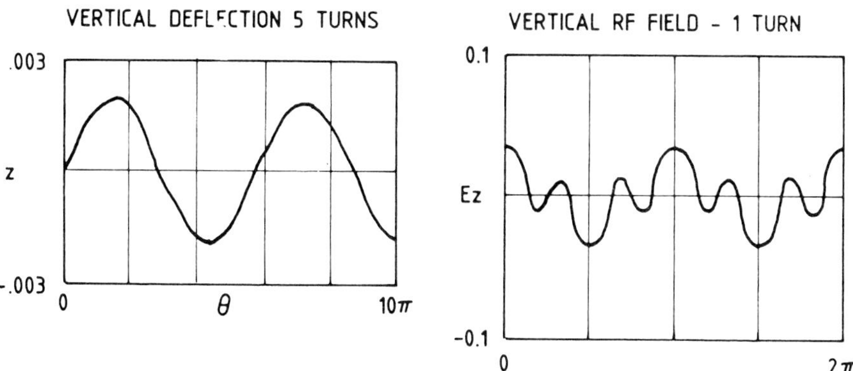

Figure 4. a) Projection of the reference trajectory on the vertical plane as a function of azimuth. This figure illustrates the effect of the vertical components in the RF field. b) Vertical RF field over one turn, as a function of azimuth.

and vertical tunes are $\nu_x = 1.006$, and $\nu_z = 0.227$, in agreement with measurement. These maps can be transformed into normal form. The *DA* code of Douglas [10] was used in this analysis. From the eigenvalues of the first order map, R, we can obtain the transformation to block diagonal form (19). We find that $\mathcal{R}_1$ and $\mathcal{R}_2$ are rotation matrices, as expected, and $\mathcal{R}_3$ is drift-like in spite of the acceleration; (this is also expected as there is no first-order longitudinal RF focusing in an isochronous cyclotron).

As was pointed out in the introduction, cyclotrons are only quasi-periodic and one should ideally work with accelerated maps. Nevertheless, much can be learned about the topology of this normal form map, up to 3rd order, by studying the normal form of a simplified autonomous Hamiltonian of the form (1) with a vector potential

$$A_\theta = -\frac{x + r_0}{2} B_0 - \frac{1}{r_0} \left\{ Q_x x^2 + Q_z z^2 \right\} - \frac{S_x}{r_0} \left\{ \frac{x^3}{3} - x z^2 \right\} \tag{27}$$

where $A_r = A_z = 0$; Q_x, Q_z, S_x, are the effective quadrupole and sextupole strengths, respectively. There is a small skew sextupole component arising from breaking midplane symmetry but we will neglect it here. Because (27) is independent of θ, the Hamiltonian (1) has the solution

$$\mathcal{M}(\theta) = \exp\left\{ -\int_0^\theta : H : d\theta' \right\} \quad \Longrightarrow \quad \mathcal{M}(\theta) = e^{-\theta : H :} \tag{28}$$

which can be factored (with some work) into the form of (17). Because we are neglecting acceleration, the energy of each particle is conserved and $p_t = (E_0 - E)/P_0 c$, where

E is the particle energy, and E_0 and P_0 are the energy and momentum of the central trajectory, respectively.

If we carry out the normalization procedure described in section 5, we find that f_3 contains 19 complicated terms. If there are no explicit resonances, the normal form contains $g_2 = -[\mu_x J_x + \mu_z J_z]$, (equivalent to the "rotation matrix" $\mathcal{R}$, eq. (19)), and

$$
\begin{aligned}
g_3 = {} & \frac{(1 - \nu_x)}{4\nu_x \beta} \, J_x p_t + \frac{(1 - \nu_z)}{4\nu_z \beta} \, J_z p_t \\[2mm]
& + \pi S_x \frac{k_x^2}{\nu_x \beta} \, J_x p_t - \pi S_x \frac{k_x k_z}{\nu_x \beta} \, J_z p_t \\[2mm]
& + \pi k_x \frac{1 - \beta^2}{\nu_x \beta} \, p_t^3 + \pi S_x \frac{2 k_x^3}{3 \nu_x^3 \beta^3} \, p_t^3
\end{aligned}
\tag{29}
$$

which contains four *"chromaticity"* terms proportional to p_t and two time-of-flight aberration terms proportional to p_t^3, a substantial simplification. Here $k_x = r_0/\nu_x$, $k_z = r_0/\nu_z$ and $\beta = v/c$, the relativistic factor. We see from (29) that if $\nu_x = 1$ (that is the system is resonant in the radial plane!) then the first term disappears. If the system is also made achromatic then $S_x = 0$ and all the *chromaticity* terms vanish except the second one involving $(1 - \nu_z)$. Under these circumstances $\mathcal{R}_1 = \mathcal{R}_x = \mathcal{I}_x$, $\mathcal{R}_2 = \mathcal{R}_z$, and we find that

$$
\begin{aligned}
\mathcal{M} &= \left\{ \mathcal{A}_3^{-1} \mathcal{A}_2^{-1} \mathcal{N} \mathcal{A}_2 \mathcal{A}_3 \right\} = \left\{ \mathcal{A}_3^{-1} \mathcal{I}_x \mathcal{R}_z e^{:g_3:} \mathcal{A}_3 \right\} \\[2mm]
&= \hat{\mathcal{A}}_3^{-1} \mathcal{R}_z e^{:\hat{g}_3:} \hat{\mathcal{A}}_3
\end{aligned}
\tag{30}
$$

where

$$
\hat{g}_3 = \frac{(1 - \nu_z)}{4\nu_z \beta} \, J_z p_t + \pi k_x \frac{1 - \beta^2}{\nu_x \beta} \, p_t^3
$$

and the $\hat{\mathcal{A}}$ now contain only terms involving z; *i.e.* all the non-linearities involving only the x coordinate cancel! If we could make $\nu_z = 1$ also, the remaining *chromaticity* term would vanish and then $\mathcal{R}_x = \mathcal{R}_z = \mathcal{I}$, i.e., $\mathcal{R}_\perp = \mathcal{I}$. Hence

$$
\mathcal{M} = \left\{ \mathcal{A}_3^{-1} \mathcal{A}_2^{-1} \mathcal{N} \mathcal{A}_2 \mathcal{A}_3 \right\} = \mathcal{I} e^{:\alpha p_t^3:}.
\tag{31}
$$

Eq. (31) tells us that under these circumstances, all of the non-linearities, except the one proportional to p_t^3 annihilate each other up to 3rd order in the map![3]

This result was originally discovered by Karl Brown [11] for achromatic beam line systems and has been analysed, using Lie-algebraic methods, in a different manner to the above by Forest *et. al.* [12].

In an isochronous cyclotron, $\nu_x \simeq 1$, $S_x \simeq 0$ (at least on the average but not necessarily locally) and $\nu_z \ll 1$ over most of the accelerating region. Hence there is a large chromaticity term in the vertical direction as is well known but all the radial plane nonlinearities will be small to 3rd order in the map. What happens at 4th order is another story! However, as we pass (adiabatically) through the resonances in the vertical plane,

---

[3] Note that the Lie transformation $e^{:g_n:}$ of order n in the Lie polynomial generates aberration (equivalent Taylor series) coefficients of order $n - 1$. Whether we are discussing the order of the Lie-polynomial or the order of the Taylor series coefficients should be apparent from the context.

this cancellation will be punctuated by terms with small denominators which must be retained (locally) in the normal map.

As a final topic we will briefly sketch the analysis of the Walkinshaw or $\nu_x = 2\nu_z$ coupling resonance. This resonance occurs near extraction in many cyclotrons and as has already been pointed out sometimes occurs in our cyclotron as a triple resonance (depending on the detailed parameters of the magnetic field). In the fringe-field region, we have large sextupole and octupole strengths and the magnetic field is no longer isochronous. We will ignore the octupole component for the purposes of this discussion, but it can have a profound effect on the stability of the orbits. Here, we will make use of *action-angle* variables which are related to $h_\pm$ and $v_\pm$ by

$$h_\pm = \sqrt{2J_x}\, e^{\mp i\varphi_x}; \qquad v_\pm = \sqrt{2J_z}\, e^{\mp i\varphi_z}. \tag{32}$$

Since $\nu_x = 2\nu_z$, (*i.e.* $(\mathbf{n}-\mathbf{m})\cdot\mu = 2\pi\ell$, with $\ell = 0$), we will have a vanishing denominator in (24) and this term must be included in the normal map. We can write the normal map, which now includes the resonance term, in factored form as follows:

$$\mathcal{N} = e^{:g_2:}e^{:g_3:}e^{:\alpha p_t^3:} = e^{:g_2:}e^{:g_3':}e^{:\hat{g}_3:}e^{:\alpha p_t^3:} \tag{33}$$

where $g_2 = -\mu\cdot J = -[\mu_x J_x + \mu_z J_z]$ and $g_3' = \mu_x' p_t J_z + \mu_z' p_t J_z$, the chromaticities (29). The factorization of $e^{:g_3:}$ into $e^{:g_3':}e^{:\hat{g}_3:}$ can be done trivially because we restrict ourselves to $3rd$ order in the map, *i.e.* the leading Poisson bracket in the CBH formula, $[g_3', \hat{g}_3]$, is of order 4. Because $[J_i^n, J_j^m] = 0$, we can combine g_2 and $\hat{g}_3$ into a single exponent without further ado as

$$-\hat{\mu}\cdot J = -\mu\cdot J + p_t\mu'\cdot J \tag{34}$$

where $\hat{\mu} = \mu - p_t\mu'$, *i.e.* the new tune is energy dependent. The tune-shifted resonance occurs when $(\mathbf{n}-\mathbf{m})\cdot\hat{\mu} = 0$. The sextupole driving term, $\hat{g}_3$ is given by

$$\hat{g}_3 = -\frac{\pi}{8}S_x k_z\sqrt{k_x}\,(2J_x)^{1/2}\left[e^{i\varphi_x} + e^{-i\varphi_x}\right](2J_z)\left[e^{2i\varphi_z} + e^{-2i\varphi_z}\right]. \tag{35}$$

In order to understand the topology of this resonance it is convenient to transform to the "co-moving" frame [13]. This transformation is defined by the expression

$$\phi_1 = \mathbf{k}\cdot\varphi; \qquad (\mathbf{n}-\mathbf{m}) = \lambda\mathbf{k}, \qquad \lambda = \{integer\}, \tag{36}$$

and

$$K_1 = \frac{\mathbf{k}\cdot J}{\|\mathbf{k}\|^2}, \qquad K_i = \frac{\mathbf{a}_i\cdot J}{\|\mathbf{a}_i\|^2}, \qquad i = 2, \ldots, N. \tag{37}$$

The $\mathbf{a}_i$ are $N-1$ linearly independent vectors satisfying the condition $\mathbf{k}\cdot\mathbf{a}_i = 0$; they generate $N-1$ invariants K_i that commute with K_1. In our case $N = 2$ and $\lambda = 1$; hence we obtain

$$\phi_1 = \varphi_x - 2\varphi_z, \quad\text{and}\quad \phi_2 = 2\varphi_x + \varphi_z \tag{38}$$

where $\mathbf{k} = (1, -2)$ is orthogonal to $\mathbf{a} = (2, 1)$. Thus

$$K_1 = \frac{J_x - 2J_z}{5}, \qquad K_2 = \frac{2J_x + J_z}{5}, \tag{39}$$

and $J_x = K_1 + 2K_2$, $J_z = K_2 - 2K_1$. The transformation to the co-moving frame is effected by the transformation [13]

$$\mathcal{N}_c = e^{:2\pi\ell K_1:}\mathcal{N} = e^{:2\pi\ell K_1:}\, e^{:-\hat{\mu}\cdot J:}\, e^{:\tilde{g}_3:}\, e^{:\propto p_t^3:} \tag{40}$$

where $2\pi\ell = \mathbf{k}\cdot\mu$. In our case $\ell = 0$, and the transformation is the identity map. We want to expand about the resonance, that is, we assume that we are not quite on resonance, and write $\mathbf{k}\cdot\mu = 2\pi(\nu_x - 2\nu_z) = \delta$. Hence

$$\mathcal{N}_c = e^{:-\delta K_1 - \mu\cdot a K_2:}\, e^{:\tilde{g}_3:}\, e^{:\propto p_t^3:} \tag{41}$$

where $\tilde{g}_3$ is in the co-moving coordinates,

$$\tilde{g}_3 = -\frac{\pi}{2}S_x k_z\sqrt{2k_x}\,[K_1 + 2K_2]^{1/2}\,[K_2 - 2K_1]\left[e^{i\phi_1} + e^{-i\phi_1}\right]. \tag{42}$$

We note that there is no ϕ_2 dependence. Thus we have reduced the apparently two-dimensional resonance into a one-dimensional form with parameter K_2. Furthermore, K_1 commutes with K_2. Hence, if we combine exponents as follows,

$$\mathcal{N}_c = e^{:-\delta K_1 - \mu\cdot a K_2:}e^{:\tilde{g}_3:}e^{:\propto p_t^3:} = e^{:h:}e^{:\propto p_t^3:} \tag{43}$$

and the "pseudo-Hamiltonian" h has the form

$$h = -\,\delta\,K_1 - \mu\cdot\mathbf{a}\,K_2$$
$$+\pi\delta S_x k_z\sqrt{2k_x}\,[K_1 + 2K_2]^{1/2}\,[K_2 - 2K_1]\left[\frac{\sin(\phi_1) - \sin(\phi_1 + \delta)}{1 - \cos(\delta)}\right] + \mathcal{O}(S_x^2). \tag{44}$$

Here we have made use of the *Campbell-Baker-Hausdorff* formula (see [8]) and summed the series with respect to δ. The fixed points of the resonance can be found in the usual way by setting $\zeta = \mathcal{M}\zeta$, which reduces to

$$\frac{\partial h}{\partial K_1} = 0 \quad\text{and}\quad \frac{\partial h}{\partial \phi_1} = 0, \tag{45}$$

the usual condition for a fixed point if h were the real Hamiltonian. This leads to

$$\tan(\phi_1) = -\frac{1 - \cos(\delta)}{\sin(\delta)} \simeq \sin(\phi_1) = 0 \tag{46}$$

and

$$\frac{2\pi k_z\sqrt{2k_x}\,S_x[6K_1 + 7K_2] - \delta\sqrt{K_1 + 2K_2}}{\sqrt{K_1 + 2K_2}} = 0 \tag{47}$$

which has the solutions

$$\phi_1 = \{0, \pi\} \quad\text{and}\quad K_1 = \pm\frac{\delta\sqrt{15K_2}}{72\pi k_z\sqrt{k_x}S_x} - \frac{7K_2}{6} \tag{48}$$

where we have neglected terms of order δ^2 in the expansion of the trigonometric functions. The topology of this resonance is rather complicated, being a function of δ, S_x, and K_2. The *level sets* [14] of the pseudo-Hamiltonian define the contours of normalized phase space and the level set that passes through the hyperbolic fixed point defined by (48) defines the *homoclinic orbit* or the *separatrix* of the resonance. In the normal form system, the phase space contours are just circles, at least up to the separatrix, if higher order non-linearities are neglected. The original canonical coordinates are defined by the

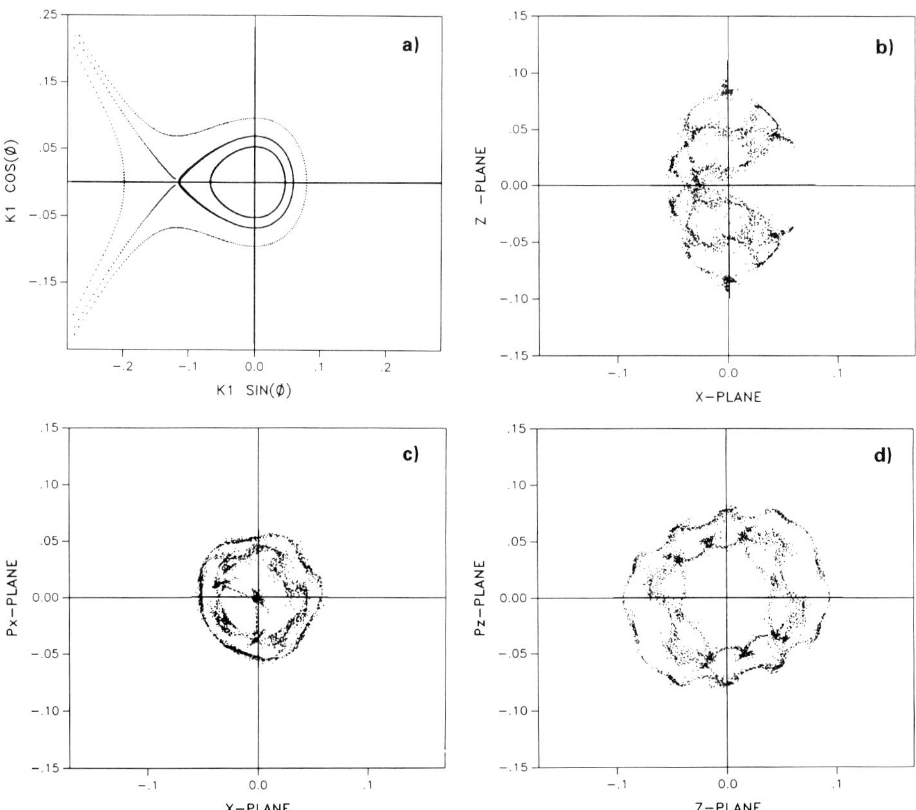

Figure 5. Phase-space plots for the $\nu_x = 2\nu_z$ resonance where $\hat{\nu}_x = 1.195$, $\hat{\nu}_z = 0.6025$ and the normalized sextupole strength is 1.700 units; $J_x = 0.0004$, $J_z = 0.0025$. a) shows the position of the fixed point at $\phi_1 = \pi$, the separatrix, and trajectories inside and outside the separatrix in order to illustrated the topology. b) x vs z projection in the original space. c) x vs p_x projection. d) z vs p_z projection. The parameters place the orbits very close to the separatrix.

relation $\zeta = \mathcal{A}^{-1}\kappa$, where $\kappa = (\phi_1, K_1, \phi_2, K_2)$ and ζ are the original coordinates. The invariants of the original map are given by $\mathcal{K}_i = \mathcal{A}^{-1}K_i$, and of course turn out to be J_x and J_z, the action invariants of the original map.

Phase-space plots for the $\nu_x = 2\nu_z$ resonance where $\hat{\nu}_x = 1.195$, $\hat{\nu}_z = 0.6025$ are shown in Fig. 5; $\hat{\nu}$ is the tune associated with the chromaticity-shifted phase advance $\hat{\mu}$, (34). Fig. 5a shows the position of the fixed point at $\phi_1 = \pi$ and the separatrix, generated by (44), along with trajectories inside and outside the separatrix in order to illustrate the topology. Here we are working in the Cartesian equivalent of K_1, ϕ_1 space. Figs. 5b–d are in the "original" coordinates defined by $\zeta = \mathcal{A}^{-1}\kappa$. However, as is conventional, the transformation to a system such that $k_x = k_z = 1$ (*i.e.* "normalized phase space" — see (19)) is used. These plots are generated by iterating the map, $\mathcal{M}$, 3000 times, which is approximately equivalent to (26)

$$\mathcal{M}^n = \mathcal{A}^{-1}\mathcal{N}^n\mathcal{A} = \mathcal{A}^{-1}e^{:nh:}\mathcal{A} \tag{49}$$

except that the iterated map generates higher order non-linearities by "feed-up" which (49) does not do.

The above results, although generated by a "continuous" Hamiltonian, could have been obtained from a *"kick-map"* with a re-definition of the sextupole strength, namely $S_x \Rightarrow S_x/\pi k_z\sqrt{k_x}$; that is, (35) can be obtained directly by substitution of (32) into the sextupole term in (27). Thus a kick-map exhibits the correct topology of the resonance and is much easier to deal with if one wishes to work by hand — the map is automatically in the factored form (33). The problem with a kick-map is that it does not automatically reflect the effects of dipole or quadrupole terms, for example the chromaticities, on the higher-order parts of the map.

7. Conclusions

In conclusion, we believe that the symbiosis of differential algebra and the Lie-algebraic formulation of optics provides a box of very powerful tools for analyzing and understanding the orbit dynamics of complex accelerators up to very high orders. Although cyclotrons are only quasi-periodic devices, *maps* still provide significant advantages over tracking in terms of speed, accuracy, and the ability to analyse many orbit-dynamical properties and in particular resonances. The analysis of resonances is accomplished very elegantly by the normal form transformations. Finally the use of differential algebra is essential for the efficient computation of the maps and the normal form transformation.

Acknowledgments

Two of the authors (W.G.D. and G.D.P.) would like to thank the Workshop Organizers for financial support.

References

[1] Schmeing H *et al.* 1989 *"Current Status of the Superconducting Cyclotron at Chalk River,"* *12th Int. Conf. on Cyclotrons and their Appl., Berlin* (Singapore: World Scientific)

[2] Davies W G, Douglas S R, Pusch G D, and Lee-Whiting G E 1991 *IEEE Part. Acc. Conf., San Francisco* **91CH3038-7** 303–5 (New York: IEEE Publishing)

[3] Berz M 1989 *"Differential Algebraic Description of Beam Dynamics to Very High Order,"* *Part. Accel.* **24** 109

[4] Forest É, Berz M and Irwin J 1989 *"Normal Form Methods for Complicated Periodic Systems,"* *Part. Acc.* **24** 91

[5] Davies W G 1991 *"Useful Forms of the Hamiltonian for Ion-Optical Systems,"* *Atomic Energy of Canada, Ltd. Report no. AECL-10364*

[6] Bulirsch R and Stoer R 1966 *"Numerical Treatment of Ordinary Differential Equations by Extrapolation Methods,"* *Num. Math.* **8** 1–13

[7] Rao S M, Wilton D R and Glisson A W 1982 *IEEE Trans. on Ant. & Prop.* **30** 409–17

[8] Dragt A J, Neri F, Rangarajan G, Douglas D, Healy L M and Ryne R D 1988 *Ann. Rev. Nucl. Part. Sci* **38** 455

[9] Forest É Private communication

[10] Douglas S R 1990 *"Automatic Differentiation of Functions,"* *Atomic Energy of Canada, Ltd. Report no. AECL-10139*

[11] Brown K L 1979 *IEEE Trans. Nucl. Sci.* **NS-26** 3490

[12] Forest É, Douglas D and Leemann B 1987 *Nucl. Instr. & Meth.* **A258** 355

[13] Forest É and Irwin J 1990 *Proc. Workshop on Non-Linear Problems in Future Part. Acc., Capri, Italy*

[14] Wiggins S 1991 *Introduction to Applied Nonlinear Dynamics and Chaos* (New York: Springer-Verlag), and references therein.

Inst. Phys. Conf. Ser. No 131
Paper presented at Int. Workshop Nonlinear Problems in Accelerator Phys. Berlin, 1992

Optics programs at TRIUMF

R. V. SERVRANCKX
TRIUMF

May 27, 1992

Abstract

This paper presents the organization of the programs that are presently used to study the optics of ring lattices at Triumf.

In particular we show how we have integrated the new technologies of Differential Algebra map analysis and Lie algebra analysis with the more traditional method of particle tracking.

A short comparison of the results obtained by the different methods is presented.

The last paragraph gives results about the first practical application of these new techniques to our racetrack booster design.

1 Introduction

Following the report of the International Committee that reviewed the substance of the Triumf Kaon Project Definition Study, the lattice study group for Kaon at Triumf decided to look into the available means to pursue the lattices analysis with more efficiency (computer speed) and more power (access to higher analysis and in particular access to resonance studies).

It was decided to use DA-maps[1] analyses (directly[2] or via Lie Algebraic techniques [3, 4]). To do so we needed to be able to create the DA-maps. This feature was introduced in the program DIMFAST, an offspring of DIMAD [6]. We also have the program COSY[7] to generate these maps. COSY however does not accept the MAD input format[8]. A routine was written within DIMAD that converts MAD input to COSY input.

2 program relationship flowchart

Figure 1 contains the flowchart describing the relationship between the various programs now in use at Triumf.

The basic philosophy was to exchange the information between the programs via the DA maps. We adopted the DA map format of the

DA package developped by M.Berz since it coincides with the format he adopted for the program COSY.

It is frustating that the many DA-map analysis programs that have been written in the last few years do not use the same format for the DA-maps. We urge the accelerator design community to choose and adopt shortly a common format for the taylor series form of transfer maps.

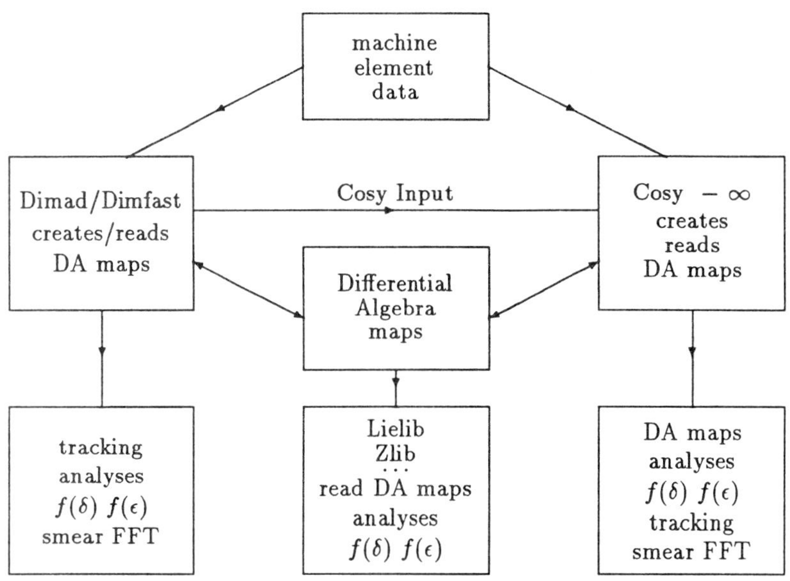

Figure 1. Flowchart of Optics programs at TRIUMF

3   Program results comparisons

In this paragraph we give, as an illustration, some comparisons between results obtained with the DA-map analysis and with the standard method of tracking.

Computations were made for two lattices. The first is a set of FODO cells (no dipoles) containing rather strong sextupoles to produce some observable geometric non linearities. The second lattice is that of the EROS ring (electron ring of Saskatoon) which is a pulse stretcher operating near a one-third resonance.

In the following tables, the momentum dependent tune coefficients are defined by the following relation ;

$$\nu = \nu_0 + \nu_1\left(\frac{\delta}{.01}\right) + \nu_2\left(\frac{\delta}{.02}\right)^2 + \cdots$$

The momentum dependent beta coefficients are given by a similar expression :

$$\beta = \beta_0 + \beta_1\left(\frac{\delta}{.01}\right) + \beta_2\left(\frac{\delta}{.02}\right)^2 + \cdots$$

The emittance dependent tune coeffcients are given by :

$$\nu = \nu_0 + \nu_x\epsilon_x + \nu_y\epsilon_y + \nu_{x2}\epsilon_x^2 + \nu_{xy}\epsilon_x\epsilon_y + \nu_{y2}\epsilon_y^2$$

The absolute difference error between DIMAD and COSY for the emittance dependent tune coefficient is computed for emittances of 10^{-4} m-rad.

Table 1. Comparison results for FODO cells

Coef	DIMAD	COSY	abs.error
	momentum dependent horizontal tune		
0	.322	.322	$< 10^{-6}$
1	$-.259\ 10^{-1}$	$-.259\ 10^{-1}$	$< 10^{-6}$
2	$0.320\ 10^{-3}$	$0.353\ 10^{-3}$	$3\ 10^{-5}$
3	$-.521\ 10^{-5}$	$-.543\ 10^{-5}$	$2\ 10^{-7}$
4	$0.112\ 10^{-6}$	$0.939\ 10^{-7}$	$2\ 10^{-8}$
	momentum dependent horizontal beta		
0	5.187	5.187	$< 10^{-5}$
1	$-.600\ 10^{-1}$	$-.600\ 10^{-1}$	$< 10^{-5}$
2	$0.129\ 10^{-2}$	$0.132\ 10^{-2}$	$3\ 10^{-5}$
3	$-.400\ 10^{-4}$	$0.316\ 10^{-4}$	$1\ 10^{-5}$
4	$0.394\ 10^{-5}$	$0.787\ 10^{-6}$	$3\ 10^{-5}$
	emittance dependent horizontal tune		
0	0.322	0.322	$< 2\ 10^{-4}$
x	$-.150\ 10^{3}$	$-.152\ 10^{3}$	$2\ 10^{-4}$
y	$0.532\ 10^{2}$	$0.559\ 10^{2}$	$2\ 10^{-4}$
x^2	$-.282\ 10^{3}$	$-.108\ 10^{5}$	$1\ 10^{-4}$
xy	$0.352\ 10^{6}$	$0.171\ 10^{6}$	$2\ 10^{-3}$
y^2	$-.669\ 10^{5}$	$-.913\ 10^{5}$	$3\ 10^{-4}$
	execution times		
	173 sec	254 sec	

Table 2. Comparison results for the EROS ring

Coef	DIMAD	COSY	abs.error
momentum dependent horizontal tune			
0	.290	.290	$< 10^{-5}$
1	$-.171$	$-.171\ 10^{-1}$	$< 10^{-5}$
2	$-.025$	$-.023\ 10^{-3}$	$2\ 10^{-3}$
3	$-.020$	$-.018\ 10^{-5}$	$2\ 10^{-3}$
4	$-.020$	$-.-16\ 10^{-7}$	$3\ 10^{-3}$
momentum dependent horizontal beta			
0	2.193	2.193	$< 10^{-5}$
1	0.330	0.310	$2\ 10^{-2}$
2	0.123	0.117	$6\ 10^{-3}$
3	0.312	0.291	$2\ 10^{-2}$
4	0.317	0.259	$6\ 10^{-2}$
emittance dependent horizontal tune			
0	0.290	0.290	$< 10^{-5}$
x	$0.158\ 10^{2}$	$0.122\ 10^{2}$	$3\ 10^{-4}$
y	$-.197\ 10^{2}$	$-.193\ 10^{2}$	$5\ 10^{-5}$
x^2	$0.170\ 10^{5}$	$0.100\ 10^{5}$	$7\ 10^{-5}$
xy	$0.600\ 10^{5}$	$-.100\ 10^{5}$	$7\ 10^{-4}$
y^2	$0.700\ 10^{5}$	$0.680\ 10^{5}$	$2\ 10^{-5}$
execution times			
236 sec	418 sec		

The cpu-times needed to obtain the same information is higher for DA-map analysis. The difference is not significant enough to base a choice of a preferred method on the execution time. The advantage for the DA approach lies in its power of analysis through the use of parameters and its ease of fitting by the use of concatenated maps for subset of lattices. This latter technique is the one used for the work presented in the next paragraph.

The numerical results obtained with the Lie Algebra analysis are in good agreement with the COSY results and the execution times are similar since the bulk of the cpu time is required for the computation of the DA-map.

4 Resonance correction of the LEB Booster and the Kaon racetrack booster

An important feature of lattice design is the implementation of a reliable control mechanism for the higher order aberrations. These aberrations sometimes need to be reduced but sometimes enhanced to reduce the risk of instabilities. One way to determine the level of higher order aberrations is to obtain the Fourier transform of the motion. This approach has the advantage that the Fourier analysis is easily obtained on the actual machines. In simulations the Fourier transform is available from the tracking of particles. So any correction scheme based on information obtained from Fourier analysis can be implemented on a real machine.

The normal form analysis available now, both in the DA analysis package and the Lie Algebraic packages provides, indirectly, informatiom on the resonance structure of a lattice.

The program COSY provides that resonance analysis and also allows fitting to be done on the results of that analysis. The use of that capability enabled us to install resonance correction schemes for the SSC LEB booster[5] and for the Triumf KAON racetrack booster.

Summarized in a few lines, here are the results for the KAON booster:

The non linear aberrations of the KAON racetrack booster are dominated by the $\nu_x - 2\nu_y$ resonance. This is clearly seen on the FFT analysis presented in figure 2 .

This resonance can be attenuated by an appropriate choice of sextupoles. After some search using the parameter analysis feature of COSY, a suitable correction scheme was found using two families of sextupoles, located in regions with low dispersion.

With the fitting capability of COSY and its resonance analysis, it was possible to obtain a solution reducing appreciably the $\nu_x - 2\nu_y$ resonance while keeping other resonances under control.

The result of this correction is shown on figure 3.

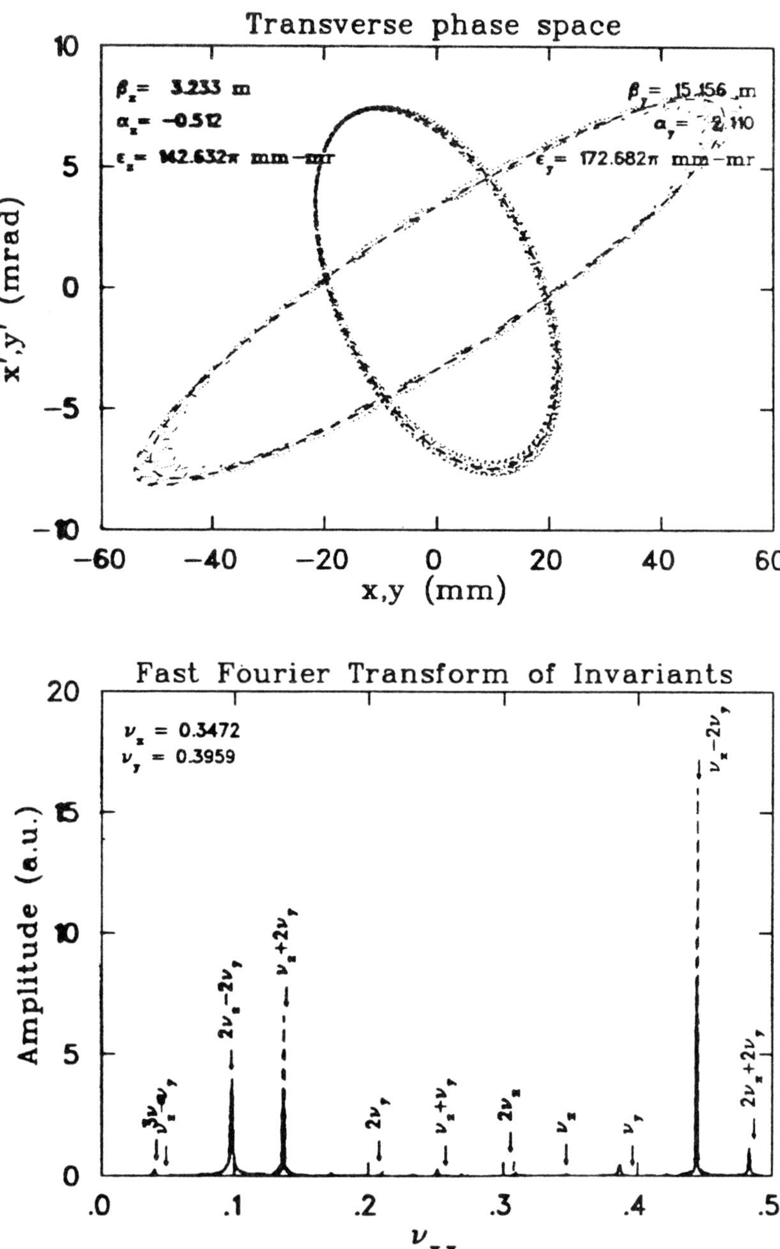

Figure 2. KAON racetrack booster - uncorrected

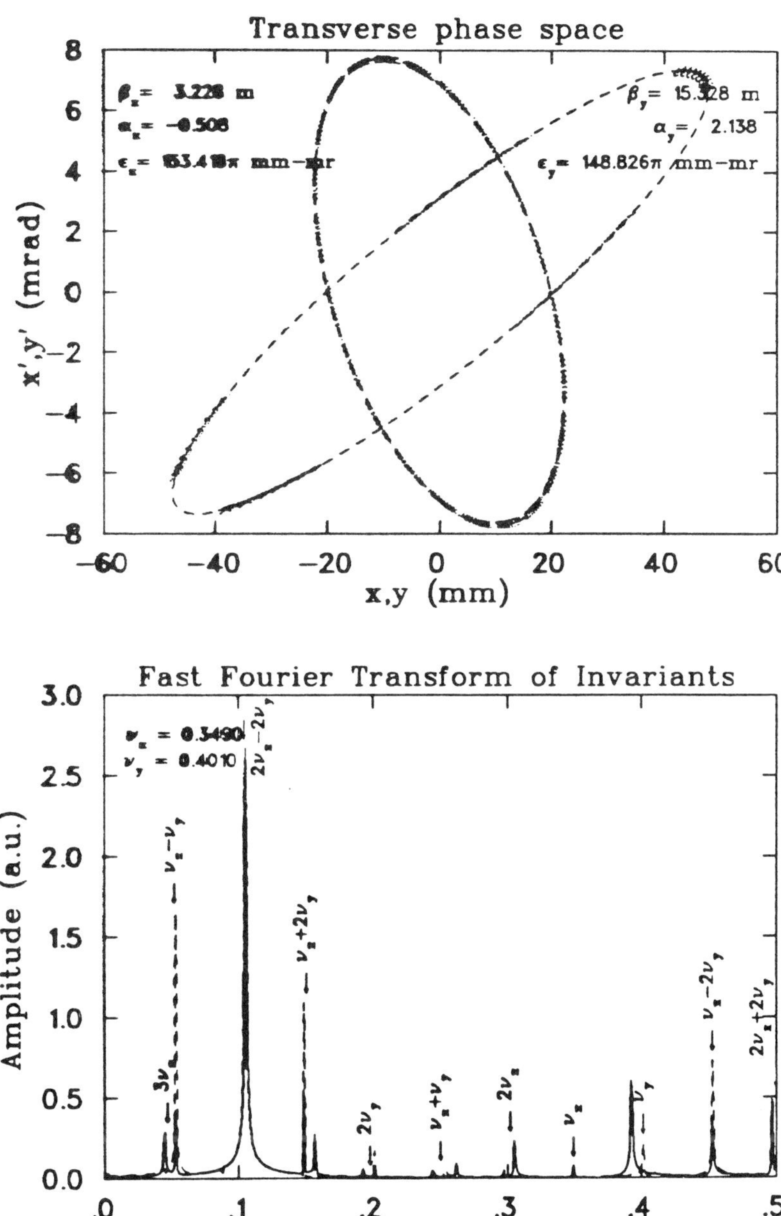

Figure 3. KAON racetrack booster - corrected

References

[1] M.BERZ Differential Algebraic description of beam dynamics to very high orders. Particle Accelerators 24(1989)109

[2] M.BERZ High Order Computation and Normal Form Analysis of Repetitive Systems, in : M.Month and M.Dienes (eds),Physics of Particle Accelerators, A.I.P. Conference Proceedings 249, p456 1992

[3] E.FOREST,J.BENGSSTON Description for LIELIB.FOR, Berkeley technical report communicated by E.FOREST

[4] Y.YAN ZLIB A Numerical Library for Differential Algebra, SSCL-300 July 1990

[5] X.WU Private communication

[6] R.SERVRANCKX,K.BROWN,L.SCHACHINGER,D.DOUGLAS Users Guide to the Program DIMAD, SLAC report 285 UC-28 May 1985

[7] M.BERZ COSY INFINITY version 5 , User's Guide and Reference Manual. Technical report MSUCL-811, December 1991

[8] D.C.CAREY F.C.ISELIN Standard input Language for Particle Beam and Accelerator Computer Programs, 1984 Summer Study on the Design and Utilization of the Superconducting Super Collider, Snowmass, Colorado, 1984

A Comparison of Methods for Long-term Tracking Using Symplectic Maps

I Gjaja[1], A J Dragt, D T Abell

Dept. of Physics, University of Maryland, College Park, MD 20742 USA

Abstract. Errors are studied for three maps that approximate the time-step-one map of the anharmonic oscillator.

1. Introduction

In accelerator physics it is common to consider the actual trajectories of particles to be near, in some not yet specified sense, to a prescribed ("design") trajectory. One then parametrizes the difference between the actual and the design trajectory by introducing deviation variables [denoted by a vector $z = (q_1, \ldots, q_n, p_1, \ldots, p_n)$] and appeals to the smallness of these to perform an expansion that facilitates computations.

In the context of one-turn maps, this approach leads to an approximate map that agrees with the exact, generally unknown, map through a given order, denoted by N, in z. This means that in the limit $\|z\| \to 0$ ($\| \cdot \|$ denotes the usual Euclidean norm), after one application of each map, the difference between the exact and approximate map results scales as $\|z\|^{N+1}$.

In practical problems, however, one is concerned with finite, and occasionally not small, values of z. In addition, when performing long-term tracking, it is necessary to iterate the map a large number of times. Under these conditions, then, when estimating the error of the approximate map, it is necessary to pay attention not only to N, but also to the total magnitude of terms of order $N + 1$, and to the symplecticity of the approximate map (assuming the actual motion is Hamiltonian). In this note we compare the behavior of four 2-dimensional maps, all of which agree with each other through order $N = 7$.

2. Exact Map

We consider the one-dimensional anharmonic oscillator described by the Hamiltonian

$$H = (p^2 + q^2)/2 - q^4/4. \tag{1}$$

Let $\mathcal{M}$ be the "time 1" map generated by H,

$$\mathcal{M} = \exp : -H : . \tag{2}$$

(In this note we use the word map to refer both to an explicit expression linking final and initial conditions, and to the operator that produces this transformation.) Figure 1 displays the phase portrait for H. It can be shown that the Taylor series for $\mathcal{M}$ has a

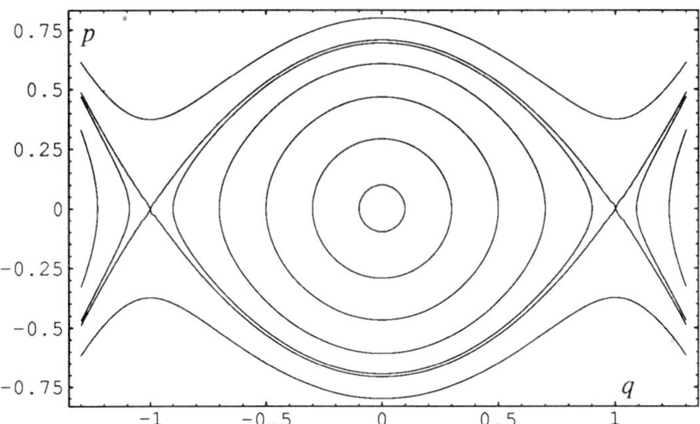

Figure 1. Phase Space Portrait of the Anharmonic Oscillator.

finite (but not infinite) domain of convergence in the q, p variables, and that the region displayed in Figure 1 is within this domain. We write $\mathcal{M}$ in the product form

$$\mathcal{M} = \mathcal{R}(\phi)\mathcal{N}. \tag{3}$$

Here $\mathcal{R}(\phi)$ denotes a linear map corresponding to a rotation by angle ϕ (in this case a simple phase advance of 1 radian) in the q, p plane, and $\mathcal{N}$ is the nonlinear portion of the map. The nonlinearity of $\mathcal{N}$ at the point z is conveniently measured by a quantity $n(z)$ defined by

$$n(z) = \|\mathcal{N}z - z\|/\|z\|. \tag{4}$$

It is easily verified that $\mathcal{N}$ differs from the identity by terms of third order in z, and hence $n(z)$ should scale as $\|z\|^2$. Figure 2 shows the value of n on various circles in phase space parametrized by polar coordinates having radius $\|z\|$ and angle θ (the circles are not curves of constant energy). The angle θ is measured counterclockwise from the q axis. For the three curves the values of $\|z\|$ are 0.2, 0.6, and 1.0, respectively. Note that the nonlinearity becomes quite large for large values of $\|z\|$. Note also that the nonlinear force arising from the $q^4/4$ term in (1) exceeds the linear force arising from the $q^2/2$ term when $|q| > 1$. In particular, as shown in Figure 1, there is a separatrix passing through the points $z = (\pm 1, 0)$.

For future comparisons, we apply $\mathcal{N}$ to the grid of initial conditions given in Figure 3. The result is displayed in Figure 4. Evidently, the lower right and upper left corners of the box are distorted the most, which is to be expected since $n(z)$ is largest there.

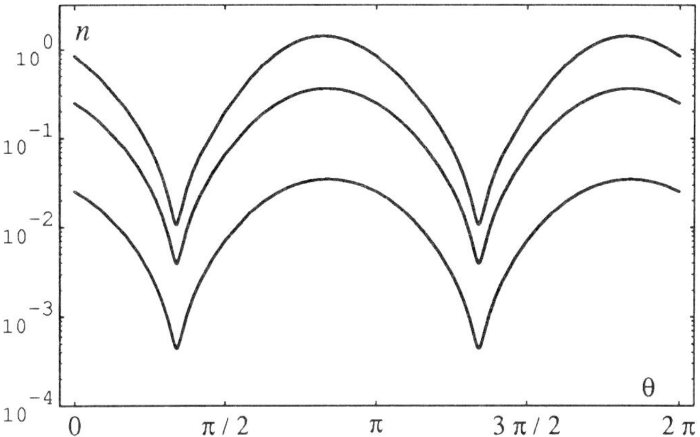

Figure 2. Nonlinearity of $\mathcal{N}$.

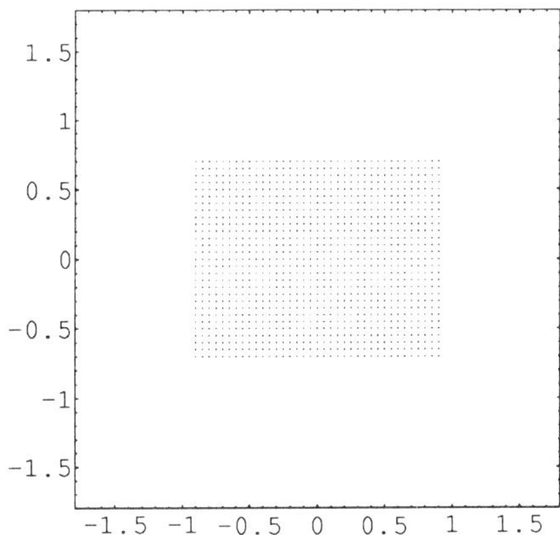

Figure 3. Set of Initial Conditions.

3. Taylor Series Map

The nonlinear map $\mathcal{N}$ is well approximated by a seventh order Taylor series map near and at moderate distances from the phase space origin. We call this map $\mathcal{T}$. Figure 5 displays the application of $\mathcal{T}$ to the initial conditions of Figure 3. The agreement between Figures 5 and 4 is very good, even for values of z for which $n(z)$ is of order one. To quantify the difference between $\mathcal{N}$ and $\mathcal{T}$ at the point z, we define a quantity ϵ^{T},

$$\epsilon^{\mathrm{T}}(z) = \|\mathcal{N}z - \mathcal{T}z\|/\|z\|. \tag{5}$$

This quantity is plotted in Figure 6, again using polar coordinates. The values of $\|z\|$ for the three curves are the same as in Figure 2. These are also the radii for the plots of various ϵ's in the subsequent sections. The map $\mathcal{T}$, because it arises from truncating a Taylor series, is not symplectic. Its departure from symplecticity at the point z is

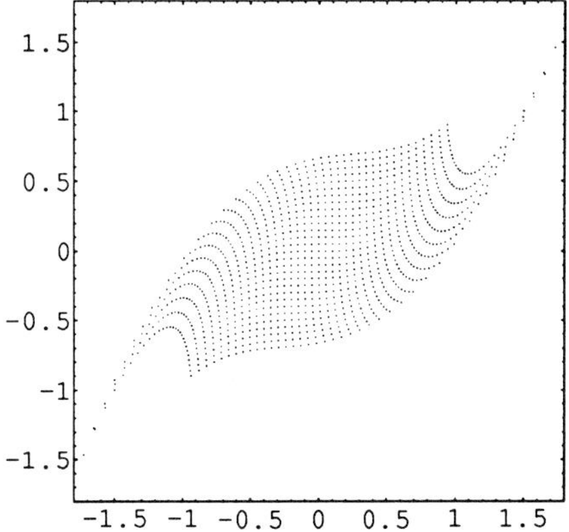

Figure 4. Transformation of Figure 3 by $\mathcal{N}$.

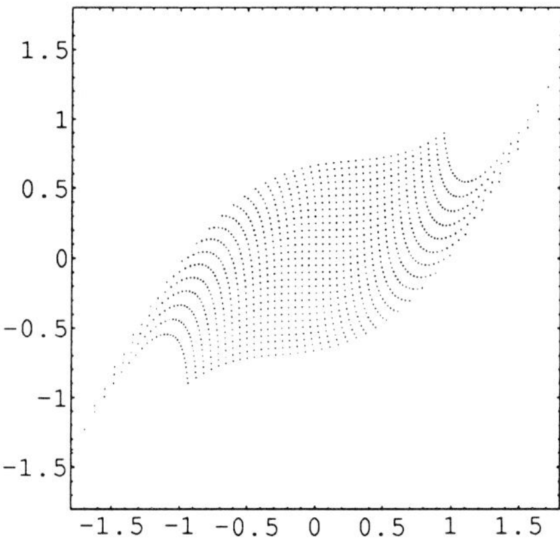

Figure 5. Transformation of Figure 3 by $\mathcal{T}$.

conveniently measured by a quantity δ^{T} defined by

$$\delta^{\mathrm{T}}(z) = |[\mathcal{T}q, \mathcal{T}p] - 1|. \tag{6}$$

Here $[,]$ denotes the usual Poisson bracket. Figure 7 shows a plot of δ^{T}. Note that ϵ^{T} and δ^{T} are comparable. Also note that both ϵ^{T} and δ^{T} scale as $\|z\|^8$, as they should. ($O(\|z^8\|)$ is absent in the exact map.)

The lack of simplecticity of $\mathcal{T}$ makes the map $\mathcal{RT}$ unfit for long-term tracking. As a way to illustrate this claim, we define the energy deviation Δ^{T},

$$\Delta^{\mathrm{T}}(n, z) = H((\mathcal{RT})^n z) - H(z). \tag{7}$$

Here n is the number of iterations of the map. In Figure 8 we plot Δ^{T} vs. n for four initial conditions, $q = 0.1, 0.2, 0.6,$ and 0.7, with $p = 0.0$ for all of them. We see that the difference in energy between the exact trajectory and that produced by repeated

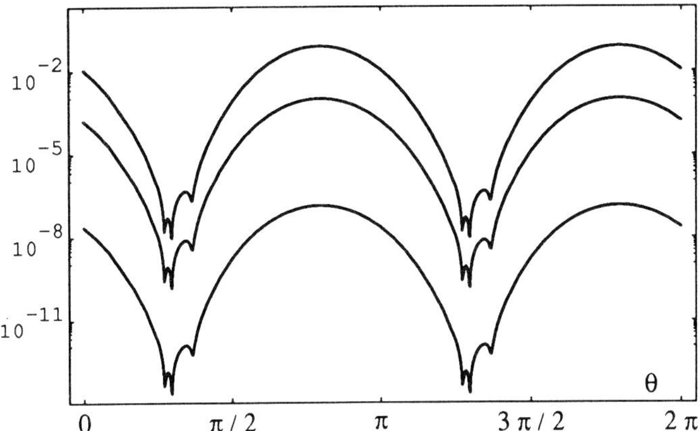

Figure 6. Accuracy of $\mathcal{T}$.

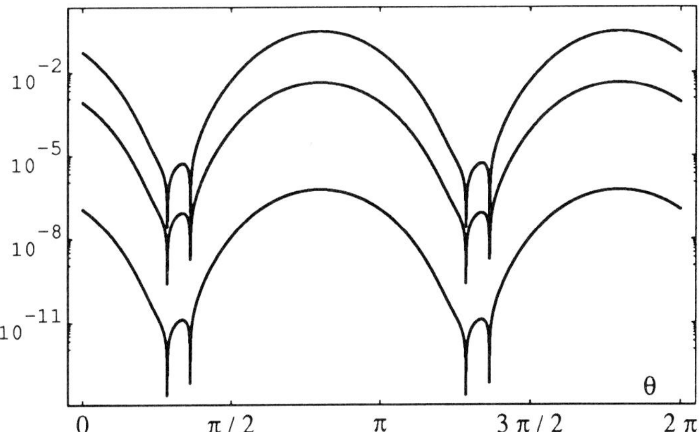

Figure 7. Nonsymplecticity of $\mathcal{T}$.

applications of $\mathcal{R}\mathcal{T}$ grows with n. (After about 240 iterations Δ^{T} for the topmost curve becomes negative and very large in magnitude.) We call attention to the interesting fact that for large n, Δ^{T} vs. n is a straight line on a semi-log scale. We are currently exploring this phenomenon further.

4. Jolt Factorization

One possible procedure for symplectifying $\mathcal{T}$ is to seek a map $\mathcal{J}$ which is a product of jolt maps [1–4]. A jolt function is one for which

$$: j :^2 z = 0, \tag{8}$$

and a jolt map is a map of the form $\exp(: j :)$. Evidently from (8), if j is a polynomial, then its associated map is a polynomial map that can easily be evaluated exactly (and

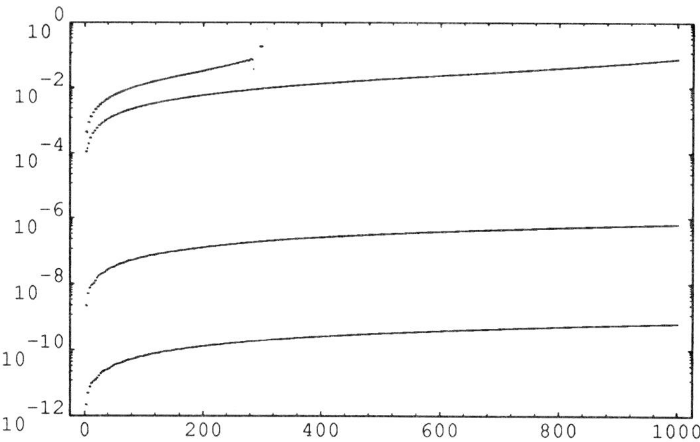

Figure 8. Change in Energy Function under $\mathcal{RT}$.

is symplectic by construction). It can be shown that any jolt function in two variables can be written in the form

$$j = \mathcal{L}k, \tag{9}$$

where $\mathcal{L}$ is a linear map and k is also a function satisfying (8). Moreover k is a function only of q. Such functions are often referred to as kicks. Thus kicks are special cases of jolts. For the problem at hand, we seek to find a sequence of 9 jolt polynomials j and their associated maps such that we have the relation

$$\mathcal{N} \simeq \mathcal{T} \simeq \mathcal{J} = \prod_{m=1}^{9} \exp : j^m : . \tag{10}$$

Each jolt j is taken to be of the form

$$j = \mathcal{R}(\phi_m)[a_4^m q^4 + a_6^m q^6 + a_8^m q^8]. \tag{11}$$

Here the ϕ_m are 9 equally spaced angles over the interval $(0, 2\pi)$, and the coefficients a_n^m are chosen to make $\mathcal{J}$ agree with $\mathcal{N}$ through terms of order 7. Figure 9 shows the application of $\mathcal{J}$ to the initial conditions of Figure 3. The agreement between Figures 9 and 4 is poorer than between 5 and 4, especially in the lower right and upper left sections of the graph, and in the tails. The difference between $\mathcal{N}$ and $\mathcal{J}$ at the point z is conveniently measured by a quantity ϵ^{J} defined by

$$\epsilon^{\mathrm{J}}(z) = \|\mathcal{N}z - \mathcal{J}z\|/\|z\|. \tag{12}$$

This quantity is plotted in Figure 10. Inspection of Figure 10 shows that ϵ^{J} scales as $\|z\|^8$, that is $\mathcal{J}$ agrees with $\mathcal{N}$ through order 7. However, the overall magnitude of ϵ^{J} is uncomfortably larger than that of ϵ^{T} and δ^{T}. We have some reason to believe, and are presently exploring whether, other jolt factorization procedures might produce a smaller ϵ^{J}.

To explore the long-term behavior of the jolt map, we repeatedly apply $\mathcal{RJ}$ to the following initial conditions: $q = 0.1, 0.3, 0.5, 0.7, 0.9$, and 1.0 with $p = 0.0$ for all of

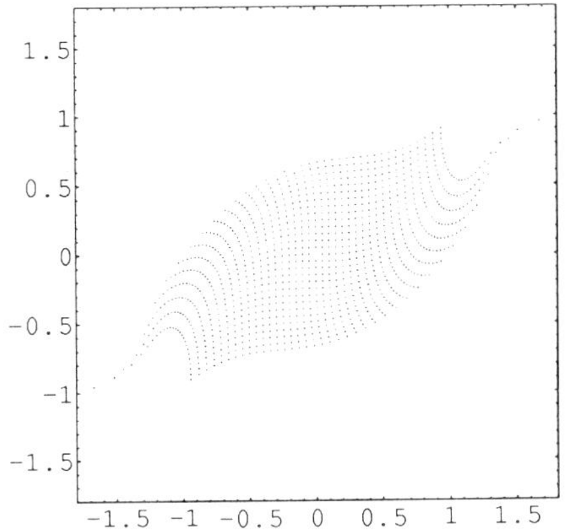

Figure 9. Transformation of Figure 3 by $\mathcal{J}$.

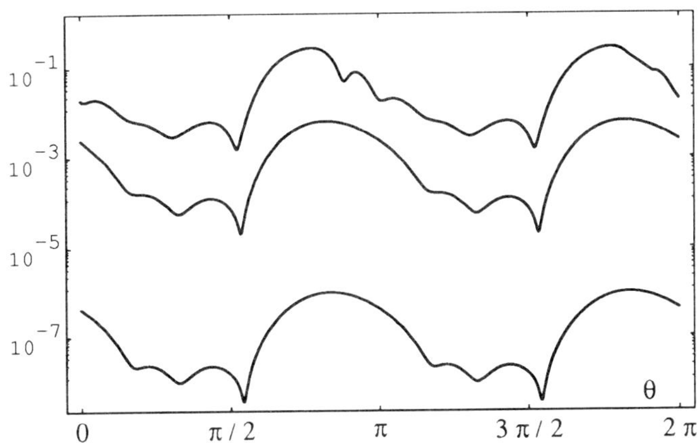

Figure 10. Accuracy of $\mathcal{J}$.

them, and $q = \pm 1.25433$, $p = \mp 0.2$. The result is displayed in Figure 11. Comparison of Figures 11 and 1 shows that for small and moderate distances from the origin, the long-term behaviors of $\mathcal{M}$ and $\mathcal{RJ}$ are very similar. Inspection of Figure 11 shows that this agreement is good even in phase-space regions where the nonlinearity is substantial. Indeed, $\mathcal{RJ}$ even reproduces some of the separatrix structure of $\mathcal{M}$. However, it should be noted that $\mathcal{RJ}$ produces small islands near the separatrix, while $\mathcal{M}$ does not, and that the unstable fixed points are moved slightly outward. These differences are consistent with the size of ϵ^{J} far from the origin.

As before, we define an energy deviation Δ^{K},

$$\Delta^{K}(n, z) = H((\mathcal{RJ})^{n} z) - H(z). \tag{13}$$

The plot of Δ^{K} vs. n is given in Figure 12 for the same initial conditions as in the previous section. Note that Δ^{K} is a complicated quasiperiodic function of n, whose

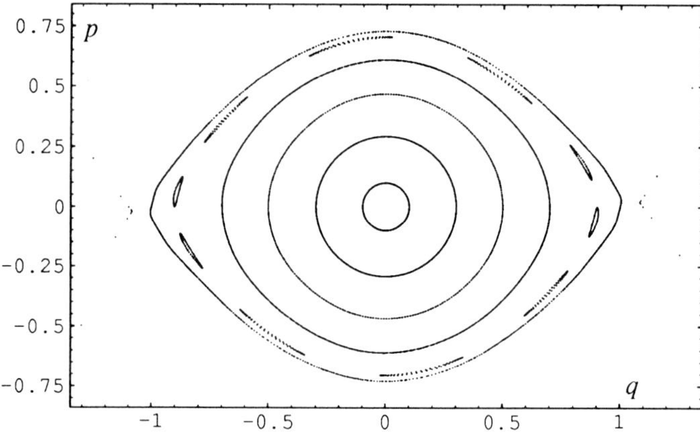

Figure 11. Long Term Jolt Tracking.

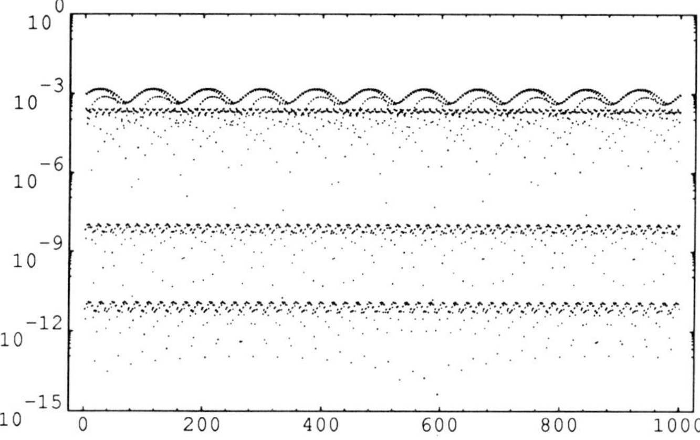

Figure 12. Change in Energy Function under $\mathcal{RJ}$.

average over many iterations is constant. This should be contrasted with the behavior displayed in Figure 8.

5. Mixed-Variable Generating Function

Finally, we examine the symplectification of $\mathcal{T}$ by means of a mixed-variable generating function of the form $q^i p^f + G(q^i, p^f)$ [5]. The superscripts i and f stand for initial and final conditions, respectively. The term $q^i p^f$ takes into account the linear transformation, which is the identity, while G, which contains only third and higher order terms in q^i and p^f, generates the nonlinear part of the transformation.

Formally, the power series for $q^f(q^i, p^i)$ and $p^f(q^i, p^i)$ can be partially inverted to obtain $q^f(q^i, p^f)$ and $p^i(q^i, p^f)$. The expressions

$$p^f + \frac{\partial G}{\partial q^i} = p^i(q^i, p^f) \tag{14}$$

and

$$q^i + \frac{\partial G}{\partial p^f} = q^f(q^i, p^f) \tag{15}$$

then yield the coefficients in the Taylor series expansion of G. Iteration of the map is accomplished by repeatedly solving (14) and (15) for q^f and p^f. Since the equations are implicit, for order higher than 4, it is necessary to use a numerical procedure. The use of the mixed-variable generating function guarantees that the transformation is symplectic.

We denote by $\mathcal{G}$ the map that is generated by G and agrees with $\mathcal{T}$ through order 7. One application of $\mathcal{G}$ to the initial conditions of Figure 3 yields Figure 13. The

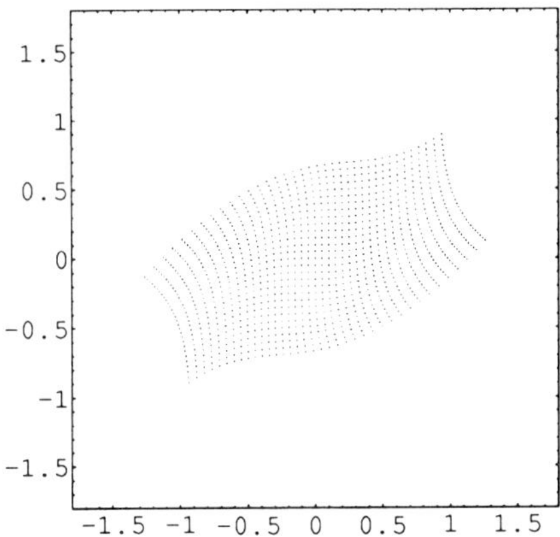

Figure 13. Transformation of Figure 3 by $\mathcal{G}$.

discrepancy between Figures 13 and 4 is significantly larger than either between 5 and 4 or between 9 and 4. To quantify the difference between $\mathcal{N}$ and $\mathcal{G}$ at the point z, we define a quantity ϵ^G,

$$\epsilon^G(z) = \|\mathcal{N}z - \mathcal{G}z\|/\|z\|. \tag{16}$$

This quantity is plotted in Figure 14. Compared to Figures 2, 6, 7, and 10, we have added the curves for $\|z\| = 0.1$ and 0.4, in order to enable the reader to verify that ϵ^G scales as $\|z\|^8$. For the map $\mathcal{G}$ the scaling relation breaks down rapidly with increasing $\|z\|$. Note that for the corresponding curves (second, fourth, and fifth from the bottom) the overall magnitude of ϵ^G is appreciably larger than that of ϵ^J, and much larger than that of ϵ^T and δ^T. This fact agrees with the observed differences between Figures 4, 5, 9, and 13.

To examine the long-term behavior of this method of symplectification, we repeatedly apply $\mathcal{RG}$ to the initial conditions $q = 0.1, 0.3, 0.5, 0.7, 0.9, 0.95,$ and 1.0, with $p = 0.0$ for all of them. The result is displayed in Figure 15. As in the case of jolt factorization, for small and moderate distances from the origin Figures 15 and 1 are very similar. It appears, however, that the region around the separatrix is not captured as well by $\mathcal{RG}$

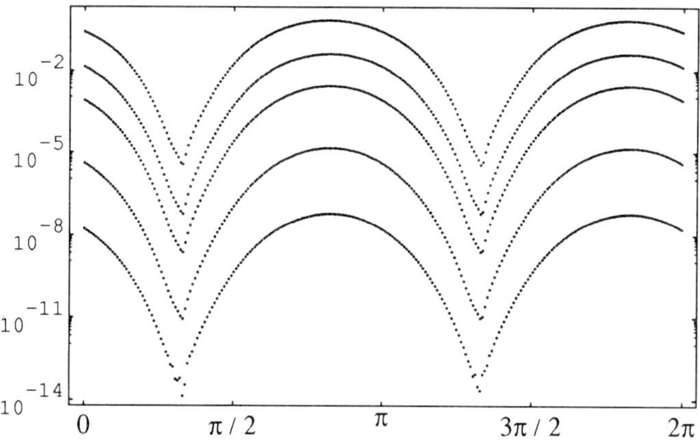

Figure 14. Accuracy of $\mathcal{G}$.

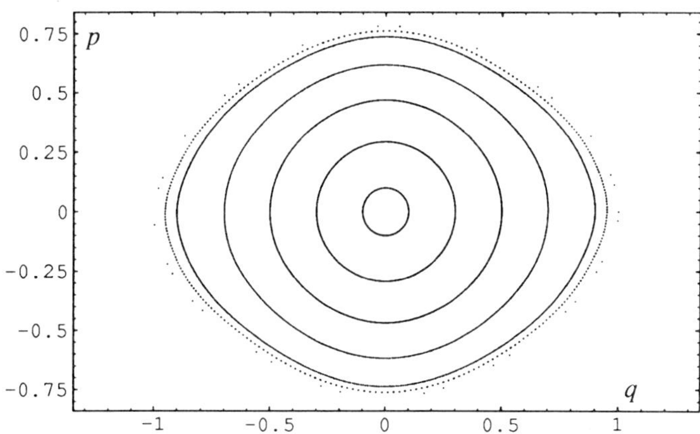

Figure 15. Long Term Tracking Using G.

as it is by $\mathcal{RJ}$. Note that for the last ring Newton's method, which is used to solve (14), fails to converge after 32 applications of the map. We have not explored whether this failure of convergence is due to the initial guess being outside the basin of attraction for the Newton fixed point, or is due to some other reason. The difference in long-term behavior of $\mathcal{RG}$ and $\mathcal{RJ}$ is consistent with the remarks made in the paragraph above. It is interesting to point out that if we do not separate the linear and nonlinear parts of $\mathcal{M}$, and use a generating function of the form $F(q^i, q^f)$ to symplectify the Taylor series for the full map (including the linear part), then the phase space portrait produced using F is rather similar to Figure 1. This approach is displayed in Figure 16 (initial conditions are the same as in Figure 15). Note that the separatrix is well reproduced. On the other hand, the figure exhibits small islands that are absent in Figure 1, and the unstable fixed points are moved slightly inward. We are currently conducting further study in order to understand the unexpected discrepancy between Figures 15 and 16.

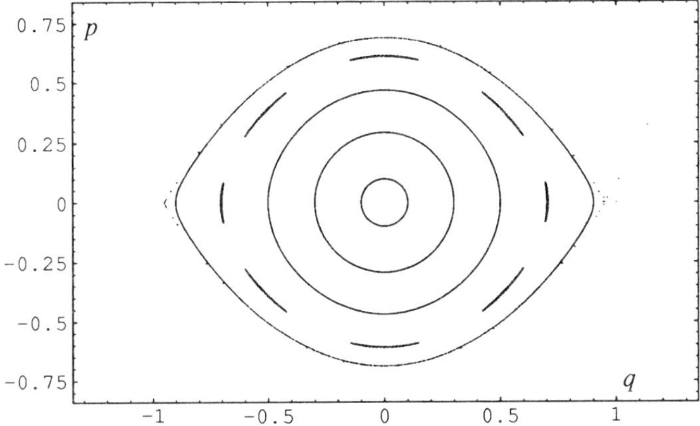

Figure 16. Long Term Tracking Using F.

In analogy with previous sections we can now define an energy deviation $\Delta^G(n, z)$ and examine its behavior as a function of n and z. Rather than displaying the figure, we merely state that the result is qualitatively similar to Figure 12.

6. Conclusions

We have studied the errors involved in the Taylor series expansion and particular cases of jolt factorization and mixed-variable generating function method for the anharmonic oscillator. While the Taylor series map is not fit for long-term tracking because it violates symplecticity, the remedy of this condition by the methods we used leads to a significant reduction in accuracy for one application of the map. Nevertheless, both jolt factorization and the mixed-variable generating function method were found to yield long-term behavior which is similar to that for the exact map, even in regions where the nonlinearity is appreciable. It is therefore worthwhile to conduct further study of these methods to see if their accuracy can be improved.

We comment on the choice of time step. We have verified that decreasing the time step improves the agreement between the exact map and any of the approximations, while increasing it does the opposite. This fact can be used in two ways. First, if the one turn map is accurate, then one can separate it into linear and nonlinear parts and symplectify the nonlinear part as follows: take the nth root of it, symplectify it, and apply the resulting map n times.[4] (The linear transformation is either already symplectic, or can be made so by a variety of procedures.) Second, rather than trying to symplectify the one turn map, one can symplectify maps for sections of the accelerator (lumps), and then apply these maps in succession. Therefore, when comparing two methods of symplectification it is worthwhile to compare not only their accuracies, but their speeds of execution as well.

Acknowledgments

We gratefully acknowledge the work of Georg Hoffstatter of Michigan State University who computed the numbers used in Figures 13–16. He performed the calculations using the code COSY INFINITY.

References

[1] Dragt A J *et al* 1988 *Ann. Rev. Nucl. Part. Sci.* **38** 455–496

[2] Irwin J 1989 *SSC Note* **228**

[3] Rangarajan G 1990 PhD Thesis, Univ. of Maryland

[4] Dragt A J, Gjaja I M and Rangarajan G 1991 *Proc. IEEE Part. Acc. Conf.* **3** 1621–1623

[5] Douglas D, Forest E and Servranckx R V 1985 *IEEE Trans. on Nucl. Sci.* **NS-32** 2279–2281

A generalization of the Hénon map: stability of the orbits, symmetries and connections to accelerator physics

Ezio Todesco

Department of Physics, University of Bologna, Via Irnerio 46, 40126 Bologna, and INFN, Section of Bologna, Italy

Abstract. A generalization of the Hènon map is discussed; such a model arises in a natural way from the extrapolation of the multipolar errors of the superconducting dipoles of the planned Large Hadron Collider in CERN. The symmetries of the map in the resonant case are studied; a qualitative analysis of the phase space structure is given. The dynamic aperture of this model is not well defined, since far from the origin one recovers a quasi linear regime; the computation of the dynamic aperture of the truncated map in its dependence on the truncation order is performed: this allows to give an estimate of the highest significant multipole which should be included in tracking simulations.

1. Introduction

In this paper we consider a map which is a generalization of the conservative Hénon mapping [1], which is obtained by replacing the quadratic term with the Taylor series of the cosine; such a map depends on two parameters: the linear frequency ω and a quantity α related to the strength and to the period of the nonlinearity. It is well known [2] that the Hénon map is the model of a cell of a hadron accelerator where the nonlinearity y is a single sextupole; indeed with the introduction of the superconducting magnets one has elements which are highly nonlinear, and contain nonlinearities of every order. Starting from the data relative to the errors of the planned dipoles for the Large Hadron Collider [3, 4], we show that t hey can be interpolated by a power law: therefore the proposed generalization of the Hénon map can approximate a LHC cell with a single nonlinearity which contains

multipoles of every order, whose values are extrapolated from the magnets data.

The discussed model has a close resemblance to the Zaslavskii map [5], which can be obtained by the Hénon map by replacing the quadratic term with the Taylor series of the sine. Generic maps with periodic nonlinearities have remarkable symmetries properties

[6]; we review the results relative to the case of resonant frequencies, showing how one has a tiling of the phase space in the case of resonances 3, 4 and 6.

Finally, we numerically analyse the stability of the orbits of the map truncated at order n for a nonresonant value of the linear frequency in its dependence on the parameter α and on the order n. We give a rule which allows to compute the order at which the truncated map has the minimum dynamic aperture; this justifies the truncation order which is commonly used in simulations. The complete map does not exhibits rapid escape to the infinity, since far enough from the origin the linear part dominates; nevertheless, there is a strong numerical evidence that in this region there are no invariant curves.

2. The model and its connections to accelerator physics

In this Section we introduce a symplectic map which is related to the conservative Hénon mapping [1]:

$$\begin{pmatrix} x' \\ p' \end{pmatrix} = R(\omega) \begin{pmatrix} x \\ p - x^2 \end{pmatrix} \qquad x, p \in \mathbf{R}. \tag{2.1}$$

A natural generalization of the Hénon map can be obtained by replacing the quadratic nonlinearity with a polynomial:

$$\begin{pmatrix} x' \\ p' \end{pmatrix} = R(\omega) \begin{pmatrix} x \\ p - \sum_{k \geq 2} a_k x^k \end{pmatrix} \qquad a_k \in \mathbf{R}; \tag{2.2}$$

(the condition $a_k \in \mathbf{R}$ is sufficient to ensure the symplecticity of the map). We assume that the coefficients have the following behaviour:

$$a_{2k} = \frac{2\alpha^{2k-2}}{(2k)!}(-1)^{k-1} \qquad a_{2k+1} = 0 \tag{2.3}$$

so that resumming the series one has:

$$\begin{pmatrix} x' \\ p' \end{pmatrix} = R(\omega) \begin{pmatrix} x \\ p + \frac{2}{\alpha^2}(\cos \alpha x - 1) \end{pmatrix}. \tag{2.4}$$

Switching to complex notation $z = x - ip$ the map reads:

$$z' = e^{i\omega} \left(z + \frac{2i}{\alpha^2} \left(1 - \cos \left(\frac{\alpha(z + z^*)}{2} \right) \right) \right). \tag{2.5}$$

We point out that according to the value of α the map exhibits different behaviours; actually, when α goes to zero we can expand the cosine and in the limit we recover the Hénon map. On the other hand, when α goes t o infinity the nonlinear part vanishes and we are left with the integrable map $z' = e^{i\omega} z$.

The map (2.4) is similar to the Zaslavskii map [5]; in fact if one scales the coordinates by α one has

$$\begin{pmatrix} x' \\ p' \end{pmatrix} = R(\omega) \begin{pmatrix} x \\ p + \frac{2}{\alpha}(\cos x - 1) \end{pmatrix}, \tag{2.6}$$

whilst Zaslavskii map is usually written in the form

$$\begin{pmatrix} x' \\ p' \end{pmatrix} = R(\omega) \begin{pmatrix} x \\ p + K \sin x \end{pmatrix}. \tag{2.7}$$

We remind the reader that the standard perturbative approach used for the Zaslavskii map is based upon an expansion of the perturbative series in powers of K, whilst the Hénon map and its generalization are usually analysed via the Birkhoff normal form [2, 3], whose small parameter is the distance t o the origin $|z| = \sqrt{x^2 + p^2}$.

Indeed, map (2.4) can be derived in a natural way from accelerator physics; we consider a magnetic lattice made up of regular and identical cells which have a linear part plus a multipole error in the one kick approximation; the one dimensional map of such a cell reads:

$$z' = e^{i\omega} \left(z + i \sum_{n=2}^{M} \sqrt{\beta}^{n+1} \frac{K_n}{n!} \frac{(z + z^*)^n}{2^n} \right), \tag{2.8}$$

where β is the value of the beta function in the nonlinearity, K_n are the integrated gradients of the multipole and $z = x - ip$ is the complex Courant–Snyder coordinate in the (x, p) plane. For a more detailed exposition of the expression (2.8) and its validity limits, we refer to [2, 3]. The coefficients K_n are computed for the magnetic elements on the base of simulations [4] and experiments, and are known up to a given order M.

Table 1. Integrated gradients of a LHC dipole and their interpolation.

Order n	K_n	$\hat{K}_n$	$K_n/\hat{K}_n$
2	$-2.90 \cdot 10^{-2}$	$-1.8 \cdot 10^{-2}$	1.6
4	$4.81 \cdot 10^{2}$	$6.0 \cdot 10^{2}$	0.8
6	$-7.73 \cdot 10^{6}$	$-2.0 \cdot 10^{7}$	0.4
8	$1.44 \cdot 10^{12}$	$6.8 \cdot 10^{11}$	2.1

We examined the values relative to the dipoles of the LHC (table 1), finding out that they roughly behave like a power law. We interpolated these coefficients according to

$$\hat{K}_{2n} = ab^{2n}(-1)^{n-1} \qquad \hat{K}_{2n+1} = 0 \tag{2.9}$$

(one has no odd multipoles, due to the symmetry of the element). A least square fit gives:

$$a = 5.4 \cdot 10^{-7} \qquad b = 1.8 \cdot 10^{2} \tag{2.10}$$

and as one can see from table 1, the ratio between the interpolated gradients $\hat{K}_{2n}$ and the real gradients K_{2n} keeps bounded between 0.5 and 2. Extrapolating eq. (2.9) one obtains:

$$z' = e^{i\omega} \left(z + i \sum_{n=1}^{+\infty} \frac{\sqrt{\beta}^{2n+1} ab^{2n}(-1)^{n-1}(z + z^*)^{2n}}{(2n)! 2^{2n}} \right), \tag{2.11}$$

i.e. we have added to the map (2.8) a tail of terms, assuming that the order that are usually neglected in all the computations behave in the same way as the orders up to M. Resumming the cosine series one has:

$$z' = e^{i\omega} \left(z + ia\sqrt{\beta} \left(\cos \frac{\sqrt{\beta} b(z + z^*)}{2} - 1 \right) \right). \tag{2.12}$$

We perform the scaling

$$w = \frac{a\beta^{3/2}b^2}{2}z \tag{2.13}$$

in order to recover the Hénon map in a neighbourhood of the origin:

$$w' = e^{i\omega}\left(w + i\frac{(ab\beta)^2}{2}\left(\cos\frac{(w + w^*)}{ab\beta} - 1\right)\right). \tag{2.14}$$

This expression is equivalent to the map (2.5) we introduced in the beginning of this Section, with

$$\alpha = \frac{2}{ab\beta}; \tag{2.15}$$

using the LHC data, we have an average $\beta = 100$ m, and having $*$ dipoles in each cell, we have

$$\alpha = 25. \tag{2.16}$$

3. Symmetries of the map in the resonant case

In this section we consider the family of maps:

$$z' = F(z, z^*) = e^{i\omega}\left(z + ig\left(\frac{z + z^*}{2}\right)\right), \tag{3.1}$$

where g is a real function of period 2π. Both the map analysed in this paper and the Zaslavskii map can be rescaled to this form. We set the linear frequency $\omega/2\pi$ on a resonant value:

$$\omega_q = 2\pi\frac{p}{q}. \tag{3.2}$$

Definition. A map is symmetric under a given group of transformations T if one has

$$F \circ T = T \circ F \tag{3.3}$$

In fact in this case T maps orbits of F in other orbits:

$$z^{on} = F^{on}(z_0) \Rightarrow Tz^{on} = F^{on}(Tz_0) \tag{3.4}$$

(where F^{on} denotes F iterated n times). One has the following

Proposition. If q=3,4 or 6, then F^{oq} commutes with a discrete gr oup of translation T_{nm} which tile the plane $\mathbf{R}^2$

Proof. We first analyse the case $q = 4$. Then one has

$$F^{o4} : \begin{cases} x^{o4} = x + g(x) + g(x + g(x)) \\ y^{o4} = y - g(-y) + g(y - g(-y)). \end{cases} \tag{3.5}$$

Clearly, F^{o4} commutes with all the translations

$$T_{mn}(x, y) = (x + 2m\pi, y + 2n\pi), \tag{3.6}$$

which tile the phase space in squares of side 2π; this is the smallest possible tiling which commutes with the map.

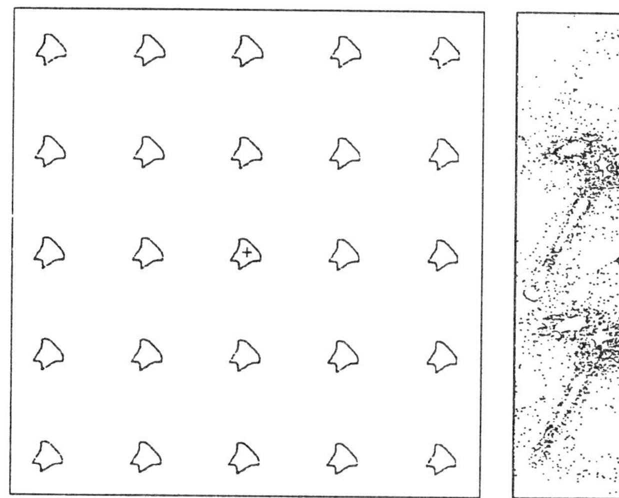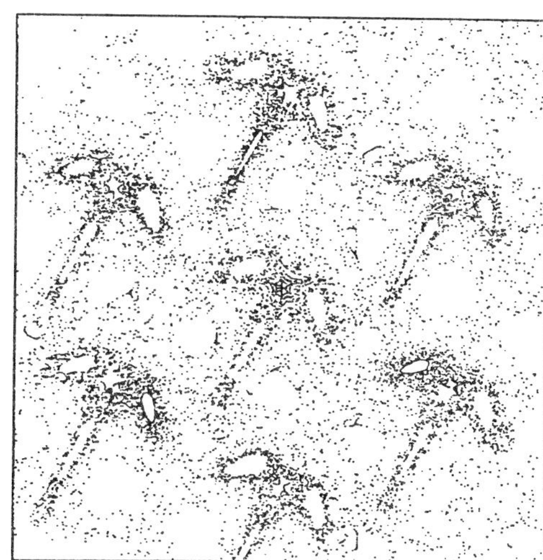

Figure 1. Phase portrait of the map (2.5) for $\alpha=1$ and $\omega/2\pi = 1/4$, scale=15 (left) and $\omega/2\pi = 1/3$, scale=10 (right).

In the case q=3 the proof is similar: one computes

$$F^{\circ 3} : \begin{cases} x^{\circ 3} = x + \frac{\sqrt{3}}{2}[g(x') - g(x^{\circ 2})] \\ y^{\circ 3} = y + g(x) - \frac{1}{2}[g(x') + g(x^{\circ 2})] \end{cases} \tag{3.7}$$

and one verifies that the maximum group of translations which commute with $F^{\circ 3}$ is

$$T_{mn}(x,y) = \left(x + 4m\pi + 2n\pi, y + 2\frac{\sqrt{3}}{3}n\pi \right). \tag{3.8}$$

In fact one has

$$F \circ T \begin{cases} x' - 2\pi(n+m) \\ y' + 2\pi\sqrt{3}m + \frac{2}{3}\sqrt{3}\pi n \end{cases} \tag{3.10}$$

$$F^{\circ 2} \circ T \begin{cases} x^{\circ 2} - 2\pi m \\ \phantom{x^{\circ 2}}... \end{cases} \tag{3.11}$$

$$F^{\circ 3} \circ T \begin{cases} x + 4\pi m + 2\pi n + \frac{\sqrt{3}}{2}[g(x' - 2\pi m) - g(x^{\circ 2} - 2\pi m)] \\ y + \frac{2\sqrt{3}}{3}\pi n + g(x + 4\pi m + 2\pi n) - \frac{1}{2}[g(x' - 2\pi m - 2\pi n) + g(x^{\circ 2} - 2\pi m)] \end{cases}$$

$$\Rightarrow F^{\circ 3} \circ T = T \circ F^{\circ 3}; \tag{3.12}$$

T_{mn} tiles the phase space in triangles of side 2π. The same argument can be used to prove the case q=6.

We checked the symmetry of the map (2.5) via a numerical computation for $q = 3$ and $q = 4$. In figure 1 we show the tiling of the phase space in triangles (left) and squares (right). Since α has been set to one, inside each tile the map is a very good approximation of the Hénon map: therefore, we

have infinitely many copies of the resonant Hénon map which tile the plane. In the case $q = 3$ the symmetry is not perfect: this is due to the impossibility of reproducing

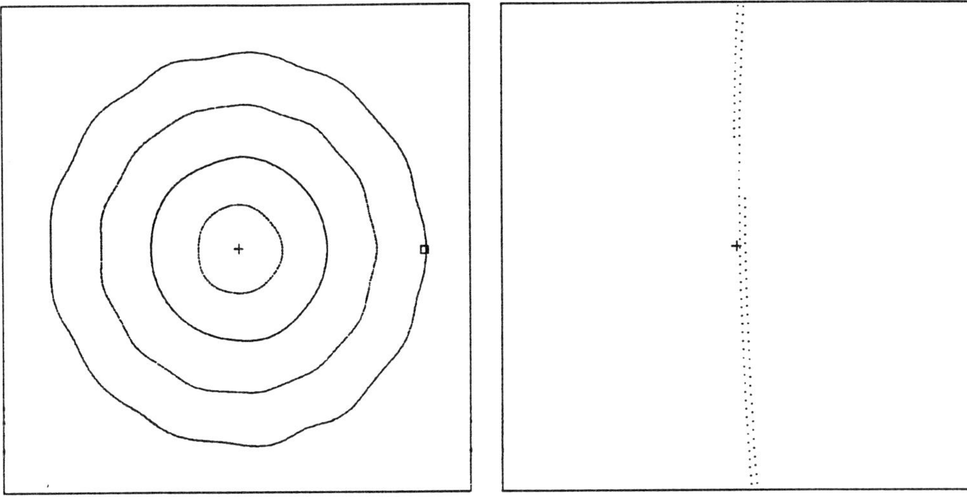

Figure 2. Phase portrait of the map (2.5) for $\alpha = 20$ and $\omega/2\pi = .61803$, scale $= 1$ (left), and magnification of the same orbit (scale$=.002$, right).

the same orbit in the different tiles when the orbit is chaotic: the rounding off error in the operation of translation $z \to T_{mn}(z)$ is amplified by the sensitivity to the initial conditions and therefore the orbits in the different tiles do not look all the same. In the other resonant cases with $q \neq 3, 4, 6$ no tiling is possible, and one can have quasi crystal symmetry.

4. Boundary of stability of the truncated map

In this Section we discuss the structure of the phase space of the map (2.4) for a generic nonresonant value of the frequency, varying the parameter α. We make the following remarks:

i) for $|z| << 2/\alpha$ the nonlinear part can be well approximated by the first significant order, i.e. we have the dynamics of the Hénon map;

ii) for $|z| >> 2/\alpha^2$ the nonlinear part is very small compared to the linear one. One has a domain of "quasi–linear" motion, where the rotation of a constant angle $e^{i\omega}z$ is perturbed by a small nonlinear term. Numerical analysis leads to the rather surprising result that there are no invariant curves, even if the perturbation is very small. In figure 2 (left) we show the orbits of the map (2.4) for $\alpha=20$; at $|z| = 1$ (i.e. in the quasi–linear domain) the orbits look like circles. Indeed, if we magnify the circle up to a scale of order $2/\alpha^2$ we see that it is not an invariant curve (figure 2, right).

iii) given a point z_0 in the phase space one can compute which is the dominant nonlinear term: since the function $x^n/n!$ reaches its maximum for $x \sim n$, the dominant order is

$$\bar{n} \sim \alpha|z_0| \tag{4.1}$$

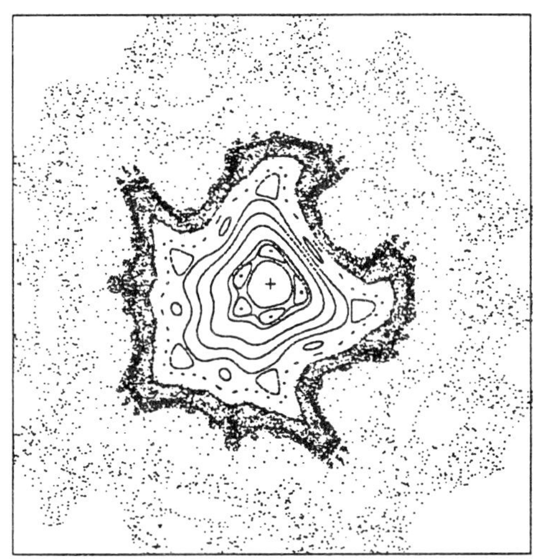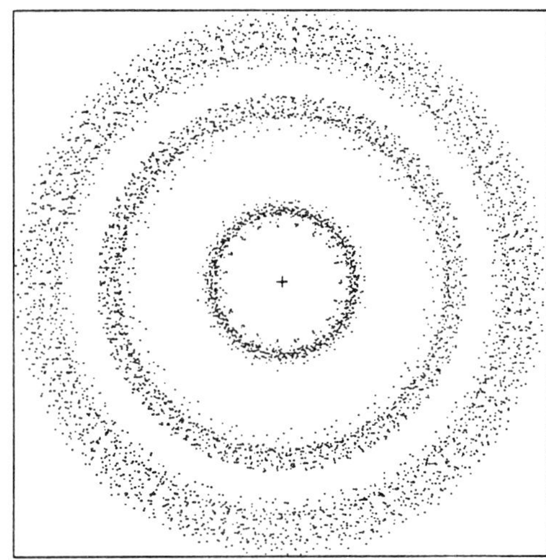

Figure 3. Phase portrait of the map (2.5) for $\alpha = 2$ and $\omega/2\pi=.21$, close to the origin (scale=5, left) and far from it (scale=100, right)

Moreover, we observe that map (2.4) does not have a region of fast escape to infinity, due to the boundedness of the nonlinearity; we can give the obvious estimate:

$$|z^{on}| - |z_0| \leq \frac{4n}{\alpha^2}; \qquad (4.2)$$

From the above discussion, we can give a qualitative description of the phase portrait of the map (2.4). For $\alpha \leq 1$ there is a linear zone close to the origin, than a region filled with KAM curves and islands equal to the corresponding region of the Hénon map; for $z \sim 2/\alpha^2$ the higher orders dominate, and one finds again KAM curves and islands up to a chaotic region. At $z > 2/\alpha^2$ one has the quasi–linear domain with a slow diffusion bounded by eq. (4.2). A typical example of this phase space structure is shown in figure 3, where the map with $\alpha = 2$ is plotted close to the origin (left) and far from it, in the quasi–linear domain (right).

For $\alpha > 1$ the phase portrait apparently consists of only circles; one has invariant circles up to $z \sim 2/\alpha$; for higher values of the amplitude one has the quasi–linear domain without invariant curves.

Since the map has orbits which far from the origin look like circles, the computation of the stability region is difficult, and strongly depends on the number of iterations. We considered the truncated map, which has a region of fast escape to infinity and therefore can be easily numerically analysed. We computed the biggest value of the amplitude $|z|$ of the initial condition which stays bounded for 1000 iteration, considering 10 different values of the argument of z. The linear frequency was fixed on the golden mean value. The results are summarized in table 2.

For $\alpha < 1$ one has a Hénon map at least up to $z \sim 2$; therefore, as the last invariant curve of the Hénon map is at .56, all computed truncations have the same stability region. For $\alpha > 1$ one has a different situation: the dynamic aperture as a function of the truncation order first reduces and then increases. The analysis of the

Table 2. Dynamic aperture versus α and truncation order.

α	2	4	8	16	32	64
0.02	0.56	0.56	0.56	0.56	0.56	0.56
0.2	0.56	0.56	0.56	0.56	0.56	0.56
2	0.56	0.70	0.80	2.5	2.3	3.1
20	0.56	0.24	0.29	0.45	0.73	1.4
200	0.56	0.044	0.040	0.052	0.080	0.14
2000	0.56	0.0092	0.0053	0.0059	0.0086	0.015
20000	0.56	0.0020	0.00071	0.00068	0.00092	0.0015

data shows that the minimum $|z|$ is reached for the order $\bar{n}$ which is the dominant order (4.1) at that amplitude:

$$\bar{n} \sim \alpha|z| \tag{4.3}$$

The orders higher than $\bar{n}$ contribute to increase the aperture.

For the LHC cell we have the estimate (2.16), then the stability region reaches a minimum when n is between 4 and 8. Therefore, in order to overestimate the aperture of the cell we can include multipoles up to order 8. It must be pointed out that this analysis is restricted to the on momentum particles (i.e. which have the nominal energy) and to a lattice model which considers only the systematic errors in the dipoles, and therefore neglects other nonlinearities such as the chromatic sextupoles and the corrector magnets [3]. Nevertheless, this analysis gives an indication on the minimum order of truncation which should be used, starting from the extrapolation of the data of the magnets. The truncation order commonly used in simulations (8, i.e up to the 18th pole) is therefore justified by this analysis.

Acknowledgments

We want to acknowledge Prof. Turchetti and Dr. Bazzani for stimulating and helpful discussions.

References

[1] Hénon M 1969 *Quarterly of Applied Mathematics* **27** 291–312

[2] Bazzani A, Turchetti G, Mazzanti P and Servizi G 1988 *Nuovo Cimento* **B 102** 51–80

[3] Scandale W, Schmidt F and Todesco E 1991 *Particle Accelerators* **35** 53–81

[4] Asner A et al. 1987 *CERN yellow report* **87-05**

[5] Zaslavskii G M et al. 1986 *Sov. Phys. JETP* **64** 294–303

[6] Yannacopoulos A N and Rowlands G 1991 *Phys. Lett.* **A 155** 133–136

Chaotic path at a nonlinear resonance

S.Y. Lee

Department of Physics, Indiana University, Bloomington, IN 47405

Abstract.
Special characteristics of particle loss at a betatron sum resonance is discussed. Particle motion at the nonlinear resonance region are decomposed into a Courant-Snyder invariant circle and a resonance curve. Intersection points in the phase space of these two curves correspond to unstable fixed points. These unstable fixed points form an unstable fixed curve, which is the resonance curve. Particles are found to stream out of the invariant surface through the nonlinear resonance line, which is the passage to unbounded motion at a sum resonance. Careful studies of the pathway may lead to understanding of the Diffusion process as well as the fast beam loss observed in particle tracking calculations.

1. Introduction

For particle motion in a circular accelerator, betatron oscillations, $x(s), z(s)$, around closed orbits satisfy Hill's equation[1]:

$$\frac{d^2x}{ds^2} + K_x(s)x = \frac{\Delta B_z}{B\rho}; \quad \frac{d^2z}{ds^2} + K_z(s)z = -\frac{\Delta B_x}{B\rho}, \tag{1}$$

with

$$\Delta B_z + i\Delta B_x = B_0 \sum_{n \geq 2}(b_n + ia_n)(x + iz)^n,$$

where b_n and a_n are respectively the normal and the skew multipole components. Here $K_x(s), K_z(s)$ are quadrupole strength functions, $B\rho = p/e$ is the momentum rigidity and s is the longitudinal particle coordinate, which advances from 0 to C($=2\pi$R), the circumference, as the particle completes one revolution of the cyclic accelerator, where R is the average radius.

Oscillations about the closed orbit due to the linear focusing force of quadrupoles, $K_{x,z}(s)$, are called betatron oscillations. The number of oscillation periods in one revolution are betatron tunes, ν_x, ν_z, which can be adjusted by varying the quadrupole strength. Both $K_{x,z}(s)$ and the anharmonic term, $\frac{\Delta B_{z,x}}{B\rho}$, are periodic functions of s with period C. The higher order anharmonic term, $\frac{\Delta B_{z,x}}{B\rho}$, which arises from higher order multipoles, is normally small. However when the condition, $m\nu_x + n\nu_z = \ell$, with m, n, ℓ as integers is satisfied, particles can encounter coherent perturbation kicks from the multipoles, $x^{|m|}z^{|n|}$, term. The number, $|m| + |n|$, is called the order of resonance. These nonlinear resonances usually lead to beam diffusion, halo, and beam loss.

Nonlinear anharmonic terms, $\frac{\Delta B_{z,x}}{B\rho}$, are especially important in high energy storage rings, where superconducting magnets generate sizable nonlinearity in the Hamiltonian. Particle tracking calculations have been used extensively in dynamical aperture studies. Considerable progress has been achieved[2].

In recent years, nonlinear mechanics has been studied in various subfields of physics. When nonlinearity is strong, chaotic motion appears. Phenomena of chaos is an interesting branch of physics. Numerical particle tracking calculations for the SSC and LHC did indicate that the Lyapunov exponent is positive near dynamical aperture boundary. However physics is sometimes hidden in numerical jungles of tracking simulation calculations.

In this paper, our goal is to study a possible pathway to chaotic motion in a 2D sum resonance model. The plan of the paper is as following. In section 2, we study general feature of particle motion near a single resonance. Tracking calculation is used to demonstrate proposition. The conclusion is given in section 3.

2. Single Resonance in 2D

Based on the theory of betatron motion[1], the betatron amplitude of particle motion is given by,

$$x = \sqrt{2\beta_x J_x}\cos(\phi_x); \quad z = \sqrt{2\beta_z J_z}\cos(\phi_z)$$

with J_x, J_z as invariant actions for unperturbed motion. Since betatron amplitude is normally small, so that the kick angles due to nonlinear magnets are of the order 10^{-3} or less than the corresponding kick angles of quadrupoles. Thus the Hamiltonian based on perturbation expansion of action angle variables is given by

$$H = H_0(J_x, J_z) + \sum_{m,n,\ell} g_{m,n,\ell} J_x^{\frac{|m|}{2}} J_z^{\frac{|n|}{2}} \cos(m\phi_x + n\phi_z - \ell\theta + \chi_{m,n,\ell}) \tag{2}$$

where $g's, \chi's$ can be obtained from the nonlinear multipoles. Depending on the operating condition of accelerators, only a few of these multipole terms are important. Assuming a single resonance dominant regime, i.e. $m\nu_x + n\nu_z = \ell$, ($m > 0$), the Hamiltonian can be approximated by

$$H \approx H_0(J_x, J_z) + g J_x^{\frac{|m|}{2}} J_z^{\frac{|n|}{2}} \cos(m\phi_x + n\phi_z - \ell\theta + \chi) \tag{3}$$

This simple model is integrable. Using a generating function,

$$F_2(\phi_x, \phi_z, J_1, J_2) = J_1(m\phi_x + n\phi_z - \ell\theta + \chi) + J_2\phi_z,$$

one obtains,

$$J_x = mJ_1; \quad J_z = nJ_1 + J_2; \quad \phi_1 = (m\phi_x + n\phi_z - \ell\theta + \chi); \quad \phi_2 = \phi_z. \tag{4}$$

The new Hamiltonian is given by

$$\tilde{H} = \tilde{H}_0(J_1, J_2) + g(mJ_1)^{\frac{|m|}{2}}(nJ_1 + J_2)^{\frac{|n|}{2}} \cos\phi_1. \tag{5}$$

Since the new Hamiltonian is independent of θ and ϕ_2, thus $\tilde{H}$ and J_2 are constants of motion. For a given particle near a 2D resonance, J_2 is a constant of motion, given by $J_2 = J_{20} = J_{z0} - \frac{n}{m}J_{x0}$ with J_{x0}, J_{z0} as the initial actions.

Let the unperturbed Hamiltonian H_0 be

$$H_0(J_x, J_z) = \nu_{x0}J_x + \nu_{z0}J_z + \frac{1}{2}\alpha_{xx}J_x^2 + \alpha_{xz}J_xJ_z + \frac{1}{2}\alpha_{zz}J_z^2. \tag{6}$$

We obtain then

$$\tilde{H}(J_1, \phi_1, J_2) = \delta_1 J_1 + \frac{1}{2}\alpha_{11}J_1^2 + g(mJ_1)^{\frac{m}{2}}(nJ_1 + J_2)^{\frac{n}{2}}\cos\phi_1 + [\nu_{0z}J_2 + \frac{1}{2}\alpha_{zz}J_2^2] \tag{7}$$

with

$$\delta_1 = m\nu_{x0} + n\nu_{z0} - \ell + m\alpha_{xz}J_2 + n\alpha_{zz}J_2, \quad \alpha_{11} = m^2\alpha_{xx} + 2mn\alpha_{xz} + n^2\alpha_{zz}.$$

Here two constants of motion, J_2, and $\tilde{H}$, determine particle orbit completely, i.e.

$$\tilde{H}(J_1, \phi_1, J_2) = \tilde{H}(J_{10}, \phi_{10}, J_2). \tag{8}$$

2.1. The third order resonance $\nu_x \pm 2\nu_z = \ell$

Let us consider $\nu_x \pm 2\nu_z = \ell$ resonance with initial horizontal kicks, i.e. $J_{10} = \mp\frac{1}{2}J_2$ (see Eq.(4)) The phase space evolution equation, Eq.(8), becomes,

$$(2J_1 \pm J_2)[\frac{\alpha_{11}}{4}(J_1) \pm g\sqrt{J_1}\cos\phi_1 + \frac{\delta_1}{2} \mp \frac{\alpha_{11}}{8}J_2] = 0 \tag{9}$$

Thus the particle trajectory follows the paths of two intersecting circles in the phase space map of $(\sqrt{J_1}\cos\phi_1, \sqrt{J_1}\sin\phi_1)$. The circle $2J_1 \pm J_2 = 0$ is called the *launching circle*, or more appropriately the Courant-Snyder invariant circle, while the circle

$$\frac{\alpha_{11}}{4}J_1 \pm g\sqrt{J_1}\cos\phi_1 + \frac{\delta_1}{2} \mp \frac{\alpha_{11}}{8}J_2 = 0$$

is the nonlinear coupling circle. *The intersections of these two circles are unstable fixed points of the Hamiltonian.*

For a difference resonance, $n < 0$, the launching circle is also the limiting circle [3]. When detuning parameter, α_{11}, is small, the coupling circle is a straight line. Varying betatron tunes, the coupling circle scans through the launching circle for a given initial J_2 value. When these two circles overlap, effect of resonance becomes strong. At condition $\delta_1 + \alpha_{11}J_{10} = 0$, the coupling circle passes through origin in phase space. This is equivalent to full coupling.

Fig. 1 shows features of Eq.(9) in phase space coordinates, $(\sqrt{J_x}\cos\phi_1, \sqrt{J_x}\sin\phi_1)$, where $\phi_1 = \phi_x - 2\phi_z + \chi$. Note that all different initial conditions pass through the same coupling line which depends on betatron tunes and the strength of nonlinear coupling resonance. Here the nonlinear coupling circle is a straight line due to $\alpha_{11} = 0$. On a difference resonance, particles move inward towards the origin of the coupling circle and reach onto the other side of the launching circle.

For sum resonance, unbounded motion may result from the coupling circle, where the radius of the coupling circle is inversely proportional to the detuning parameter

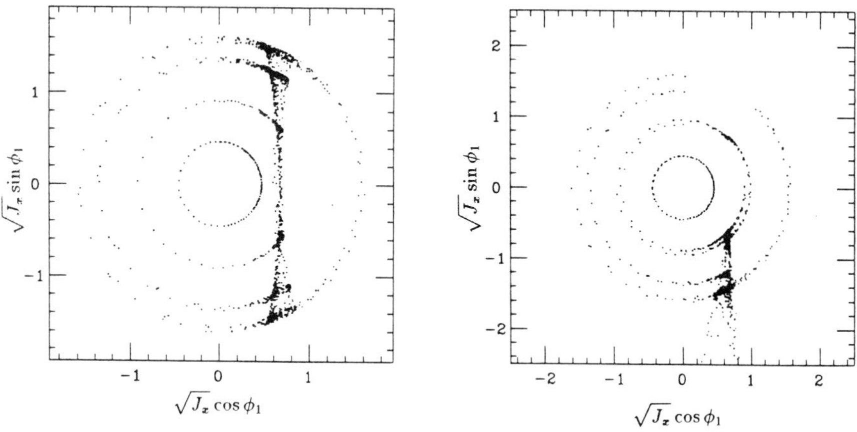

Figure 1. Phase space maps, obtained from a simple tracking program of linear lattice plus a single sextupole at $\nu_x - 2\nu_z = \ell$. The initial condition is given by $J_{z0} = 0$ with varying, J_{x0}, ϕ_{10}. Similar plot for the third order sum resonance, $\nu_x + 2\nu_z = \ell$, is shown on the right

α_{11}. Particles, starting out from a launching circle, will reach outward in phase space along the coupling circle. The right hand part of Fig. 1 shows a phase space map of $(\sqrt{J_x}\cos\phi_1, \sqrt{J_x}\sin\phi_1)$ for a simple linear lattice with a single sextupole at the $\nu_x + 2\nu_z = \ell$ resonance, where $\phi_1 = \phi_x + 2\phi_z + \chi$.

If the initial condition is only vertical motion, i.e. $J_{x0} = 0$, then the vertical motion will stay on the Courant-Snyder invariant circle until it encounters a coupling resonance line. When the initial horizontal and vertical actions are equal, the motion will not stay on a simple Courant-Snyder curve. But the specific unstable fixed point location is not changed greatly. Fig.2 shows a similar plot at the $\nu_x + 2\nu_z = \ell$ resonance with $x_0 = z_0, x_0' = z_0'$. Note here that the feature is similar to that on the right part of Fig.1. Particles that move along these invariant surfaces become chaotic when the invariant curve intersects the resonance curve. *The chaoticity is due to the nonintegrability of the tracking model*, which differs from the simple one resonance model of Eq. (3). The particle survival time to reach the resonance curve depends on the initial condition and the property of unstable fixed points.

Note also that there are two unstable fixed points on Figs. 1 and 2. Particles from the top unstable fixed point move downwards along separate paths depending on the initial phase space locations. The Courant-Snyder invariant circle is sliced into two halves. Particles that move along these two sections in opposite directions reach the other unstable fixed point in the lower part and stream out of the invariant region. The motion becomes chaotic due to nonintegrability of the tracking model.

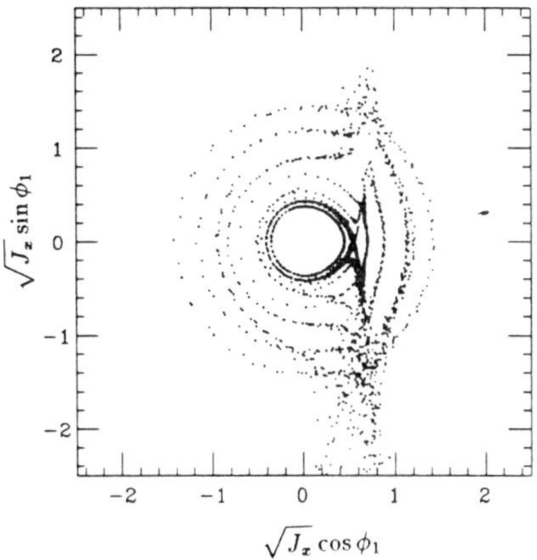

Figure 2. Similar to that of Fig.1, for the reduced phase space coordinates, is shown for the $\nu_x + 2\nu_z = \ell$ resonance at initial conditions, $x_0 = z_0, x'_0 = z'_0$.

2.2. $2\nu_x + 2\nu_z = \ell$ resonance

For $2\nu_x + 2\nu_z = \ell$ resonance with an initial horizontal kicks, similar equation of motion can be derived, i.e.

$$(2J_1 + J_2)[\frac{\alpha_{11}}{8}(2J_1) + g(2J_1)\cos\phi_1 + \frac{\delta_1}{2} - \frac{\alpha_{11}}{8}J_2] = 0. \tag{10}$$

Let us now define the phase space as $(\sqrt{2J_1}\cos\frac{\phi_1}{2}, \sqrt{2J_1}\sin\frac{\phi_1}{2})$. Eq.(10) is composed of a launching circle $2J_1 + J_2 = 0$ and *an ellipse or a hyperbola* described by

$$\frac{\alpha_{11}}{8}(2J_1) + g(2J_1)\cos\phi_1 + \frac{\delta_1}{2} - \frac{\alpha_{11}}{8}J_2 = 0,$$

for the nonlinear resonance curve. Locations where the nonlinear resonance curve intercepts the launching circle are *unstable fixed points of the Hamiltonian*. Particles follow the Courant-Snyder invariant circle until they encounter the unstable fixed point, which then leads to unstable motion. The left hand side of Fig. 3 shows a phase space plot of $(\sqrt{J_x}\cos\frac{\phi_1}{2}, \sqrt{J_x}\sin\frac{\phi_1}{2})$ for a simple tracking with linear lattice and a single octupole at $2\nu_x + 2\nu_z = \ell$ resonance. Clearly, particles follow the Courant-Snyder invariant curve until they encounter the unstable fixed point. Particle motion appears chaotic afterwards. Thus particle loss is located at a unique phase space curve in (J_1, ϕ_1). We have presented our tracking data for $2\nu_x + 2\nu_z = \ell$ resonance in the phase space as a function of $(\sqrt{J_x}\cos\frac{\phi_1}{2}, \sqrt{J_x}\sin\frac{\phi_1}{2})$ on Fig.3. One can also present them in the phase space of $(\sqrt{J_x}\cos\phi_1, \sqrt{J_x}\sin\phi_1)$. The result is a curved resonance line that intercepts the Courant-Snyder invariant circles.

When betatron tunes are such that the nonlinear resonance coupling curve does not intercept the launching circle, particle motion is stable.

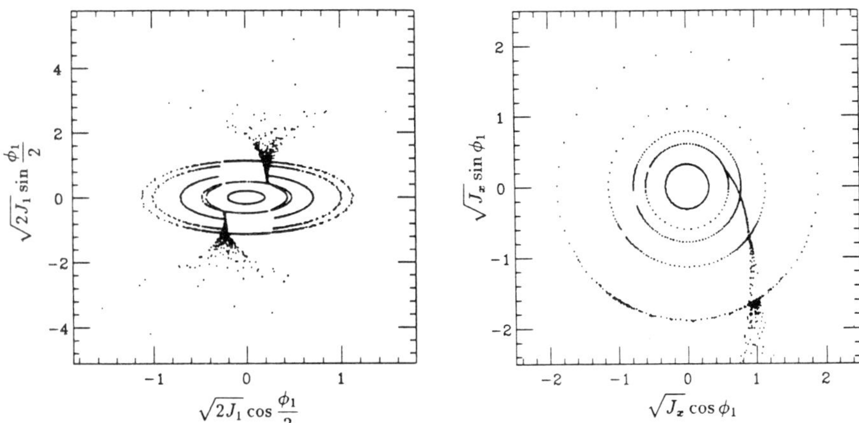

Figure 3. The Poincaré maps (see text for explanation) are shown for a Simple tracking calculation with a single octupole at a $2\nu_x + 2\nu_z = \ell$ resonance.

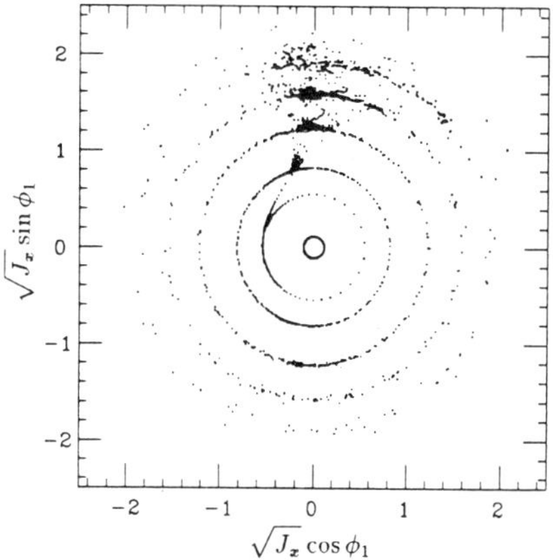

Figure 4. The Poincaré maps (see text for explanation) are shown for a Simple tracking calculation with a single decapole at a $3\nu_x + 2\nu_z = \ell$ resonance.

2.3. General Discussion on $m\nu_x + n\nu_z = \ell$ resonance

General characteristics of resonance behavior remain valid for $m\nu_x + n\nu_z = \ell$ resonance. The right side of Fig. 4 shows Poincaré maps of $(\sqrt{J_x}\cos\phi_1, \sqrt{J_x}\sin\phi_1)$, with $\phi_1 = 3\phi_x + 2\phi_z$ and $J_x = 3J_1$ of tracking calculations with a single decapole at the $3\nu_x + 2\nu_z = \ell$ resonance. Note that three resonance coupling curves can be seen crossing through the Courant-Snyder circle. Particles spill out of the Courant-Snyder invariant circle through unstable fixed points.

At higher order multipoles, the betatron amplitude growth rate is usually very large. Thus the reconstruction of invariant curves becomes more difficult.

3. Conclusion

The characteristic feature of the betatron motion at a resonance condition can be described as a crossing of the Courant-Snyder invariant curve and resonance curve. The unstable fixed points of the Hamiltonian are located at intersections of these two curves. *These unstable fixed points form an unstable fixed curve in 2D.* Particles that move along the Courant-Snyder invariant curve stream out of the invariant region through one of these unstable fixed points.

Simple particle tracking calculations (symplectic but non integrable) are used to demonstrate these characteristic features. Experimental measurement of these feature will be interesting. Another possible application of this feature can be used to identify particle loss mechanisms in a particle tracking calculation.

Particle beam bunch in accelerator is an ensemble of particles with many different betatron tunes embedded in a multinet of higher order nonlinear resonances. Most particles in this ensemble move along Courant-Snyder invariants. A few particles located near important sum resonances will stream out of the unstable phase space locations. Studies of these phase space locations is important in the determination of dynamical apertures.

acknowledgement

I am grateful to beneficial discussions with Dr. Y. Yan.

References

[1] E.D. Courant and H.S. Snyder, Ann. Phys. (NY) $\underline{3}$,1 (195 8).

[2] See papers in these proceedings.

[3] J. Liu, et al., Particle Accelerators, to be published;

Inst. Phys. Conf. Ser. No 131
Paper presented at Int. Workshop Nonlinear Problems in Accelerator Phys. Berlin, 1992

Third Order Achromats Based on Mirror Symmetries

W. Wan, E. Goldmann, and M. Berz

Department of Physics and Astronomy and National Superconducting Cyclotron Laboratory, Michigan State University, East Lansing, MI 48824

Abstract.
A method is presented that allows the design of third order achromats consisting of a sequence of cells, in which subsequent cells agree with this base cell up to symmetry reflections. By correcting certain suitably chosen aberrations of the cell, the sequence of cells can be made free of all aberrations. Using methods of computational theorem proving, a complete classification of all n-cell systems requiring the least number of corrections in the base cell is obtained for $2 \leq n \leq 8$. It turns out that four cell systems utilizing mirror symmetry are particularly suitable for the design of achromats, and that there is a class of 32 different designs all requiring the same number of corrections.

1. Introduction

The search for systems without any aberrations up to a certain order, so-called achromats, is an old and fundamental problem of optics.While it is not very difficult to design first order achromats in mid plane symmetric systems, the correction of all second order aberrations already requires substantial thought [1]. It was recognized that in order to eliminate aberrations, it is very advantageous to impose certain symmetry conditions on the systems, and so most second order achromats consist of four identical cells which are connected by various mirror symmetries.

The past years have seen an interest in the design of third order achromats, which was mostly stimulated by the application of normal form techniques [2, 3, 4]. Various third order achromatic systems consisting of only repetitive arrangements of identical cells have been found using these ideas. The first such system, which is to be considered a proof of principle only, consisted of thirty identical cells and a total of 300 bending magnets [5]. By using techniques of resonance suppression, recently Neri [6, 7] produced much more realistic and potentially practically useful systems consisting of as little as five identical cells placed in series.

For reasons directly connected to normal form theory, it is not very practical to try to reduce the number of cells below five. But since current normal form methods are only applicable directly to systems made of completely identical cells, it seems worthwhile to study the use of additional symmetries, which have been recognized in the past as being useful for the correction of aberrations.

So instead of using identical cells as building blocks like in the normal form approach, we allow cells which are obtained from the original one through mirror imaging about the x-y plane, which corresponds to a reversion, cells which are obtained through mirror imaging about the x-z plane, which corresponds to a switching of the bending direction, as well as those that are obtained through both of these mirror imaging processes from the base cell. We denote the base forward cell with F, the reversed cell with R, the switched cell with S, and the combined switch-reversed cell with C.

Fig. 1 shows symbolically the four different allowed configurations of cells. To obtain achromatic systems, we use combinations of these cells which are derived from a forward cell F in which suitable aberrations have been removed.

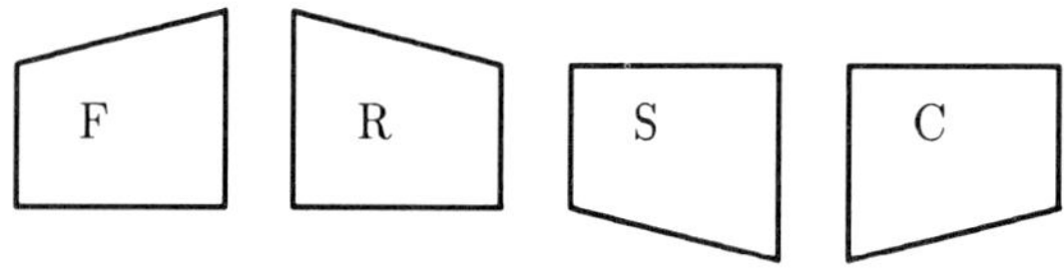

Figure 1. Forward (F), Reversed (R), Switched (S), and Combined Switched-Reverse (C) Cells

2. A Method to Classify Symmetry-Based Achromats

In order to study the nonlinear behavior of the combined systems, we have to take into account the inter-correlations between the aberrations of the map which are connected to the condition of symplecticity [8, 9], a direct consequence of the Hamiltonian structure of particle optical systems. To this end, to order three, the map $\mathcal{M}$ of the forward cell is represented in the form

$$\mathcal{M}_F = M \circ \exp(: H :) \tag{1}$$

where M is the linear part of the map and H is a polynomial of exact order 3 and 4 describing the nonlinear part [10, 11]. Using the Taylor representation of $\mathcal{M}_F$, and letting $\mathcal{A}_1$ denote the map describing a switch of the sign of the x and a coordinates, we obtain for the map $\mathcal{M}_S$ of the switched system

$$\mathcal{M}_S = A_1^{-1} \circ \mathcal{M}_F \circ \mathcal{A}_1. \tag{2}$$

Similarly, if we let A_2 denote the map exchanging the sign of the a and b coordinates, we obtain for the map $\mathcal{M}_R$ of the reversed system

$$\mathcal{M}_R = \mathcal{A}_2^{-1} \circ \mathcal{M}_F^{-1} \circ \mathcal{A}_2, \tag{3}$$

and we also obtain for the map $\mathcal{M}_C$ of the combined switched and reversed system:

$$\mathcal{M}_C = A_1^{-1} \circ \mathcal{M}_R \circ \mathcal{A}_1. \tag{4}$$

To construct aberration free systems, the task is now to find linear maps M as well as polynomials H such that a suitable combination of F, R, S, or C cells produces a system that is completely free of aberrations.

This search was performed in an automated and complete way using the code COSY INFINITY [10]. The first order map M was chosen to have rotations in the horizontal and vertical planes, where for an n cell system, the angles were all possible combinations of multiples of $2\pi/n$. Furthermore, the chromatic properties of the linear map were required to satisfy either $(x|\delta) = 0$ or $(a|\delta) = 0$.

For each of these choices of first order maps, all possible combinations of arrangements consisting of the four elementary cells F, R, S, and C were studied. For each of these combinations, it was determined in an analytical way which terms in H of the forward cell have to be removed to obtain a third order achromat.

To find these terms, first each monomial in the polynomial H is studied separately by setting all the other terms to zero. For this case, the maps of the four cells and the n cell system under consideration are computed, and it is checked if any aberrations are produced. If this is the case, then the term certainly has to be corrected since it cannot be expected that other monomials of H cancel its contribution.

After all monomials have been considered in this way and those which did not produce any aberrations individually have been determined, it is checked if the remaining terms collectively produce any aberrations due to coupling effects. To check this, all monomials that passed the initial test are assigned 16 digit random number coefficients. Then the aberrations of the resulting system are computed again. Since the probability of producing an accidental zero with floating point operations of 16 digit random numbers is in the order of 10^{-16}, if still all aberrations of the system vanish, then the system is actually aberration free with a probability of about $1 - 10^{-16}$, i.e. beyond any reasonable doubt.

3. A Classification of Achromats

After extensive computer searches extending over all the millions of possible combinations of systems requiring hundreds of hours of CPU time, the following results were found. First of all, among the systems consisting of any number of cells from two to eight, four cell systems required the least number of corrections.

$$
\begin{array}{ll}
xa^2 & a^3 \\
xb^2 & ayb \quad ab^2 \\
a^2\delta & b^2\delta
\end{array}
$$

$$
\begin{array}{llll}
x^2a^2 & a^4 \\
xayb & x^2b^2 \quad a^2y^2 \quad a^2b^2 \\
y^2b^2 & b^4 \\
a^2\delta^2 & b^2\delta^2
\end{array}
$$

Table 1. Second and third order monomials that have to be removed to obtain a system free of position aberrations of up to third order

Of all the possible four cell systems characterized by either $(x|\delta) = 0$ or $(a|\delta) = 0$ as well as different rotation angles in the horizontal and vertical parts of linear map and different combinations of the four cells, a small family of systems proved to be particularly useful. As the first case, we study the goal of correcting all aberrations of positions, which because of the condition of symplecticity [9] also entails a correction of time of flight terms except those that depend only on energy. In this case, there is a family of sixteen systems, all of which require the correction of the same terms in H. These terms are listed in table 1.

Moreover, for each choice of horizontal and vertical rotation angles, there is exactly one combination of the F, R, S, and C maps that yields a system free of position aberrations. These combinations are shown in table 2 for the first order condition $(x|\delta) = 0$ and in table 3 for the first order condition $(a|\delta) = 0$.

θ_x	$\theta_y = 0$	$\theta_y = 90$	$\theta_y = 180$	$\theta_y = 270$
0	F C S R	F C S R	F C S R	F C S R
90	F C F C	F C F C	F C F C	F C F C
180	F C S R	F C S R	F C S R	F C S R
270	F C F C	F C F C	F C F C	F C F C

Table 2. Systems without Position Aberrations with $(x|\delta) = 0$

If the goal is to also remove the angle dependent aberrations, i.e. to produce a true achromat, then additional terms in the polynomial H have to be removed. The family of sixteen systems is the same as before, and the terms in M that have to be removed are shown in table 4.

Altogether, both in the case of systems that are free of position aberrations as well as in the case of systems that are achromatic, the classification of the systems requiring the minimal number of corrections follows a very systematic pattern. The monomials which have to be removed are always the same, and for each rotation combination, there is exactly one minimal system, which in the case of the four cell systems is even independent of the rotation angle in the vertical plane. Apparently the mirror symmetries are essential

θ_x	$\theta_y = 0$	$\theta_y = 90$	$\theta_y = 180$	$\theta_y = 270$
0	F R S C	F R S C	F R S C	F R S C
90	F R F R	F R F R	F R F R	F R F R
180	F R S C	F R S C	F R S C	F R S C
270	F R F R	F R F R	F R F R	F R F R

Table 3. Systems without Position aberrations with $(a|\delta) = 0$

$$
\begin{array}{lllll}
x^3 & xa^2 \\
xy^2 & xb^2 & ayb \\
x^2\delta & x\delta^2 & a^2\delta & y^2\delta & b^2\delta
\end{array}
$$

$$
\begin{array}{lllll}
x^4 & x^2a^2 & a^4 \\
x^2y^2 & xayb & x^2b^2 & a^2y^2 & a^2b^2 \\
y^4 & y^2b^2 & b^4 \\
x^2\delta^2 & a^2\delta^2 \\
y^2\delta^2 & b^2\delta^2
\end{array}
$$

Table 4. Second and third order monomials that have to be removed to obtain a third order achromatic system

in that there are no minimal systems consisting of only F cells. Furthermore, there are systems consisting of only F and R cells, i.e systems that produce a net deflection of four times that of the F cell. In case there are S or C cells, they always occur in pairs, which means that the net deflection of these systems is always zero.

So obviously the employed mirror symmetry has merits and drastically reduces the number of fitting conditions. To obtain a system free of position aberrations, one has to correct only seven out of nineteen third order terms and ten out of thirty-seven fourth order terms in H. To obtain a third order achromat requires the removal of ten third order terms and fifteen fourth order terms.

To close our discussion, we note that the technique employed here could also be used in the environment of a formula manipulator, in which case completely rigorous results could be obtained. However, this would come at a substantially increased computational expense which in our estimate would make the extensive searches described here impossible in practice. While the search process does not seem feasible in such an environment, the reader is encouraged to cross check the aberration free systems presented here; we refrain from such efforts.

So far, the techniques described here have been used to design both a system free of position aberrations as well as a third order achromat, which are presented in [12]. Several more designs are under consideration. Obviously the method can be generalized to the correction of fourth order aberrations, and we are planning to perform their classification as well. Furthermore, the method can also be used for the correction of only certain sub classes of aberrations.

Acknowledgement

For several discussions and comments in the early stages of this work, we would like to thank Prof. Hermann Wollnik.

References

[1] R. V. Servranckx and K. L. Brown. Circular machine design techniques and tools. *Nuclear Instruments and Methods*, A258:525, 1987.

[2] A. J. Dragt and J. M. Finn. Normal form for mirror machine Hamiltonians. *Journal of Mathematical Physics*, 20(12):2649, 1979.

[3] E. Forest, M. Berz, and J. Irwin. Normal form methods for complicated periodic systems: A complete solution using Differential algebra and Lie operators. *Particle Accelerators*, 24:91, 1989.

[4] M. Berz. Differential algebraic formulation of normal form theory. *Nonlinear Effects in Accelerators*, 1993.

[5] A. J. Dragt. Elementary and advanced Lie algebraic methods with applications to accelerator design, electron microscopes, and light optics. *Nuclear Instruments and Methods*, A258:339, 1987.

[6] F. Neri, in: M. Berz and J. McIntyre (Eds.). Proceedings of the 1990 workshop on high order effects in accelerators and beam optics. Technical Report MSUCL-767, National Superconducting Cyclotron Laboratory, Michigan State University, East Lansing, MI 48824, 1991.

[7] Filippo Neri. Private communication.

[8] A. J. Dragt. Lectures on nonlinear orbit dynamics. In *1981 Fermilab Summer School*. AIP Conference Proceedings Vol. 87, 1982.

[9] H. Wollnik and M. Berz. Relations between the elements of transfer matrices due to the condition of symplecticity. *Nuclear Instruments and Methods*, 238:127, 1985.

[10] M. Berz. COSY INFINITY Version 6 reference manual. Technical Report MSUCL, National Superconducting Cyclotron Laboratory, Michigan State University, East Lansing, MI 48824, 1993.

[11] M. Berz. Arbitrary order description of arbitrary particle optical systems. *Nuclear Instruments and Methods*, A298:426, 1990.

[12] W. Wan and M. Berz. In *Third Computational Accelerator Physics Conference*. AIP **Conference Proceeings, 1993.**

Design of modern high resolution magnetic spectrometers

A F Zeller[1]

National Superconducting Cyclotron Lab, Michigan State University, East Lansing, MI 48824

Abstract. The choice of correcting nonlinear aberrations in high resolution magnetic spectrometers with software or hardware is examined. The ability of raytracing methods, using realistic focal plane detector resolutions, is demonstrated for the S800 spectrograph under construction at the NSCL. Furthermore, Differential Algebraic methods are shown to reproduce the results for accurately known fields at a considerable savings in design time.

1. Introduction

High resolution, large acceptance magnetic spectrographs offer considerable opportunities for nuclear physics research. With the trend towards higher beam energies the resulting increase in dipole mass, and hence cost, dictates that the design be as efficient as possible. Correction of aberrations with edge profiles and multipole elements for large solid angle devices leads to induced higher order effects which are impossible to correct for the largest solid angles. If one needs to measure positions at the focal plane for physics reasons, then the use of ray reconstruction methods is natural. Additionally, the difficulty in predicting the three dimensional behavior of complicated aberration correcting edge curvatures on dipole entrances and exits have led many modern spectrometer designers to the method of software corrected solutions. This does result in more complicated and expensive detector systems, but has many advantages. As detector development progresses, the spectrograph resolution may be increased by simply replacing the detector. A traditional spectrometer with machined edge contours is difficult to upgrade. Likewise, a change in the optical mode can be more readily accommodated. The S800 spectrometer being built at the NSCL is an example of a spectrograph that is entirely software corrected [1,2]. Spectrographs that uses both hardware and software correction are those being built at CEBAF [3]. For comparison of the S800 to other spectrographs, a figure of merit, which is composed of solid angle, resolution, magnetic rigidity and the energy range, *versus* year of construction is shown in Fig 1. The S800, labeled MSU, is among the best in terms of its combination of large solid angle, energy range and

[1] Supported by the U.S. National Science Foundation, PHY89-13815

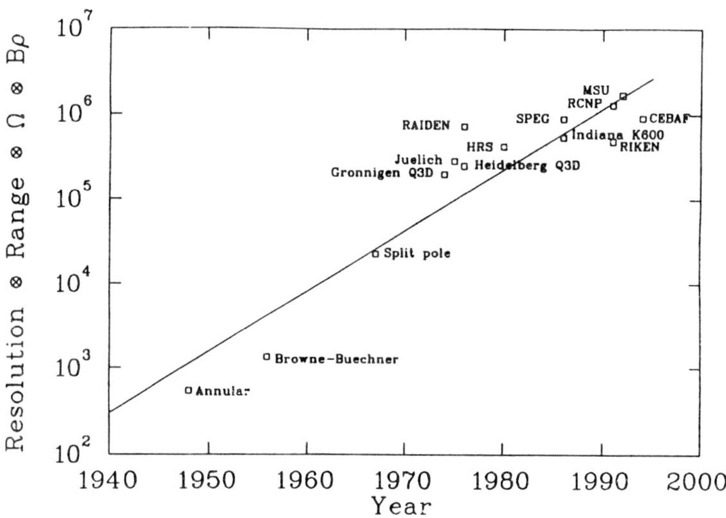

Figure 1. Figure of merit versus year of construction for spectrograph developement. The figure of merit is the product of the solid angle in msr, the resolution (assuming a 1mm object size), the energy range in percent, and the magnetic rigidity in T-m.

good energy resolution. This result is based on using software to correct aberrations and assumes a 1 mm object at the beginning of the beam analysis system. The parameters of the S800 are given in Table 1. A schematic view of the magnetic elements is given in Fig 2. It should be noted that the spectrograph disperses in the vertical plane while the scattering angle is measured in the horizontal plane, thus a two dimensional detector system is required. It turns out that two detector systems are required to measure the angles at which the particle enters the focal plane so the complete trajectory can be reconstructed.

Spectrometers may also reduce aberrations with active multipoles incorporated into the quadrupoles. The large solid angle of the S800 results in large contributions to the resolution in the dispersive plane from the spherical aberrations (x/θ^2 and x/ϕ^2). This is shown in the top part of Fig 3, calculated with TURTLE [4]. The results of correcting the two terms by placing sextupoles successively in the first quadrupole and then in the second are shown in the middle and lower illustrations of Fig 3. However, TURTLE is only a second order code, so when the corresponding calculations are made with RAYTRACE [5], the results are not so impressive. Shown in Fig 4 is a calculation with a sextupole designed to correct (x/θ^2). It becomes obvious that while second order (symmetric in theta) has been corrected, higher order terms have been introduced. Attempting to correct other second order terms with a sextupole requires sextupole field strengths comparable to the quadrupole field strength at the pole tip.

Table 1.

PARARMETERS OF THE MSU 1.2 GeV/c SPECTROGRAPH

ENERGY RESOLUTION:	$\Delta E/E = 10^{-4}$ WITH 1mm RADIAL OBJECT SIZE FOR BEAM ANALYSIS SYSTEM
ENERGY RANGE:	$\Delta E/E = 10\%$
SOLID ANGLE:	$\Omega = 10\text{-}20$ msr
RESOLVING POWER:	$D/M = 12.3$
RADIAL DISPERSION:	$D = 9.1 \,\text{cm}/\%$
RADIAL MAGNIFICATION:	$M = 0.74$
AXIAL DISPERSION:	$R_{34} = 0.88$ mm/mr
ANGULAR RESOLUTION:	$\Delta\theta \leq 2$ mr (TOTAL OF BEAM PLUS SPECTROGRAPH CONTRIBUTIONS)
FOCAL PLANE SIZE:	50 cm (RADIAL) X 15 cm (AXIAL)
FOCAL PLANE TILT:	$28.5°$
MAGNETIC RIGIDITY:	$B\rho = 4\text{T-m}$
DIPOLE FIELDS:	$B = 1.5\text{T}(\rho = 2.7$ m)
DIPOLE GAP:	$D = 15$ cm
DIPOLE SIZE:	3.5 m LONG X 100 cm WIDE (75° BEND) QTY OF 2
WEIGHT OF DIPOLES:	70 TONS EACH
QUAD SIZES:	#1) 20 cm ID X 40 cm LONG
	#2) 35 cm X 17 cm X 40 cm
DETECTOR REQUIREMENTS:	TWO 2-DIMENSIONAL DET., 1m SEPARATION
	#1) 50 cm X 15 cm
	#2) 62 cm X 16 cm
	RESOLUTION: RADIAL 0.2 mm
	AXIAL 0.4 mm

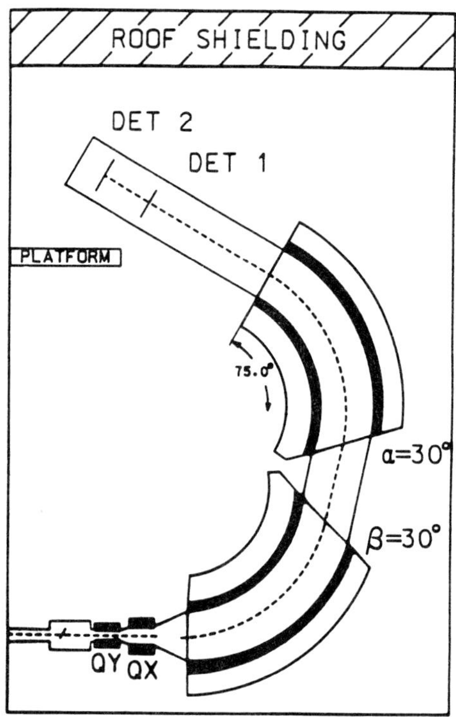

Figure 2. The magnetic elements of the S800 spectrometer.

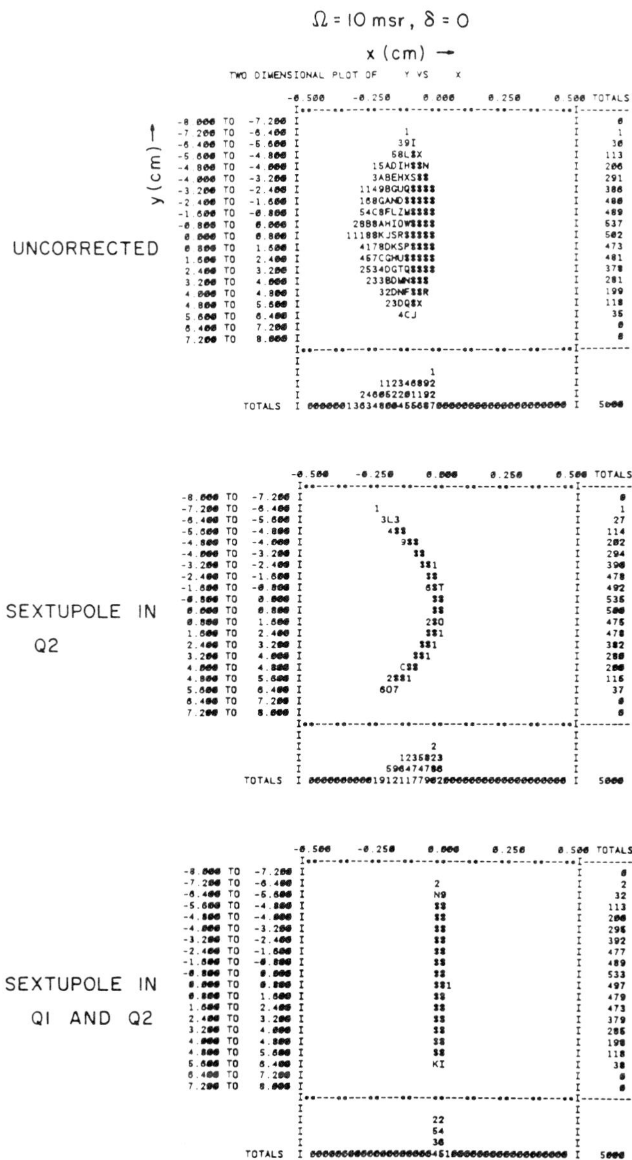

Figure 3. TURTLE calculations showing the effect of sextupoles placed in the quadrupole elements.

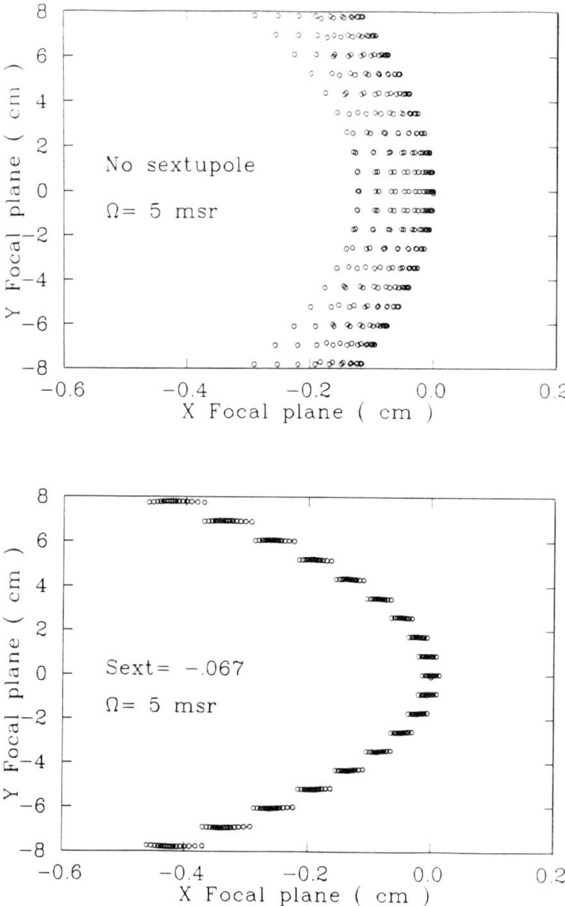

Figure 4. RAYTRACE calculation showing the effect of a sextupole in one of the quadrupoles.

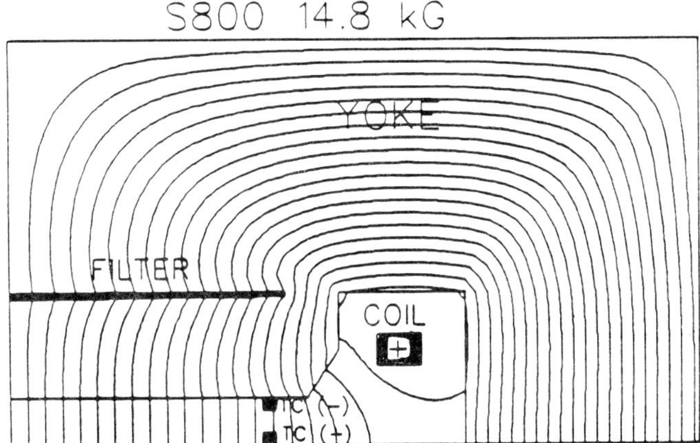

Figure 5. Magnetostatic calculation for one quarter of the S800 dipole showing the field profile producing elements.

2. Dipole design

In a spectrometer like the S800 where focal plane measurements of the positions and angles are required to fulfill both the physics requirements and for aberration corrections, the degree of dipole field flatness can be reduced. From simple constant first order considerations it can be shown that a gradient in the field of up to 1 G/cm can be tolerated [6], as long as the gradient is known. This greatly reduces the size of the dipoles and the cost of the spectrograph. The S800 dipoles are designed to provide the required field uniformity from 0.5 to 1.6 T with sufficient width to achieve the following combinations of solid angle and momentum acceptance: a 10 msr solid angle with a 5% momentum acceptance and a 15-20 msr solid angle varying smoothly over the momentum acceptance range with about a 1% momentum acceptance. The exact combination will depend on the measured field profile and focal plane detector resolution. From a number of considerations [6] it was decided to construct the dipoles and the quadrupoles using superconducting coils.

The goal of meeting the required dipole field uniformity over the whole operating region is accomplished by introducing several features to the design; namely a tapered air gap (filter) behind the pole tip and an auxiliary trim coil which goes along the pole tip and returns along the median plane. These features are shown in Fig 5, which is a cross sectional view of one quarter of the magnet with the field lines also shown. The effects of the filter and the trim coils are shown in Fig 6. Note that the filter affects the finite permeability solution (ie, the nonlinear saturation properties of the iron) and the trim coils affect the infinite permeability solution. The trim coils run at about one percent of the main coil amp-turns and are not superconducting.

3. Resolution calculations

Determination of whether a software corrected spectrometer design meets the expected goals is not as easy as verifying that a hardware corrected spectrograph has enough

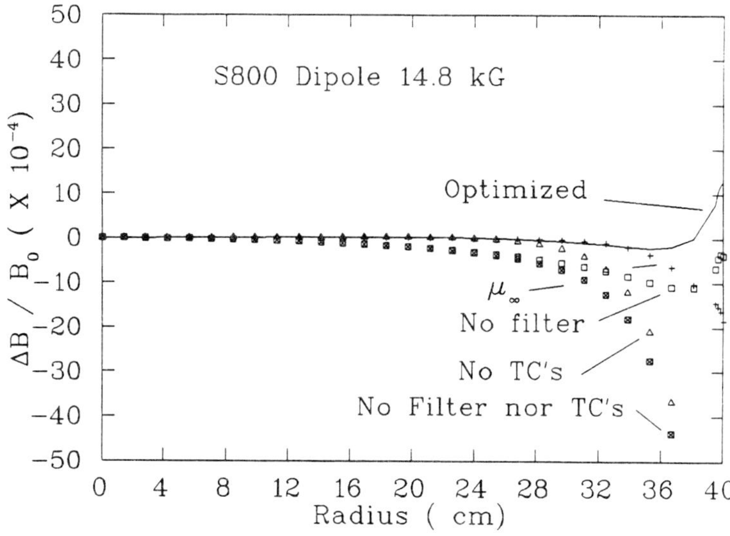

Figure 6. Gradient profile showing the effect of the elements illustrated in the preceeding figure.

higher order features. With a hardware spectrometer all that is required is tracing a large number of trajectories, which fill out the expected solid angle and momentum acceptance, to the focal plane. If the resulting image is small enough, then the spectrometer works. Detector resolution enters into the picture only to the extent that it must be sufficient so that it doesn't add to the spot size. With a software spectrometer, such as the S800, a raytracing code such as MOTER [7], is used. MOTER is an optimizing code which adds the assumed detector resolutions to a set of random rays. One then determines a set of coefficients, that look like the standard aberration terms, which MOTER uses to determine the resolution. The accuracy of measurements at the focal plane then determines how well you can correct the coefficients. After determining the required set of coefficients, the resulting histogram of residuals is the resolution.

A big concern is how to determine which coefficients are important. Due to limitations on the number of coefficients which can be accommodated it is vital that all of the significant terms be found. Some terms are easily found from second and third order codes, and some are readily deduced from a knowledge of optics. For example, in large solid angle devices, if the second order spherical terms are important, then so too are the higher order opening terms. The next step is to run sets of rays through RAYTRACE that have different combinations of the phase space. From the resulting patterns of X *versus* Y at the focal plane, many other aberration terms can be deduced. With this set of coefficients, and a set of assumed detector resolutions, MOTER is used to evaluate the resolution. The resulting histograms are shown in Figs 7 and 8 for the momentum resolution and scattering angle resolution, respectively. The sets of curves are labeled with the assumed detector resolutions; eg, 443 means the detector has a resolution in X of 0.4mm, a resolution in Y of 0.4mm and resolutions in the determined focal plane angles of 0.3mr. All resolutions are expressed in FWHM. Illustrated are the results for

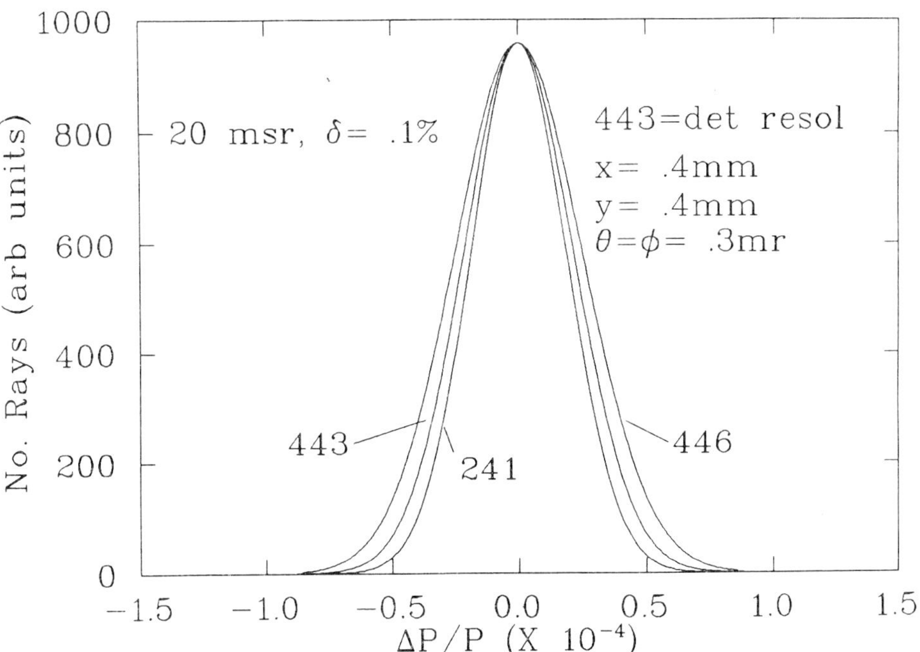

Figure 7. MOTER results showing the resolution function for momentum.

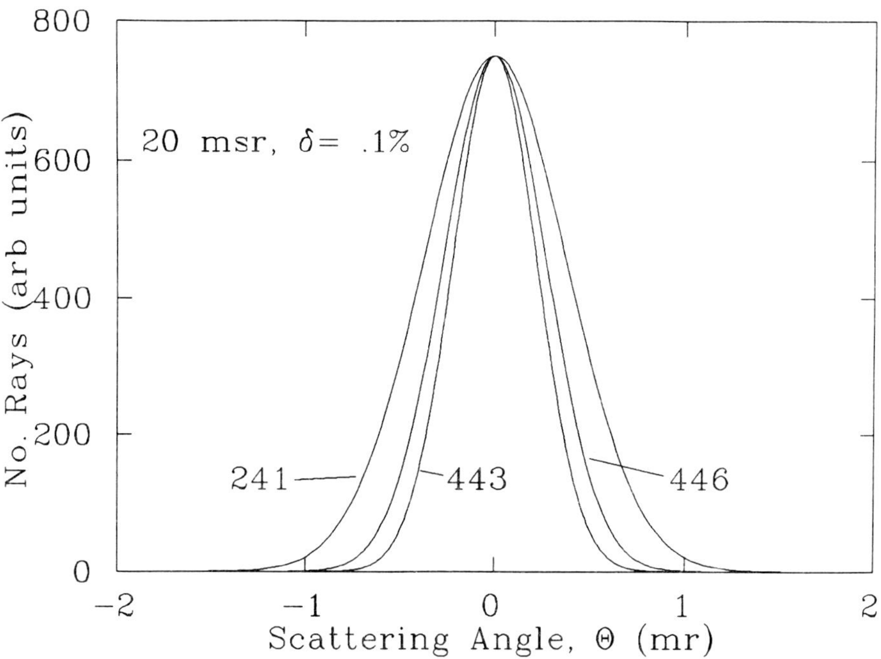

Figure 8. MOTER results showing the resolution function for scattering angle.

Table 2.

S800 RESOLUTION
1 G/cm gradient in dipoles

		Detector Requirements			$\Delta P/P$ ($\times 10^{-4}$)		θ (mr)	
Ω(msr)	$\delta(\pm\%)$	x(mm)	y(mm)	$\theta=\phi$(mr)	FWTenthM	FWHM	FWTenthM	FWHM
3.6	.5	0	0	0	.32	.18	.12	.066
3.6	.5	.4	.4	.6	.91	.50	.95	.52
20	.01	.4	.4	.6	.85	.47	1.16	.64
20	.8	.2	.4	1.	.82	.45	1.70	.90
20	.1	0	0	0	.41	.22	.13	.072
20	.1	.2	.4	1.	.82	.45	1.61	.88
20	.1	.4	.4	.6	1.09	.60	1.18	.64
20	.1	.4	.4	.3	.90	.49	.93	.51
10	2.5	0	0	0	.30	.16	.12	.072
10	2.5	.2	.4	1.	.86	.47	1.15	.63
10	2.5	.4	.4	.6	1.00	.55	1.00	.55
10	2.5	.4	.4	.3	.90	.49	.93	.51

the large solid angle, small momentum acceptance limit. Similar results are obtained for other combinations of solid angle and acceptance. The results are summarized in Table 2. The results are achieved by using about 25 coefficients each for momentum and scattering angle.

If you have been very lucky then after a few hours of CPU time, your calculations will verify that the spectrometer will meet the required resolutions, if you build it properly. However, what will likely happen is that the set of coefficients will not be sufficient and/or the detector requirements will be impossible to achieve. Now you will have to start to find coefficients by trial and error - putting in terms and evaluating the results. Since for even a relatively simple spectrograph such as the S800, this takes two to four hours of CPU time, the process is very slow. For cases like the CEBAF high resolution spectrometer the calculational time goes up to twelve to twenty hours, and progress is painfully slow.

The advent of Differential Algebraic (DA) methods of calculations of particle optic systems to arbitrary order by codes such as COSY INFINITY [8] has aided considerably in reducing the time needed to find aberration coefficients. One needs to run the code up to whatever order is required, then transfer the important terms to MOTER. The results for the S800 are given in Table 3, which compares the results for trial and error deduced terms (labeled as "old") with those deduced from COSY. It should be pointed out that COSY was only run to fifth order and that the results for momentum resolution *vis-a-vis* scattering angle resolution depend on the weighting function and the choice of initial rays, so small differences in achieved resolutions are not significant.

The next step in the evolution of spectrograph design is to skip the raytracing procedure altogether and do all the calculations with the DA method. This is now possible with COSY INFINITY [9]. Detector resolution assumptions are built into the calcula-

Table 3.

S800 spectrograph resolutions

	No gradients in dipoles				Gradients			
	$\Omega=20,\delta=.1$		$\Omega=10,\delta=2.5$		$\Omega=20,\delta=.1$		$\Omega=10,\delta=2.5$	
	$\Delta p/p$	$\Delta\theta$	$\Delta p/p$	$\Delta\theta$	$\Delta p/p$	$\Delta\theta$	$\Delta p/p$	$\Delta\theta$
Old	0.49	0.53	0.46	0.50	0.49	0.51	0.46	0.51
Old+COSY	0.49	0.53	0.45	0.50	0.49	0.51	0.45	0.50
COSY only	0.50	0.62	0.45	0.63	0.50	0.61	0.45	0.64

Units are $\Delta p/p = 1 \times 10^{-4}$, $\Delta\theta =$ mr, $\Omega =$ msr, $\delta = \pm\%$

Table 4. The resolution of the S800 for various correction orders. The resolutions are based on detector resolutions of 0.4 mm and 0.3 mrad FWHM (from ref [9]).

	Energy Resolution $E/\Delta E$	Angle Resolution $mrad$
Linear Resolution	9,400	-
Uncorrected Nonlinear	100	40
Order 2 Reconstruction	1,500	6.7
Order 3 Reconstruction	6,300	1.7
Order 4 Reconstruction	8,700	1.4
Order 5 Reconstruction	9,100	0.8
Order 8 Reconstruction	9,100	0.8

tions so accurate assessments can be made. In contrast to the raytracing calculations the forward going map is inverted to give the coordinates at the target, which is just what is needed for a spectrograph operating in the energy loss mode. Further details are given in [9]. Table 4, from [9] shows the results as a function of the order of the calculation. It should be noted that the solid angle and momentum acceptance is the extremum of the proposed operating range, and that the dipoles are assumed to be as wide as needed to accept the particles. In these calculations the fringe field has been assumed to have a simple analytical shape, but the calculations will be repeated for the mapped fields as they become available.

4. Field mapping and status

The success of a software corrected spectrometer depends on the quality of the magnetic field measurements, since both raytracing and matrix techniques require field maps or an analytic description of the field. In particular, the shape of the fringe fields must be well described because of their contributions to the multipole content of the system [10]. To successfully use the DA method the higher order derivatives of the system must also be known. The dipole field mapping procedure adopted for the S800 consists of search

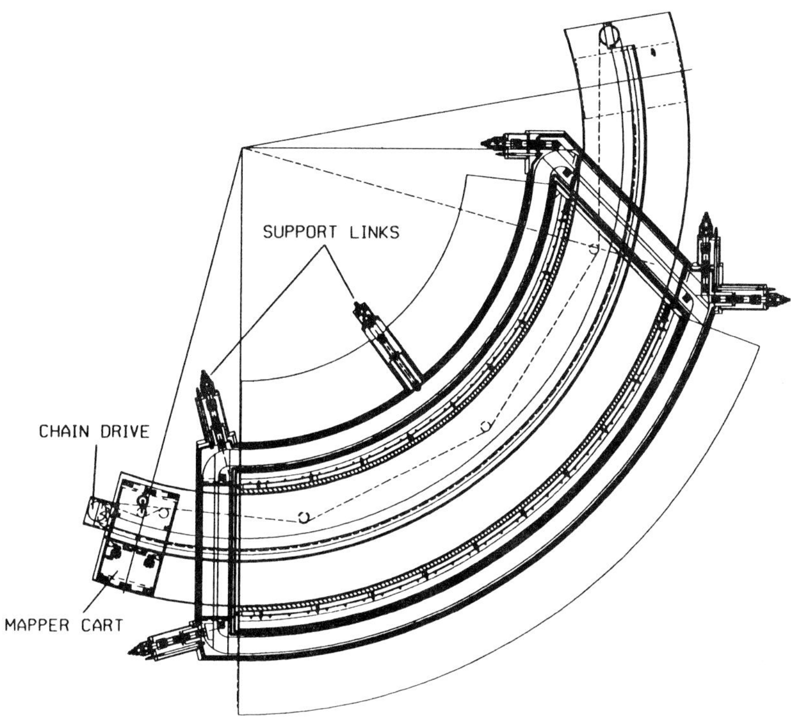

Figure 9. Field mapping system for the dipoles.

coils, which register the changes in the magnetic field, relative to an absolute measure of the field with an NMR. It is similar to the system used to map the magnetic field of the K1200 superconducting cyclotron [11]. The mapping was so successful that new beams can be readily extracted by setting the calculated values for the 28 trim coils and like number of magnetic extraction elements. The complication of field measuring on the S800 dipoles is that the magnets are designed to operate only in a vertical orientation, so system has been designed to track through the magnet along the arc. The system is illustrated in Fig 9. Because the search coil method is inherently a differential method, we will have derivatives of the field as well as the field itself. We plan to make several measurements off the median plane to get derivatives in the Y direction.

By minimizing the dipole mass with a sector magnet, we have required the construction of a superconducting coil having one negative curvature side. This presents complications in the construction process, since each turn must be individually placed in its insulated channel. The winding process for the negative curvature side is shown in Fig 10. Further details of the construction are given elsewhere [12].

The use of DA methods in the design of software corrected spectrometers will significantly speed up the process of verifying that the proposed design meets the resolution specifications when different assumptions of focal plane detector resolutions are made. Additionally, if the magnetic field measurements are of sufficient quality then the same methods are applicable for determining the desired quantities of actual particles in the

Figure 10. Winding the negative curvature side of the dipole.

spectrometer. This will alleviate the necessity of doing calibration runs at the start of each experiment. We plan on using this method when the S800 spectrometer is completed within the next three years.

References

[1] Nolen J A, Zeller A F, Sherrill B, Dekamp J C and Yurkon J 1989 *Technical Report MSUCL-694*, National Superconducting Cyclotron Lab

[2] Zeller A F, Nolen J A, Harwood L H and Kashy E 1982 *Workshop on High Resolution, Large Acceptance Spectrometers*, ANL/PHY-81-2, Argonne National Lab

[3] Conceptual design report - basic experimental equipment, Technical report, CEBAF April 13, 1990

[4] Carey D C 1978 *NAL-64*, Fermi National Accelerator Lab report

[5] Kowalski S and Enge H A 1987 *Nucl. Instrum. Methods* **A258** 407

[6] Nolen J A 1982 *Workshop on High Resolution, Large Acceptance Spectrometers*, ANL/PHY-81-2, Argonne National Lab

[7] Thiessen H A and Klein M 1972 *Proc. IV Inter. Conf. on Magnet Technology*

[8] Berz M 1990 *Nucl. Instrum. Methods* **A298** 426

[9] Berz M, Joh K, Nolen J A, Sherrill B M and Zeller A F 1992, *Phys. Rev.* C, in press

[10] Neri F 1991 *Proc. PILAC Optics Workshop*, ed Thiessen H A, LA-UR-91-3436, Los Alamos National Lab

[11] Harwood L H and Nolen J A 1984 *Proc Tenth Inter. Conf on Cyclotrons and their Applications* ed Marti F, p 101

[12] Zeller A F, Nolen J A, Swanson R, DeKamp J C and Laumer H 1992 *Advances in Cryogenic Engineering* **37** 417

Alternating-Phase Focusing: A Model to Study Nonlinear Dynamics

L Sagalovsky and J R Delayen

Argonne National Laboratory, 9700 S Cass Avenue, Argonne, IL 60439, USA

Abstract. We discuss a new model to study alternating-phase focusing (APF). Our approach is based on representing the accelerating electric field with a continuous phase modulated traveling wave. The resulting nonlinear equations of motion can be solved analytically to predict the regions of stable APF motion. We also identify the key parameters which adequately describe the physics of APF. The model is believed to be applicable to low-β ion linacs with short independently-controlled superconducting cavities being developed at ANL.

1. Introduction

The basic idea of APF is to achieve stable beam transport in both axial and radial planes of motion by alternating the sign of equilibrium phase of the accelerating electric field. The advantages of realizing a three-dimensional beam focusing without the use of solenoids or quadrupole magnets must be weighed against the compromises in longitudinal acceptance one is forced to make. Previous works in the field [1, 2, 3, 4, 5] addressed APF in the context of a linac with a discrete number of accelerating gaps spaced in a predetermined manner to achieve a particular value of the synchronous phase in each gap (such as the case in the π-mode Wideroe linac and the Alvarez DTL). In contrast, the application we have in mind is a linac with superconducting accelerating cavities of the type described in [6]. These low-β cavities are short, can be independently controlled in adjusting both the phase and the amplitude of the electric field, and were shown to produce very high accelerating gradients [7]. The model presented in this paper is thought to be a good description of essential APF physics for the superconducting linacs and a starting point in trying to determine the practical limits of APF.

2. Beam dynamics in APF linacs

2.1. Analytical model and assumptions

We assume that the electric field is described by a cylindrically symmetric traveling wave with a continuous phase modulation. Here, we choose the modulation to be sinusoidal:

$$E_z(r, z; t) = E_0 \cos \left[\omega t - \int_0^z k(z') dz' + \phi_0 + \phi_1 \sin \left(2\pi z/\Lambda \right) \right], \tag{1}$$

where ω is the angular velocity and k is the wave number of the rf field, ϕ_0 is the equilibrium phase in the absence of APF, and Λ and ϕ_1 are the APF period and phase modulation amplitude respectively. For the central reference trajectory z_c, we choose

$$\omega t - \int_0^{z_c} k(z')dz' = 0 \tag{2}$$

In subsequent analysis, we will neglect the effect of the velocity change in one APF period; the reference particle is assumed to travel with a constant β.

We can compute the average accelerating gradient by integrating eq. 1 for the reference trajectory:

$$< E >= \frac{E_0}{\Lambda} \int_0^{\Lambda} \cos\left[\phi_0 + \phi_1 \sin\left(2\pi z/\Lambda\right)\right] \, dz = E_0 \cos\phi_0 J_0(\phi_1) \tag{3}$$

where J_0 is a Bessel function.

2.2. Equations of motion

The equations of motion are

$$\frac{d^2 z}{dt^2} = \frac{q}{m} E_z(r, z; t), \tag{4}$$

$$\frac{d^2 r}{dt^2} = \frac{q}{m} E_r(r, z; t), \tag{5}$$

with E_z given by eq. 1 and E_r given to the first order in r by Maxwell's equations,

$$E_r(r, z; t) = -\frac{r}{2} \frac{\partial E_z}{\partial z}. \tag{6}$$

For an arbitrary longitudinal deviation from the reference trajectory given by $\Delta z = z - z_c$, the equation of motion becomes

$$\frac{d^2 \Delta z}{dt^2} = \frac{q E_0}{m} \left\{ \cos\left[\phi_0 - k\Delta z + \phi_1 \sin\left(2\pi(z_c + \Delta z)/\Lambda\right)\right] - \cos\left[\phi_0 + \phi_1 \sin\left(2\pi z_c/\Lambda\right)\right] \right\}. \tag{7}$$

We will first look at the APF linear motion.

2.3. Linear stability

Let us define some dimensionless parameters which we are going to use throughout this paper,

$$\Delta\phi \equiv -k\Delta z = -2\pi \frac{\Delta z}{\beta\lambda}, \tag{8}$$

$$\tau \equiv \frac{z_c}{\Lambda} = \frac{\beta c t}{\Lambda}, \tag{9}$$

$$\nu \equiv \frac{\Lambda}{\beta\lambda}, \tag{10}$$

$$\eta \equiv \frac{q E_0 \beta\lambda}{\frac{1}{2}mv^2}. \tag{11}$$

The linearized equations of motion are

$$\frac{d^2\Delta\phi}{d\tau^2} + \pi\eta\nu\left[(-\nu + \phi_1\cos 2\pi\tau)\sin(\phi_0 + \phi_1\sin 2\pi\tau)\right]\Delta\phi = 0, \qquad (12)$$

$$\frac{d^2 r}{d\tau^2} - \frac{\pi}{2}\eta\nu\left[(-\nu + \phi_1\cos 2\pi\tau)\sin(\phi_0 + \phi_1\sin 2\pi\tau)\right]\Delta r = 0. \qquad (13)$$

By expanding the linear coefficients in eq. 13 in a Fourier series, we get the familiar Mathieu-Hill equations:

$$\frac{d^2\Delta\phi}{d\tau^2} - 2\left[B + \sum_{n=1}^{\infty} C_n\sin(2\pi n\tau + \theta_n)\right]\Delta\phi = 0, \qquad (14)$$

$$\frac{d^2 r}{d\tau^2} + \left[B + \sum_{n=1}^{\infty} C_n\sin(2\pi n\tau + \theta_n)\right]r = 0, \qquad (15)$$

where

$$B = \frac{\pi}{2}\eta\nu^2 J_0(\phi_1)\sin\phi_0, \qquad (16)$$

$$C_n = -\pi\eta\nu|J_n(\phi_1)|\begin{cases}\sqrt{\nu^2\cos^2\phi_0 + n^2\sin^2\phi_0} & \text{if } n \text{ odd}\\[2mm]\sqrt{\nu^2\sin^2\phi_0 + n^2\cos^2\phi_0} & \text{if } n \text{ even}\end{cases}, \qquad (17)$$

$$\tan\theta_n = -\tan\phi_0\begin{cases}\frac{n}{\nu} & \text{if } n \text{ odd}\\[2mm]\frac{\nu}{n} & \text{if } n \text{ even}\end{cases}. \qquad (18)$$

The equations are analogous to those obtained in ref. [5] using a discrete thin-lens approximation and a standing wave approach. Here, however, the beam dynamics variables B and C_n depend only on four independent parameters: ϕ_0, ϕ_1, ν, and η; moreover, the dependence is given in an explicit analytic form. Keeping just the $n = 1$ term, we obtain a well-known Mathieu equation for which we can compute the stable region boundaries. Fig. 1 shows the linear stable region. Fig. 2 shows stability boundaries

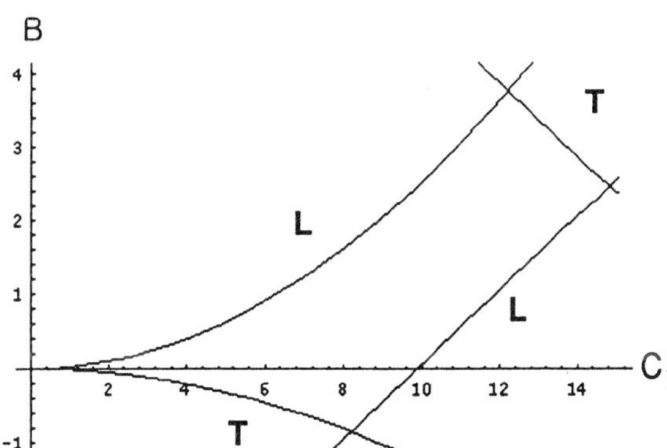

Figure 1. Transverse (**T**) and longitudinal (**L**) boundaries for the linear stability region.

in the $\phi_1 - \nu$ space for APF phase advances of less than 90° and $\phi_0 = 5°$, $\eta = 0.05$; fig. 3 shows the effect of increasing the "acceleration parameter" η to 0.25.

We next turn our attention to the nonlinear problem of calculating the longitudinal acceptance for the APF linac.

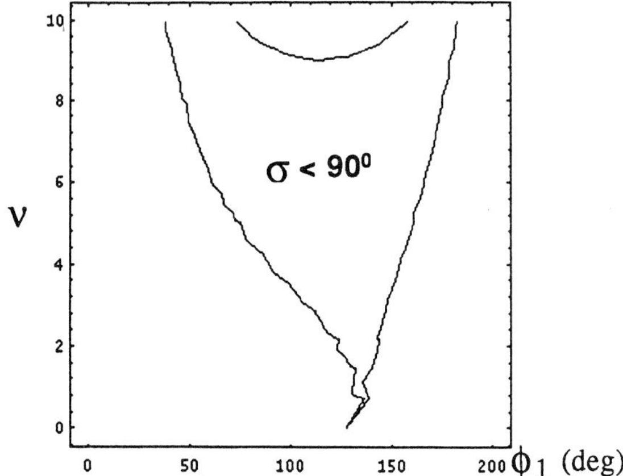

Figure 2. Stability boundaries for trajectories not exceeding 90° in either transverse or longitudinal phase advance with $\phi_0 = 5°$ and $\eta = 0.05$.

2.4. Longitudinal acceptance

The full nonlinear equation of longitudinal motion is given by

$$\frac{d^2\Delta\phi}{d\tau^2} = -\pi\eta\nu^2 \left\{ \cos\left[\phi_0 + \Delta\phi + \phi_1 \sin\left(2\pi\tau - \frac{\Delta\phi}{\nu}\right)\right] - \cos\left[\phi_0 + \phi_1 \sin 2\pi\tau\right]\right\}. \quad (19)$$

We can find the effective potential for eq. 19 by using the averaging method given in ref. [8] and applied to the problem of longitudinal acceptance in ref. [5]. We review the method here for completeness.

Consider an equation of the form

$$\frac{d^2x}{dt^2} = -\frac{\partial U_0}{\partial x} + f, \quad (20)$$

where

$$f = \sum_{n=1}^{\infty} \left[u_n \sin\left(n\Omega t\right) + v_n \cos\left(n\Omega t\right)\right]. \quad (21)$$

If the period of the motion caused only by the potential U_0 is T and $\Omega \gg 1/T$, the particle motion can be described by

$$x(t) = X(t) + \xi(t), \quad (22)$$

where $X(t)$ and $\xi(t)$ are caused by the slowly varying potential U_0 and the rapidly oscillating force f respectively. In this case, eq. 21 can be averaged to yield

$$\frac{d^2X}{dt^2} = -\frac{\partial U_{\text{eff}}}{\partial X}, \quad (23)$$

where the effective potential U_{eff} is given by

$$U_{\text{eff}} = U_0 + \frac{1}{4\Omega^2} \sum_{n=1}^{\infty} \frac{u_n^2 + v_n^2}{n^2}. \quad (24)$$

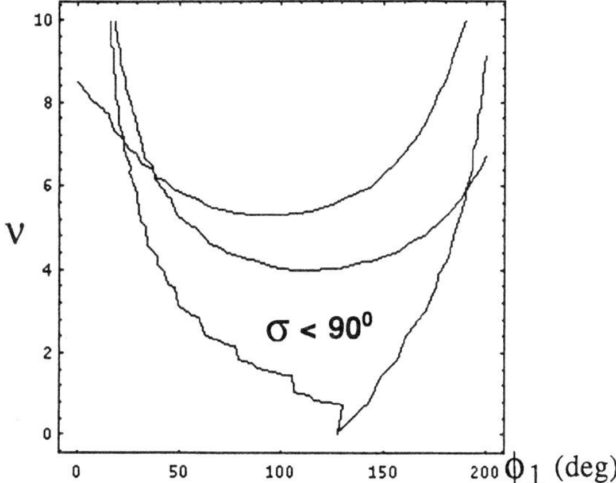

Figure 3. Stability boundaries for trajectories not exceeding 90° in either transverse or longitudinal phase advance with $\phi_0 = 5°$ and $\eta = 0.25$.

In the problem at hand, eq. 19 can be transformed to the canonical form of eq. 21 by using well-known Fourier expansions:

$$\cos\left(x\sin\theta\right) = J_0(x) + 2\sum_{n=1}^{\infty} J_{2n}(x)\cos(2n\theta); \tag{25}$$

$$\sin\left(x\sin\theta\right) = 2\sum_{n=1}^{\infty} J_{2n-1}(x)\sin\left((2n-1)\theta\right). \tag{26}$$

We find the effective potential to be given by

$$U_{\text{eff}} = U_0 + \sum_{n=1}^{\infty} U_n \tag{27}$$

with

$$U_0 = \pi\eta\nu^2 J_0(\phi_1)\left[\sin\left(\phi_0 + \Delta\phi\right) - \Delta\phi\cos\phi_0 - \sin\phi_0\right], \tag{28}$$

$$U_n = \left(\frac{\eta\nu^2}{2}\right)^2 J_n^2(\phi_1)\frac{S_n}{n^2}, \tag{29}$$

where

$$S_n = \begin{cases} \sin^2\left(\phi_0 + \Delta\phi\right) + \sin^2\phi_0 - 2\sin\left(\phi_0 + \Delta\phi\right)\sin\phi_0\cos\left(\frac{n}{\nu}\Delta\phi\right) & \text{if } n \text{ odd} \\ \cos^2\left(\phi_0 + \Delta\phi\right) + \cos^2\phi_0 - 2\cos\left(\phi_0 + \Delta\phi\right)\cos\phi_0\cos\left(\frac{n}{\nu}\Delta\phi\right) & \text{if } n \text{ even} \end{cases}. \tag{30}$$

Checking the validity of the assumption that $\Omega \gg 1/T$, we find $\Omega = 2\pi$, $2\pi/T = \sqrt{|\pi\eta\nu^2 J_0(\phi_1)\sin\phi_0|}$, and the assumption is satisfied if $\eta\nu^2|J_0(\phi_1)\sin\phi_0| \ll 16\pi^3$. For any practical application, the requirement is $\nu \ll 150$.

Given the effective potential, we can calculate the equation for the separatrix in the $(\Delta\phi, \Delta W/W)$ space and the total longitudinal acceptance. The separatrix is given by

$$\frac{\Delta W}{W} = \pm\frac{1}{\pi\nu}\sqrt{2\left[\Delta U - U_{\text{eff}}(\Delta\phi)\right]}, \tag{31}$$

where

$$\Delta U = U_{\text{eff}}(\Delta\phi_c) \tag{32}$$

and $\Delta\phi_c$ is the unstable fixed point of the motion given by

$$\left.\frac{\partial U_{\text{eff}}}{\partial\Delta\phi}\right|_{\Delta\phi=\Delta\phi_c} = 0, \quad \left.\frac{\partial^2 U_{\text{eff}}}{\partial^2\Delta\phi}\right|_{\Delta\phi=\Delta\phi_c} < 0. \tag{33}$$

Fig. 4 illustrates the relationship between the potential well $U_{\text{eff}}(\Delta\phi)$ and the stability boundaries in the $(\Delta\phi, \Delta W/W)$ plane of motion. The width of the separatrix Ψ is simply

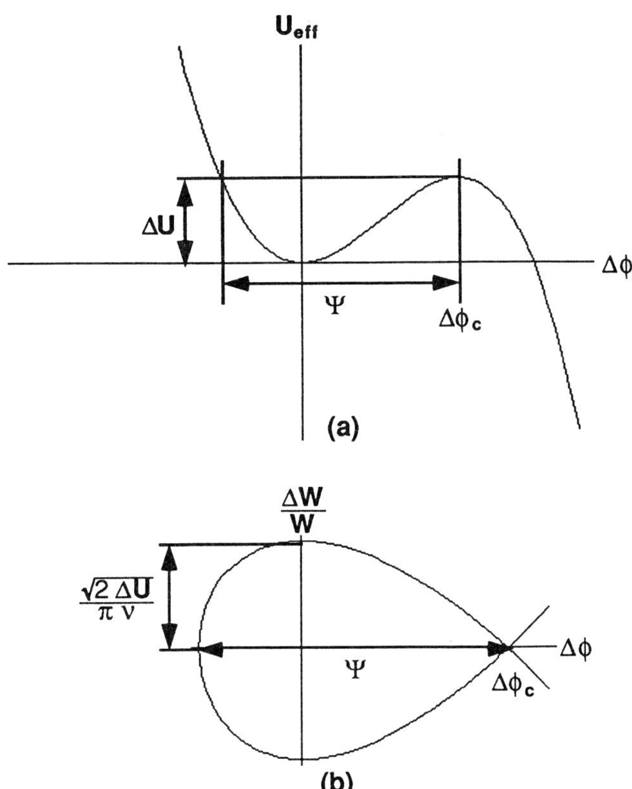

Figure 4. Relationship between (a) the effective potential $U_{\text{eff}}(\Delta\phi)$ and (b) the stable region in the $(\Delta\phi, \Delta W/W)$ phase space.

the distance between the values of $\Delta\phi$ at which $U_{\text{eff}}(\Delta\phi) = \Delta U$ (cf. fig. 4). The height of the separatrix is given by

$$\left(\frac{\Delta W}{W}\right)_{\text{max}} = \frac{\sqrt{2\Delta U}}{\pi\nu}. \tag{34}$$

The area of the stability region in the $(\Delta\phi, \Delta W/W)$ phase space (longitudinal acceptance) is

$$\alpha_L = 2\left(\frac{\Delta W}{W}\right)_{\text{max}} \int_{\Delta\phi_c-\Psi}^{\Delta\phi_c} \sqrt{1 - \frac{U_{\text{eff}}}{\Delta U}}\, d\left(\Delta\phi\right). \tag{35}$$

Below we give an explicit solution for α_L accurate to the second order in $\Delta\phi$, i. e. we consider terms up to $O\left(\Delta\phi^3\right)$ only in the Taylor expansion of U_{eff}.

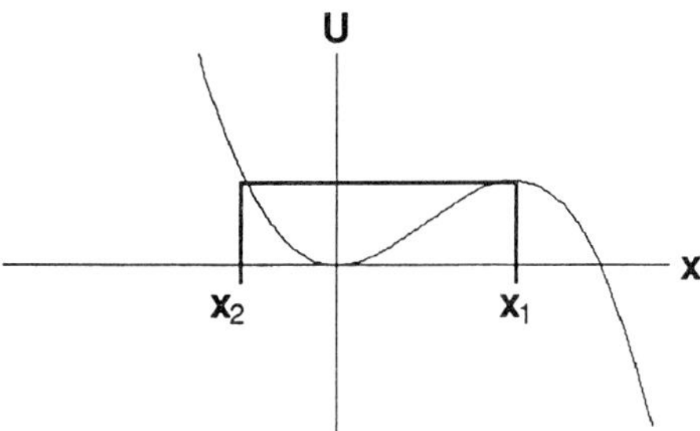

Figure 5. A general potential $U(x)$ described by a cubic polynomial in x.

2.4.1. Second-order solution Consider a potential function $U(x)$ described by

$$U(x) = \frac{a}{2}x^2 - \frac{b}{3}x^3 \tag{36}$$

and the stable motion confined to $x_2 < x < x_1$ as shown in fig. 5. The turning points x_1, x_2, potential well depth $U(x_1)$, and the area of the stable (x, x') region are calculated to be

$$x_1 = \frac{a}{b}, \tag{37}$$

$$x_2 = -\frac{1}{2}\frac{a}{b}, \tag{38}$$

$$U(x_1) = \frac{1}{6}\frac{a^3}{b^2}, \tag{39}$$

$$\alpha = 2\int_{x_2}^{x_1}\sqrt{2\left[U(x_1) - U(x)\right]} = \frac{6}{5}\frac{a^{5/2}}{b^2}. \tag{40}$$

In the problem at hand, the effective potential U_{eff} given in eq. 27 can be Taylor expanded to $O\left(\Delta\phi^3\right)$ to yield

$$U_{\text{eff}}(\Delta\phi) = \frac{a}{2}\Delta\phi^2 - \frac{b}{3}\Delta\phi^3 + \cdots, \tag{41}$$

where a is the square of the linear motion's phase advance σ_L,

$$a = \sigma_L^2 = 2B + \frac{1}{2\pi^2}\sum_{n=1}^{\infty}\left(\frac{C_n}{n}\right)^2 \tag{42}$$

and b is given by

$$b = \frac{\pi}{2}\eta\nu^2 J_0(\phi_1)\cos\phi_0 + \frac{3}{8}\eta^2\nu^2\sin 2\phi_0\sum_{n=1}^{\infty}(-1)^n J_n^2(\phi_1)\left(1 - \frac{\nu^2}{n^2}\right). \tag{43}$$

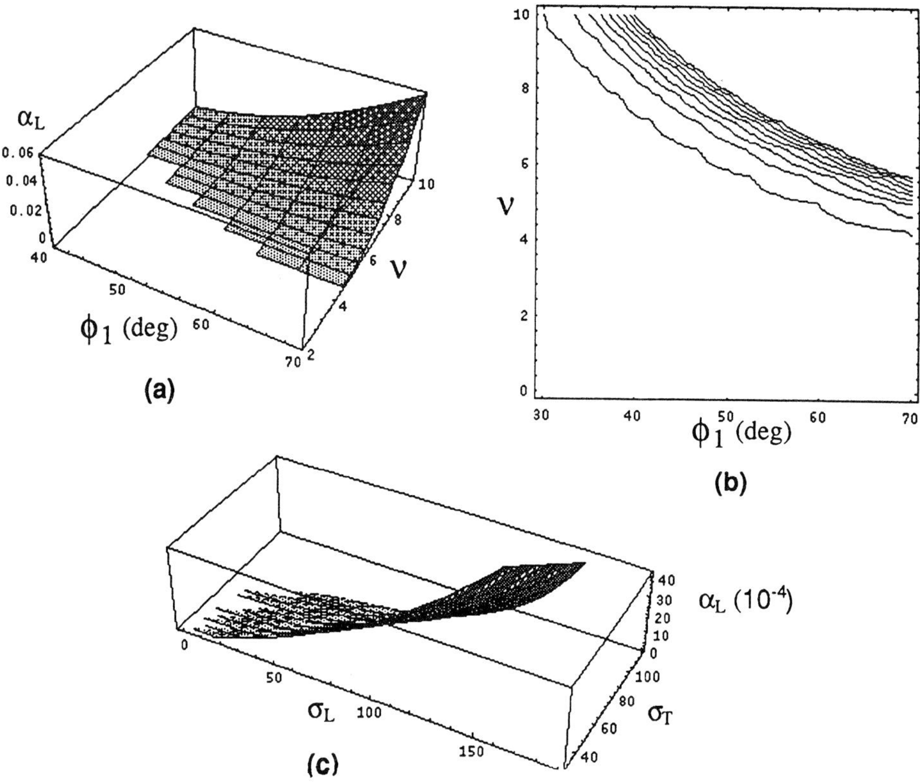

Figure 6. Plots of the longitudinal acceptance α_L for $\phi_0 = 5°$, $\eta = 0.1$. (a) Plot of α_L as a function of ϕ_1 and ν. (b) Contour plot representation of (a). (c) Plot of α_L as a function of σ_L and σ_T, the longitudinal and transverse phase advances respectively.

Then, the width of the separatrix Ψ and the $(\Delta\phi, \Delta W/W)$ acceptance α_L are given by

$$\Psi = \frac{3}{2}\frac{a}{b}, \tag{44}$$

$$\alpha_L = \frac{6}{5\pi\nu}\frac{a^{5/2}}{b^2}. \tag{45}$$

Fig. 6 shows the results of acceptance calculations for $\phi_0 = 5°$, $\eta = 0.1$ using eq. 45 and keeping just the $n = 1$ term in eqs. 42, 43. Computer simulations indicate that for most practical cases the second-order acceptance approximation is accurate with an error of less than 10%.

3. Conclusions and future work

The model of the traveling wave with continuous phase modulation presented in this paper gives quantitative predictions to the problem of longitudinal stability in APF linacs. The model describes the physics of APF with four parameters and yields analytic solutions for the effective potential and the acceptance for the longitudinal motion to any order in $\Delta\phi$.

Future work on the model will include investigations of practical limits in linacs with independent superconducting cavities, space-charge current limits, and ways to improve the acceptance by modulating both the phase and the amplitude of the accelerating field.

4. Acknowledgments

This work was supported by the U. S. Army Strategic Defense Command under the auspices of the U. S. Department of Energy.

References

[1] Kushin V V 1970 *Atomnaya Energiya* **29** (3)
    Kushin V V and Mokhov V M 1973 *Atomnaya Energiya* **35** (3)

[2] Beley A S, Kaplin S S, Knizhnyak N A and Shulika N G 1981 *Zh. Tekh. Fiz* **51** 656

[3] Baev V K and Minaev S A 1981 *Zh. Tekh. Fiz* **51** 2310
    Baev V K, Gavrilov N M, Minaev S A and Shal'nov A V 1983 *Zh. Tekh. Fiz.* **53** 1287

[4] Garashchenko F G, Kushin V V, Plotnikov S V, Solokov L S, Strashnov I V, Fedotov P A, Kharchenko I I and Tsulaya A V 1982 *Zh. Tekh. Fiz.* **52** 460

[5] Okamoto H 1989 *Nucl. Instrum. Methods* **A284** 233–247

[6] Delayen J R 1987 *Nucl. Instrum. Methods* **A259** 341–357
    Delayen J R, Bohn C L and Roche C T 1989 *Nucl. Instrum. Methods* **295** 1–4

[7] Delayen J R, Bohn C L, Kennedy W L, Nichols G L, Roche C T and Sagalovsky L 1991 *Proceedings of the 5th Workshop on RF Superconductivity*, DESY

[8] Kapitsa P L 1951 *Zh. Eksper. Teor. Fiz.* **21** 588

Review of the Dynamic Aperture Experiment at the CERN SPS

J. Gareyte, W. Scandale, F. Schmidt

CERN, CH-1211 Geneva 23, Switzerland

1. Introduction

Since 1986 experiments are performed in the CERN SPS to study the beam dynamics in presence of strong non-linear fields intentionally introduced along the accelerator circumference. They are motivated by the need to refine the aperture and field quality criteria for the design of future large hadron accelerators like LHC, SSC, or RHIC. The experimental procedure consists in exciting already existing sextupoles in order to introduce in a controlled fashion non-linearities in an otherwise linear lattice. To probe large amplitudes, we give a large coherent deflection to a pencil beam of small emittance and momentum spread, or we create a large emittance beam with repetitive small amplitude coherent kicks. In the first case, a few hundred turns are sufficient to create a 'hollow' distribution of charges around the central orbit due to non-linear filamentation, in the second case, few seconds are required to spread-out the particles all around the available physical aperture of the accelerator. The beam behaviour is observed with several instruments: current transformers record lifetime, Schottky noise detectors give tune and tune-spread, flying wires provide transverse profile, and orthogonal pairs of position monitors are able to produce a phase space portrait. More recently, sinusoidal and random tune modulations have been added to simulate the effect on the beam stability of the unavoidable ripples in power supplies of the guiding magnets and the speed of the induced diffusion process has been measured using collimators and loss detectors. A review of the experimental results and of their interpretation is presented below.

2. Experimental Conditions

Well understood and clear experimental conditions, with most of the spurious effects eliminated and the phenomena under study carefully isolated, are required in order to make meaningful comparisons between the experimental results and their numerical simulations with computer tracking programs. For these reasons the experiments are performed at an energy where the SPS is very linear. Between 100 and 250 GeV/c the remanent fields effects can be neglected, as are the space charge effects, whilst the saturation effects in the magnets are not yet perceptible. An energy of 120 GeV/c is finally selected to benefit from the maximum strength of the added non-linear fields. Beam intensities not in excess of $(1-2) \times 10^{12}$ protons allow to minimise the resistive wall instabilities. The chromaticities are corrected using four families of sextupoles in order to minimise the tune variation with momentum deviation. Intentional non-linearities are provided by powering independently eight sextupoles foreseen for the resonant extraction of the beam at 450 GeV/c. Currents of 140 A are routinely used which correspond to sextupolar strengths at 120 GeV/c of:

$$B''\ell/B\rho = 0.28m^{-2}$$

The sign of the sextupole gradients alternates around the ring in such a way as to suppress the third integer resonance - except in an early experiment where that resonance was strongly excited - whilst the chromaticities on central orbit are kept unchanged. It has been soon discovered that residual closed orbit deviations and linear coupling complicate the understanding of the experiments. Currently the closed orbit is corrected down to an r.m.s value of less that 0.5 mm and the linear coupling down to a minimum tune approach of 0.005. Most of the experiments are performed with RF off and debunched beams.

3. Short Term Dynamic Aperture Experiment [1, 2]

Studies of the short term dynamic aperture were started in 1986 while the SPS was operating in fixed target mode. The beam was injected at 14 GeV/c, accelerated up to 120 GeV/c and allowed to circulate at this energy for 8 sec on a magnetic flat top. The full ring was filled with bunches and the RF was kept on. At the beginning of the 120 GeV/c flat top intentional sextupoles were energised and the beam emittance increased until losses occurred. This was done by repetitively deflecting the beam with the fast kicker normally used to measure the tunes. In absence of intentional non-linearities, it was sufficient to pulse the kicker twice at its maximum deflection of 5.9 mm at a β of 100m to reach the machine aperture in the vertical plane, which was of the order of $\pm 15mm$. In these conditions the intensity losses occur abruptly. When the sextupoles were energised, the acceptance was reduced. The amplitude of the two kicks was therefore lowered as to produce losses not in excess of 10 to 20% of the circulating intensity, which were starting just after the second kick and lasting for a few seconds. Following that, in all cases there was another period of a few seconds before the end of the flat top with no visible losses. The size of the surviving beam measured with a high gain of the wire scanner was assumed to be the dynamic aperture of the machine. In the first experiments the distribution of sextupoles was such as to excite the 1/3 order resonance, while this resonance was suppressed in the later ones, in which several working points around the fifth order resonances were explored and an horizontal dynamic aperture of 13 to 21 mm was measured. Once the synchrotron motion and closed orbit distortions were taken into account, computer tracking simulations performed over up to 10^5 turns, i.e. a few seconds of real machine time, were able to provide estimates of the dynamic aperture in reasonable agreement with measured values. The loss mechanism was clearly related to the crossing of the fifth order resonances, whilst the upper limit of the detuning and the smear of stable trajectories were found to be of the order of 0.01 units and 0.035 respectively. In 1988, another experimental session was devoted to the study of the short term stability, with the same machine conditions as in 1986. At that time, an instrument was made available to record the beam oscillation over several thousand turns in two position monitors separated by about $\pi/2$ in betatron phase. Using it, phase portraits could be drawn and tunes could be precisely measured both in pulsed and in storage mode [3]. To explore the effect of the intentional non-linearities, a pencil beam of 1.3 mm r.m.s. size was used, which was obtained by scraping the larger injected beam on a collimator. Hollow beams of amplitudes up to 17 mm at $\beta = 100m$ could be produced by firing once one module of the fast kickers normally used for fast ejection at 450 GeV/c. Two working points were explored, one near the fifth and the other near the seventh order resonances. The detuning was measured as a function of the oscillation amplitude, proportional to the kick voltage, and compared to the numerical simulations. The agreement was found to be excellent, as in Fig. 1 where the observations near the fifth order resonances are reported. The dynamic aperture was measured by adding the amplitude of the coherent oscillation to twice the r.m.s. beam size for which losses were just visible. For comparison, the boundary between regions of regular and chaotic motion was evaluated by numerical simulations. For this two particles with slightly different initial conditions were tracked with the program SIXTRACK [4] and their distance in phase space recorded as a function of time. From the evolution in time of this distance the boundary between regions with stable and chaotic motion could be obtained, a linear increase corresponding to regular motion, and an exponentially growth corresponding to chaotic motion. The agreement between experimental data and simulations was excellent, as for instance for the working point near the fifth order resonances: the measured aperture was between 15.8 and 20.2 mm, the numerical estimate was 17.6 mm. With the previous experiments the concept of short term dynamic aperture was clarified to a large extent. In presence of strong non-linearities particles can be detuned onto high order resonances and lost after multiple crossings. In these conditions, the Lyapunov coefficient, easily obtainable by numerical simulations, is a good predictive indicator of the chaoticity of the trajectories and of the instability of the motion.

4. Sextupole Induced Diffusion [2]

In the 1988 experiment, diffusion induced by sextupoles was also investigated following earlier experience at the $Sp\bar{p}S$, which had revealed a transverse diffusive behaviour in presence of localised beam-beam collisions. In the context of the studies for the design of large superconducting hadron colliders, an important issue is whether the unavoidable non-linearities in the guide fields are likely to create a diffusion mechanism similar to that induced by beam-beam. In both cases non-linearities create tune-spread and resonance excitation, although with quite a different dependence with particle amplitude. In this experiment the beam was stored at 120 GeV/c with a working point near the fifth order resonances, and the beam emittance was slowly increased by repeatedly firing the Q-kicker, until a few percent losses were noticed. The eight sextupoles were powered as to avoid excitation of the third order resonance. In general, the RF was switched off and the beam was debunched. The beam intensity was recorded, see Fig. 2, while certain actions were performed. When a scraper was introduced at an amplitude of 15.4 mm, about 1% of the circulating intensity was lost. Subsequently the 1/e decay time stabilised at about 40'. When the scraper was retracted by 2.7 mm, the intensity remained constant for 30 seconds, which is the time it took for the particles to fill the gap, and then decayed with characteristic time of 65'. When the scraper was moved back at its initial position, 1% of the beam was lost again and the decay time became 36', a value close to the initial

40'. We see here a clear signature of a diffusion process. The diffusion rate evaluated in this way was 3 mm/minute at an amplitude of 12.6 mm and 6 mm/minute at 15.4 mm. At 18.1 mm the diffusion was so fast that it could not be measured with the procedure described above: this value corresponds to the short-term dynamic aperture (between 15.8 and 20.2 mm). Similar measurements were done in the working point near the seventh order resonances. In this case the largest amplitude at which no significant diffusion could be detected was 9.2 mm. The measurement was repeated in absence of the added sextupoles. There was no sign of diffusion up to an amplitude of 22 mm, well outside the dynamic aperture measured in the experiment. Numerical simulations performed for up to 10^6 turns did not reveal any chaotic motion inside the short-term dynamic aperture. This changed dramatically when a tune modulation of 3×10^{-3} units was introduced in the tracking to take into account the realistic situation of the main power supplies optimised at that time for the operation in fixed target mode. The value of the Lyapunov coefficient evaluated by tracking varies in this case with the particle amplitude, in qualitative agreement with the experimental observations of a strongly amplitude dependent diffusion rate. This is to be expected from the application of the Chirikov-Courant criterion for overlap of satellite resonances. In view of evaluating the predictive power of the detuning and smear as indicators of stable motion, Fig. 3 was drawn which included all the experimental situations considered in 1986 and 1988. The rectangular box represents the criteria used in 1988 for acceptance of the LHC lattice: it was supposed that in this rectangle the machine would be sufficiently linear to assure a good, comfortable operation. It appeared clear from the experimental data that the LHC criteria were sufficiently conservative as far as the short-term dynamic aperture was concerned, however they were certainly insufficient to ensure a good lifetime in storage mode, due to the magnetic imperfection induced diffusion.

5. Influence of a Sinusoidal modulation of the Tune On the Dynamic Aperture [5, 6]

In 1989, intensive tracking studies were performed in order to estimate the effect of a sinusoidal tune modulation on the dynamic aperture. The parameters of the modulation were selected having in mind the shape of the power spectrum of the power supplies ripple observed in the 1988 experiment: the modulation amplitude was varied from zero to 3×10^{-3} units in tune, and the frequency between a few Hz and 600 Hz. Using the SPS model [7], the value of the stability border evaluated from Lyapunov coefficients was found to be strongly dependent on the frequency of the tune modulation, see Fig. 4. However, the particles with an amplitude range between the short term dynamic aperture and the stability border, although chaotic, show no important increase in amplitude in the time range accessible to tracking with the full SPS lattice. To speed up simulation, a simple 4-D Henon map with tune modulation was studied. In Fig. 5 the survival plot for 9 Hz and 200 Hz modulation frequency is shown and compared to the situation with no modulation. The stability border is strongly affected by tune modulation, the lowest frequency having the strongest effect. One finds however that up to a high number of turns, of the order of 10^5, the plotted survival curves lie close to each other. It takes about 10^6 turns to see the influence of the tune modulation. After 10^7 turns the variation with frequency starts to be visible. However, even in the worst case of 9 Hz, the tracked particles still circulate after 5×10^7 turns. Note that only 2×10^4 turns are sufficient to estimate the Lyapunov coefficient in this simple case. The purpose of the experiments performed at the SPS in 1989, 1990 and 1991 was to get quantitative results which could be compared with the previous simulations. To achieve this it was necessary to provide a tune modulation with a depth larger than that of the natural power supply ripple and with variable frequency. We found using the continuous tune-measurement system [8] that, after an improvement of the main power supply control similar to that used in collider operation, the peak to peak variation of the tune due to ripple was lowered to $\pm 2.4 \times 10^{-4}$ units. Therefore an external modulation depth of $\pm 8.0 \times 10^{-4}$ tune units was chosen. In setting up the machine, care was taken to compensate the closed orbit distortions and the linear coupling. The tune was chosen very carefully to stay away from the lowest order resonances. Octupoles were used to decrease the amplitude dependent tune shift so as to better locate the working point in between the most dangerous resonances. The diffusion speed was measured by the scraper retraction method, however now scrapers were used in both horizontal and vertical planes both before and after the retraction to identify in which plane the diffusion had taken place. The measurement of the amplitude-dependent tune shift was found to be again in good agreement with tracking. Fig. 6 shows that the experimental data of 1989 are very close to that of 1988. Moreover Fig. 6 shows clearly how efficiently the detuning can be reduced by the use of octupoles. By adequately powering them in the SPS we succeeded in reducing the non-linear detuning by roughly an order of magnitude, which at the same time improved the dynamic aperture by 30%. This gave us confidence that correcting the non-linear tune shift in the LHC could be an efficient method to improve its performance. In storage mode experiments in 1989 and 1990, the beam size was increased by a single large amplitude coherent kick of 17.6 mm at $\beta = 100m$. The tune was adjusted to stay away from fifth and seventh order resonances: QH between 0.623 and 0.640, QV between 0.529 and 0.540. Measurement with no tune modulations and with modulations of 9 Hz, 40 Hz, and 180 Hz and the same depth of $\pm 8.0 \times 10^{-4}$ tune units were performed. As the scrapers were retracted from the periphery of the beam, there was the usual period with no noticeable losses. When the first scraper was reached a drop in lifetime appeared, followed by a second drop when the second scraper was reached. This is indeed what we can see in Fig. 7 for the experiment with no external modulation: the lifetime during the gap-filling time if of the order of 80 hours, the first drop appears after 5' and brings the lifetime to 14.5 hours, the second drop occurs after 14.2' and produces a lifetime of 6.3 hours. The diffusion process is dominant in the horizontal plane, since the plateau with large lifetime is only visible after the horizontal scraper retraction. Putting back the aperture limiters to their original positions produced about 3% losses with the horizontal scraper and about 1% losses with the vertical scraper. Diffusion speeds of about 0.5 mm/minute were recorded without tune modulation. With tune modulation they reached 2.5 to 5 mm/minute. The diffusion process, however, was almost independent on the modulation frequency. In a similar experiment performed in 1991, the importance of the modulation depth was revealed. By varying from 1.1×10^{-3} to 2.2×10^{-3} the amplitude of the sinusoidal modulation an increase of the diffusion speed by a factor of ten was observed for the three modulation frequencies, 9 Hz, 40 Hz, 180 Hz, which sample the spectral range of interest. This is not yet understood in a quantitative way. In the tracking studies it was found that introducing a second modulation frequency leads to a sizeable reduction of the dynamic aperture. This effect was tested in the 1991 experiment. The results are shown in Fig. 8. Initially there is no ripple and no measurable diffusion. With a tune modulation of 1.65×10^{-3} at 9 Hz the beam lifetime drops to seven hours. The losses stop immediately when the ripple is switched off. Two combined sinusoidal modulations at 9 and 180 Hz, each with a modulation amplitude of 0.8×10^{-3} produce a severe drop of the lifetime to two hours. In the two situations the same total modulation amplitude was used, however the effect on beam stability is dramatically larger with two frequencies than with only one frequency. This is still to be elucidated quantitatively. From the above results it is clear that in a collider like the LHC the total ripple amplitude which can be tolerated in the main power supplies is well below 10^{-3} tune units.

6. Numerical Studies of the Diffusion Induced by Tune Modulation

Also in 1991, a substantial effort was devoted to numerical simulations of the experiment in order to obtain a quantitative understanding of the various phenomena observed. In particular we tracked single particles up to 2.6×10^7 turns, corresponding to about 10' storage time in the SPS, with the machine conditions of the 1991 experiment. The dependence of particle stability on the modulation depth is shown in Fig. 9. The three dashed curves in the upper part of the plot indicate loss before 20 sec for the particles with zero and maximum momentum deviation respectively. The lower curves indicate the onset of chaos estimated with the Lyapunov coefficient. Both sets of curves show a decrease of the particle stability as a function of the modulation depth, however not yet in quantitative agreement with the experimental observations. The dependence on modulation frequency is shown in Fig. 10. The chaotic border for the particles with three different momentum deviations in the lower part of Fig. 10 show the decrease of the dynamic aperture with the decrease of the modulation frequency. Particle loss before 20 sec in the upper part of Fig. 10 do not show such a dependency, which is probably due to the fact that the simulation has not been performed for a large enough number of turns. On the other hand, when an additional sinusoidal modulation is added at 50 HZ with an amplitude as low as 2.5×10^{-4} tune units, to partially simulate the effect of the natural ripple of the SPS power supplies, a sizeable reduction of the chaotic border of about 4 mm is found, and the amplitudes at which particle loss takes place before 20 sec are reduced as well, see Fig. 10b. Furthermore, in these conditions, the strong dependence on frequency is levelled out and the particle loss occurs at a much lower amplitude. When the tracking is extended up to 10' losses continue to occur. The strong effect of the second frequency can be explained at least qualitatively in terms of sidebands, making use of the Chirikov-Courant criterion. In fact, the Fourier transform of the betatron motion has a wider distribution and contains more peaks,which tend to make the particle motion more chaotic.

7. Conclusion

The experiment on diffusion and dynamic aperture at the CERN SPS is well mastered, but requires a long and meticulous preparation to obtain clear machine conditions. However its interpretation is far from being complete. As far as short-term phenomena are considered, the predictive power of the Lyapunov coefficient has been well assessed. It has also been proven that the detuning is a parameter which strongly influences beam stability, and it is recommended to carefully minimise it in future colliders. Long-term effects are still not fully understood. The diffusion mechanism itself does not seem to be governed by simple Fokker-Plank equations [9]. On the other hand, the diffusion process is inherently related to ripple, and empirical criteria are now available for the design of the power supplier of future colliders, which have to be stabilised in current to much better than one part per 10^{-5}. This kind of studies are likely to be continued, both with new experiments and with numerical simulations in order to improve our quantitative understanding of the diffusion and the loss mechanisms. The aim is to find criteria for the field tolerances of the superconducting magnets for future colliders.

References

[1] L.Evans, J. Gareyte, A. Hilaire, W. Scandale and L. Vos, *The non-linear dynamic aperture experiment in the SPS*, EPAC-88, Rome,(1988).

[2] J. Gareyte, H. Hilaire and F. Schmidt, *"Dynamic aperture and long-term stability in the presence of strong sextupoles in the CERN SPS*, PAC-89, Chicago,(1989).

[3] A. Burns et al.,*The BOSC project*, EPAC-90, Nice, (1990).

[4] F. Schmidt, CERN SPS/88-22 (AMS), (1988).

[5] D. Brandt, E. Brouzet, A. Faugier, F. Galluccio, J. Gareyte, W. Herr, A. Hilaire, W. Scandale, F. Schmidt, A. Verdier, *Influence of the power supply ripple on the dynamic aperture of the SPS in the presence of strong non-linear fields*, EPAC-90, Nice, (1990).

[6] X. Altuna, et al, *The 1991 dynamic aperture experiment at the CERN SPS*, Particle and Field Series 48, Conf. Proc. No. 255, Corpus Cristi (1991).

[7] M. Furman, F. Schmidt, CERN/SPS 89-1 (AMS), (1989).

[8] T. Linnecar, W. Scandale, *Continuous tune measurement using the Schottky detector*, PAC-83, Santa Fe, (1983).

[9] A. Gerasimov, CERN/SL/92-30 (AP).

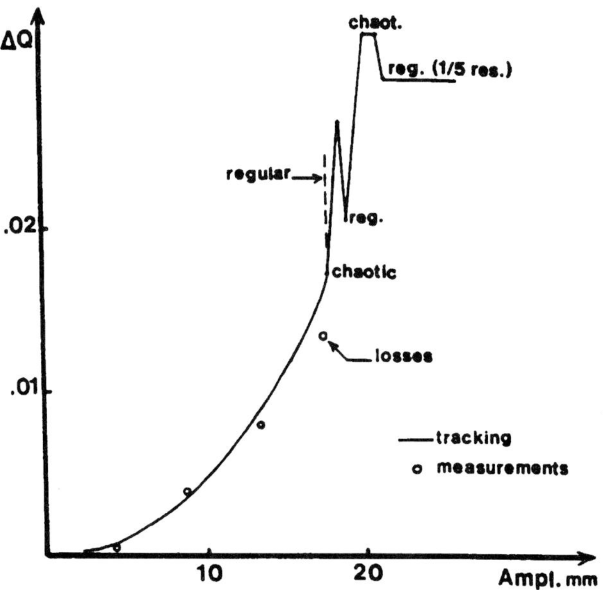

Figure 1. Tune-shift and stability

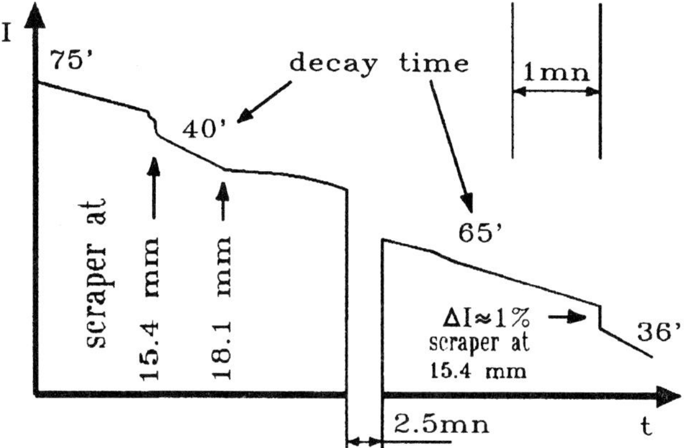

Figure 2. Measurements of diffusion rates

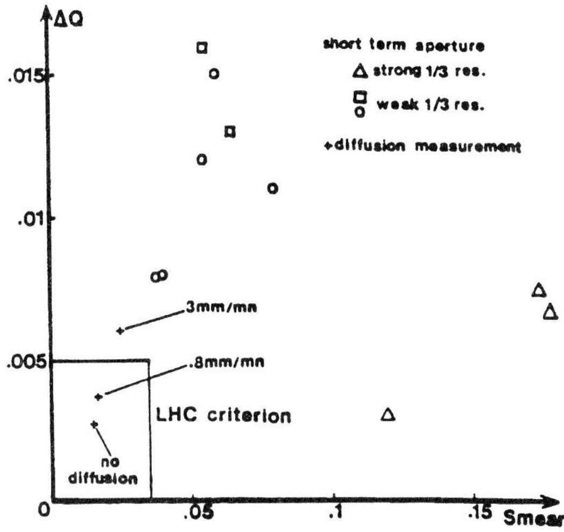

Figure 3. Tune-shift and smear at the dynamic aperture

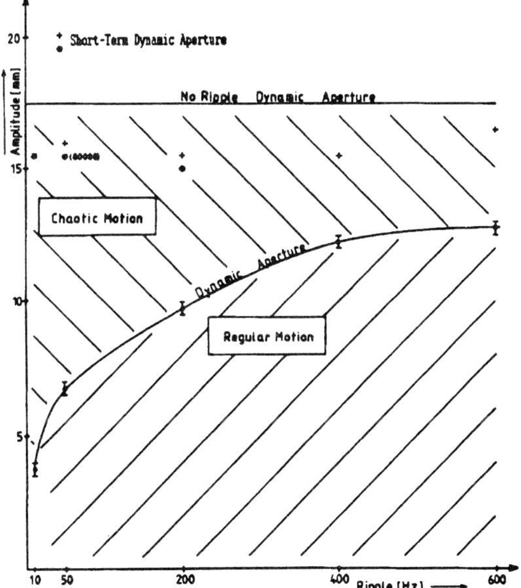

Figure 4. Influence of tune modulation

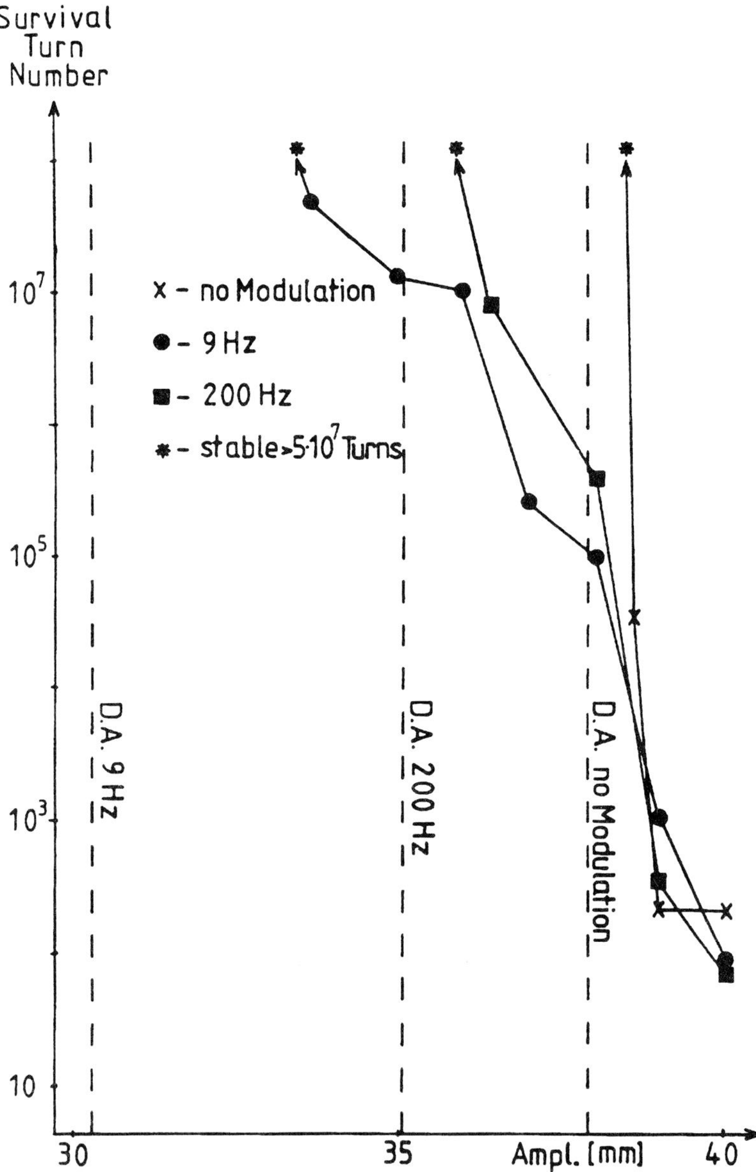

Figure 5. Long-term behavior due to tune modulation

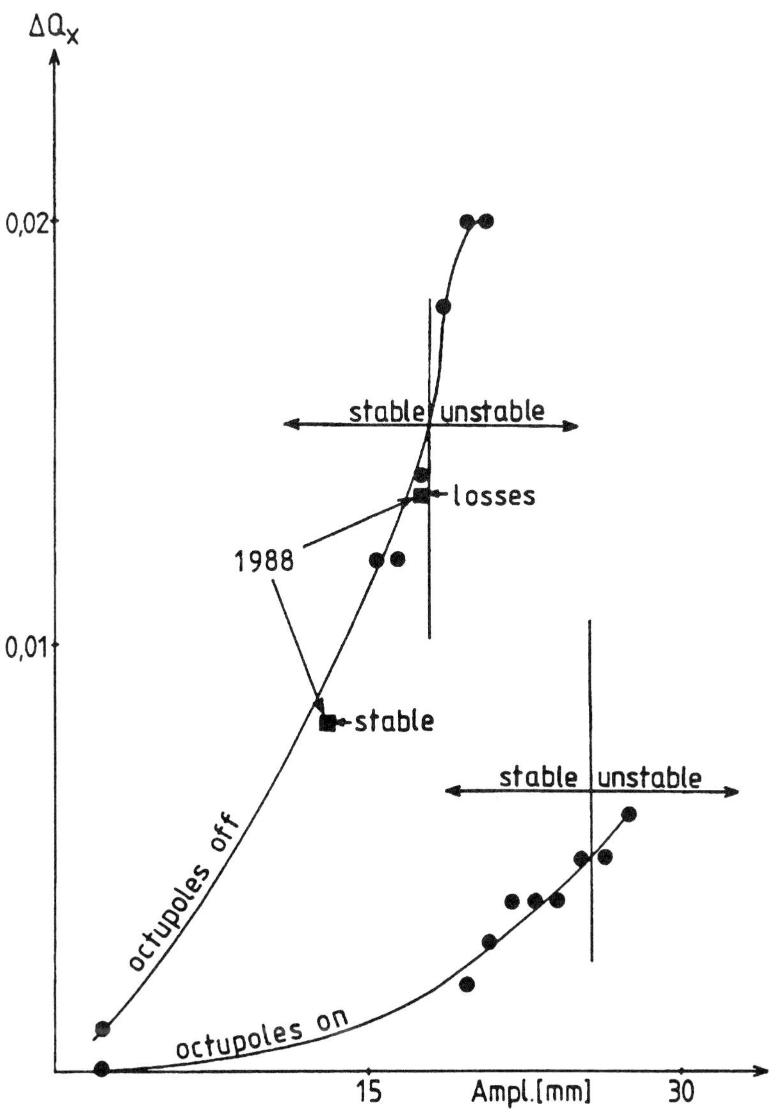

Figure 6. Measured detuning and dynamic aperture

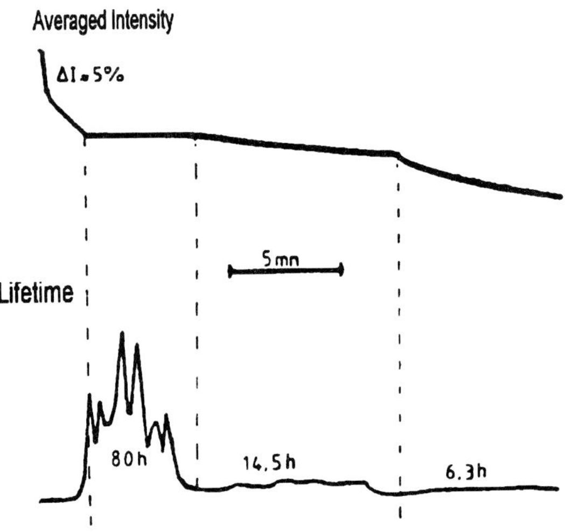

Figure 7. Diffusion measurement using scrapers

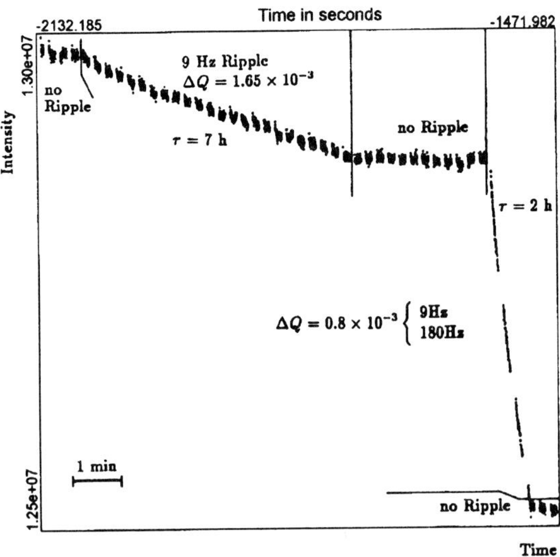

Figure 8. Lifetime for one and two frequencies

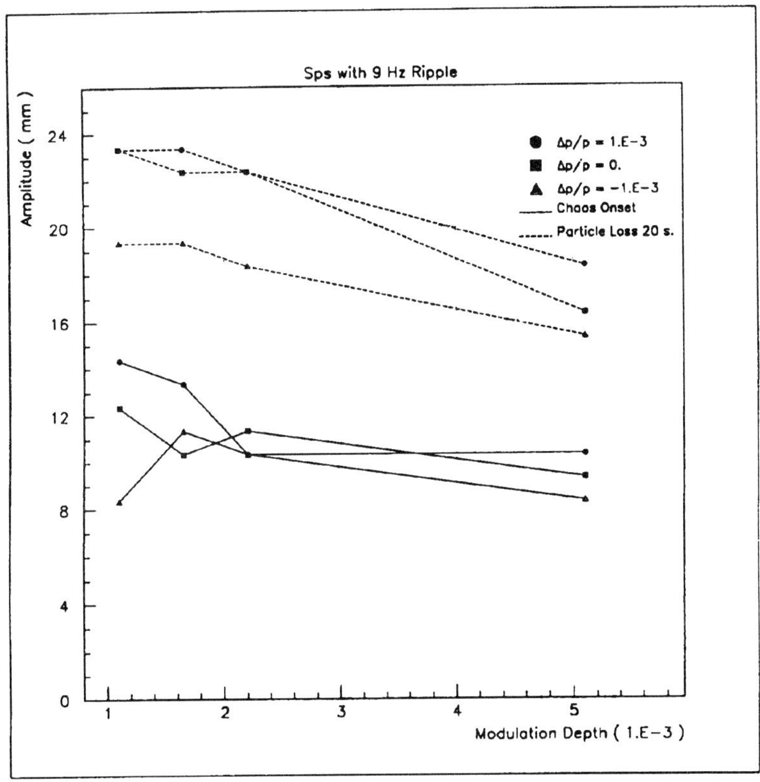

Figure 9. Effect of modulation depth on particle stability

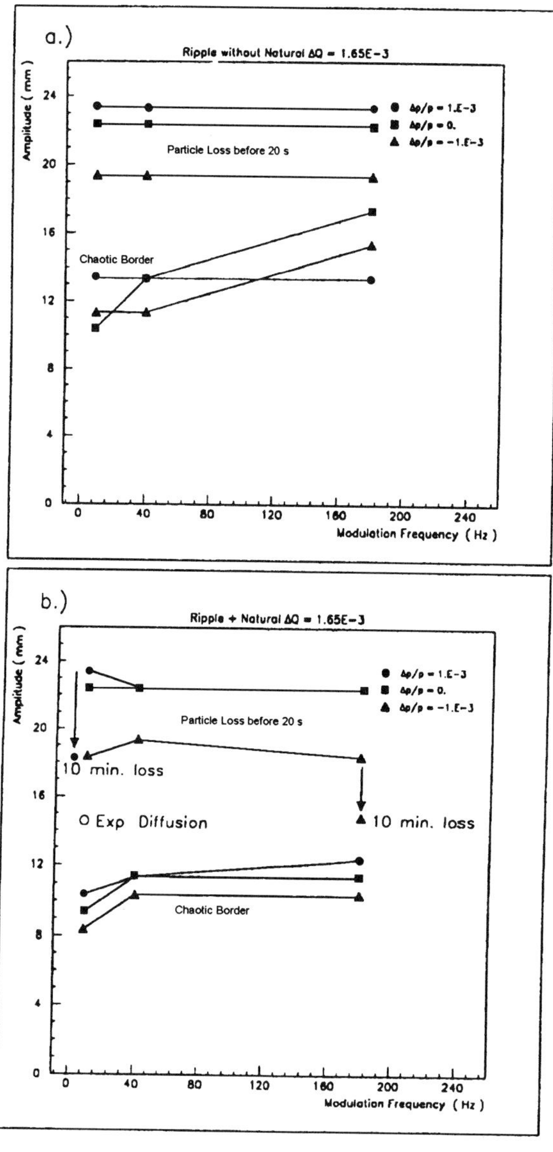

Figure 10. Effect of modulation frequency on particle stability

Review of Nonlinear Beam Dynamics Experiments

S.Y. Lee

Department of Physics, Indiana University, Bloomington, IN 47405

Abstract.
Nonlinear beam dynamics experiments, which measure Poincaré maps of nonlinear Hamiltonian in synchrotrons, are reviewed. Particle motion in resonance islands are studied in great detail. These experimental results can improve our understanding of nonlinear dynamics as well as the dynamical aperture problems in the Superconducting Super Collider.

1. Introduction

In recent years, nonlinear mechanics has been studied in various subfields of physics. Nonlinear beam dynamics studies have been especially important in the design of future colliders such as the Superconducting Super Collider (SSC) and the Relativistic Heavy Ion Collider (RHIC), since the higher order multipoles in superconducting magnets are considerably greater than those in conventional iron magnets. Theoretical studies of nonlinear fields have been used to predict both the long and the short term behavior of orbiting particles in an accelerator. In order to better understand the approximations used in theoretical predictions, experimental studies of resonant behavior are essential. This paper reviews some recent results of nonlinear beam dynamics experiments performed at the Wisconsin light source, Aladdin, the Fermilab Tevatron, and the IUCF Cooler Ring. These experiments have different characteristic features.

For particle motion in a circular accelerator, the betatron oscillations, $x(s), z(s)$, around the closed orbit satisfy Hill's equation[1]:

$$\frac{d^2x}{ds^2} + K_x(s)x = \frac{\Delta B_z}{B\rho}; \quad \frac{d^2z}{ds^2} + K_z(s)z = -\frac{\Delta B_x}{B\rho},\tag{1}$$

with $\Delta B_z + i\Delta B_x = B_0\sum_{n\geq 2}(b_n + ia_n)(x + iz)^n$, where b_n and a_n are respectively the normal and the skew multipole components. Here $K_x(s), K_z(s)$ are quadrupole strength functions, $B\rho = p/e$ is the momentum rigidity and s is the longitudinal particle coordinate, which advances from 0 to C($=2\pi$R), the circumference, as the particle completes one revolution of the cyclic accelerator, where R is the average radius. The higher order anharmonic term, $\frac{\Delta B_{z,x}}{B\rho}$, which arises fr om higher order multipoles, is normally small. Oscillations about the closed orbit due to the linear focusing force of quadrupoles, $K_{x,z}(s)$, are called betatron oscillations. The number of oscillation periods in one revolution are betatron tunes, ν_x, ν_z, which can be adjusted by varying the quadrupole strength. Both $K_{x,z}(s)$ and the anharmonic term, $\frac{\Delta B_{z,x}}{B\rho}$, are periodic functions of s with period

C. In proton accelerators, the damping of the phase space motion due to synchrotron radiation is negligible, hence the phase space area of particle motion is conserved.

Neglecting the small anharmonic term in the Hamiltonian, the betatron motion is linear. Hill's equation (1) can be solved[1] using the Floquet transformation to obtain the solution $y = \sqrt{2\beta_y J_y}\cos\phi_y$, where J_y and ϕ_y are action-angle variables, and y stands for either x or z. Here $2J_y$ is the phase space area (called the Courant-Snyder invariant or the emittance) of the betatron motion and β_y is the betatron amplitude function of the Floquet transformation (β_y is periodic in s with period C). For each turn around the accelerator, the angular variable ϕ_y increases by $2\pi\nu_y$, where ν_y is the betatron tune. The conjugate phase space coordinate, $y' = dy/ds$, can be determined using two beam position monitors (BPMs). The turn-by-turn tracking of motion in (y, y') phase space as observed at a given location in the cyclic accelerator is called the *Poincaré map*. Betatron oscillations resulting from a linear force produce ellipses in the Poincaré map.

Nonlinear perturbations in the accelerator include sextupole fields in dipoles, chromaticity correction sextupoles, octupoles, and some small higher order random error multipoles. These anharmonic terms usually do not significantly perturb the particle motion in phase space except when the betatron tunes are near to a resonance condition, which occurs at $m\nu_x + n\nu_z = \ell$, where m, n, ℓ are integers. The Poincaré map deviates from an ellipse at a resonance condition, where stable particle motion around fixed points (a stable solution to the equation of motion) in phase space bounded by invariant surfaces may occur for nearly integrable Hamiltonian systems. These stable phase space ellipses (called islands) around fixed points, are separated by the unstable fixed points. The particle phase space trajectory passing through unstable fixed points is called the separatrix.

Accelerator physicists have been interested in the nonlinear mechanics[2] in the past. Experiments at Novosibirsk VEPP2, determined the effect of beam-beam interaction under various resonance conditions. Due to higher order resonances in the beam-beam interactions, experiments at the SPS have observed larger background in the detector area. More recently, due to advances in the computer technology, large amounts of data can be recorded for post analysis. Usage of Poincaré maps in nonlinear dynamics becomes an important tool.

2. Review of Experimental Results

The experimental procedure started with a single bunch being kicked with various angular deflections, θ_K, by a pulsed deflecting magnet. The subsequent beam-centroid displacement (the betatron motion) was measured by two BPMs (four BPMs for both x and z degrees of freedom tracking). The turn-by-turn beam positions were digitized and recorded in transient recorders. Here important issues[3-9] are (1) the stability of beam closed orbit, (2) the resolution of beam position measurement, (3) linearity and dynamical range of the amplifier, and (4) digitization bandwidth for the time resolution. Depending on physics issues, the available memory can be also important. For most of electron storage rings, the damping time is of the order of milliseconds and the betatron amplitude decoheres fast. Thus the amount of memory buffer is not important. For the study of diffusion process in the hadron storage ring, available memory becomes an important issue.

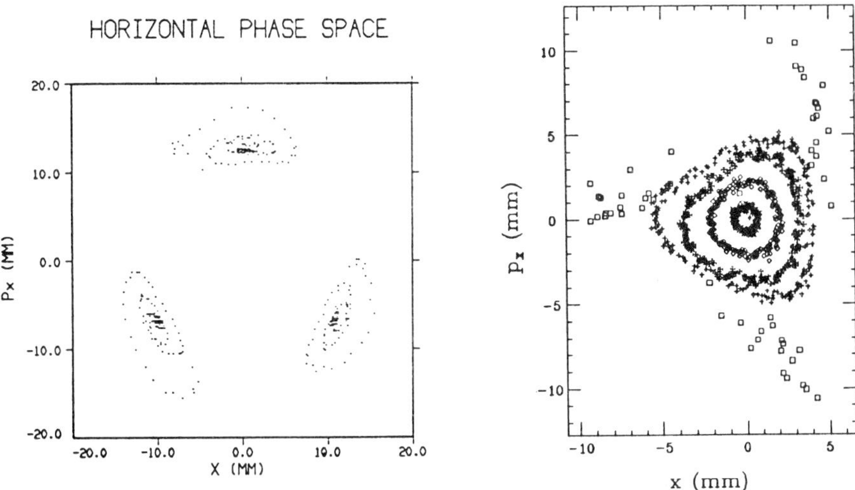

Figure 1. Poincaré maps at third order resonance are shown for the Aladin (left) and the IUCF Cooler Ring

Besides hardware issues, beam properties in storage rings are also very important in nonlinear beam dynamics experiments. To better simulate a single particle motion, nonlinear beam dynamics studies prefer small emittance beam. The BPM measures the centroid of the charge distribution. With a smaller beam size, dynamics of resonance islands can be explored. The oscillation frequencies inside the island can be measured.

The effect of betatron decoherence is smaller for smaller beam size also. When the bunch of particles is kicked to a large betatron amplitude, particles with different betatron tunes decohere in the the betatron phase space. Although each particle may remain in a large betatron amplitude of a hollow beam, the centroid of the bunch becomes zero due to decoherence. Decoherence limits the number of turns that a nonlinear beam dynamics experiment can be observed. Appendix A shows a way to disentangle the decoherence effect.

Another important issue is the linear coupling. The linear coupling messes up the interpretation of nonlinear experiments. Thus linear coupling correction is very important. Linear coupling is also an important topic in nonlinear beam dynamics experiments. For example, careful study of $\nu_x + \nu_z = n$ resonance remains to be seen.

2.1. Single Resonance Dominant regime in 1D

In a single resonance dominant regime, most experiments[3-9] studied low order resonances. The beam was kicked onto a resonance island and the Hamiltonian near the resonance island was derived. With a small emittance beam, details of island motion could be studied.

Fig. 1 shows the characteristic difference of Poincaré maps at the third order resonance condition obtained from the Aladdin[5] and the IUCF Cooler[9] (See Table 1). The Aladdin is a synchrotron radiation center at Madison, Wisconsin. For Aladdin, the beam kicked onto resonance island decoheres rapidly (the radiative damping time

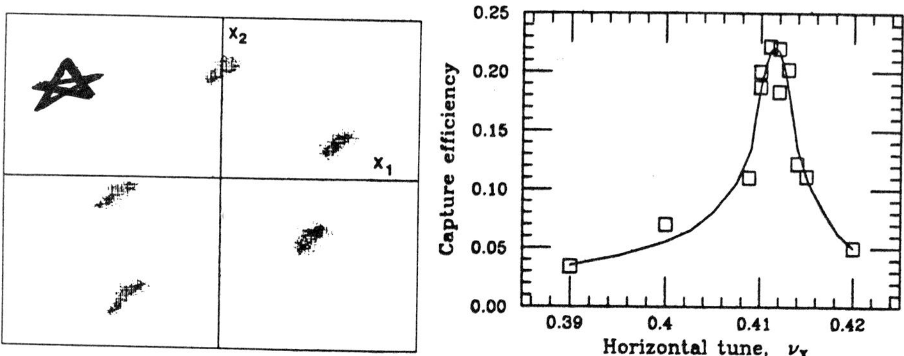

Figure 2. The raw Poincaré map and the capture efficiency for the fifth order resonance in E778 at TEVATRON

is about 89,000 turns). The decoherence is due to betatron tune spread. On the other hand, the third order island for the IUCF Cooler Ring is located at a very large betatron amplitude, so particles kicked outside the separatrix are unstable (see the right hand graph of Fig.1).

The Hamiltonian for third order resonance is given by,

$$H = \delta J_x + \frac{1}{2}\alpha_{xx}J_x^2 + \frac{(2J_x)^{3/2}}{48\pi}F\cos(3(\phi_x + \chi)) \tag{2}$$

where ϕ_x is the betatron phase, $\delta = \nu_x - \frac{\ell}{3}$, with ℓ integer, and $Fe^{3i\chi} = \int \beta_x^{3/2}\frac{B''}{B\rho}e^{3i\phi_x}ds$ with B'' as the sextupole strength. Eq.(2) shows that the third order resonance islands can be observed when the condition, $|\frac{F}{\alpha_{xx}J_x^{1/2}}| \leq 1$, is satisfied. Thus a weak third order resonance with large detuning gives rise to the third order resonance island shown in the Aladdin experiment. For a third order slow extraction process, control of the detuning parameter is important in achieving a good efficiency.

The experiments at FNAL[6][7], studied the fifth order resonance island generated by sextupoles. Due to a relatively large beam size and detuning parameter α_{xx}, the island width is small. The capture efficiency is only about 24%. The resonance detuning coefficient is about[7] $\alpha_{xx} = 47,000 \ \frac{1}{\pi m}$. The location of resonance islands can not be measured accurately due to the fact that the BPM measures only the centroid of the charges. Fig.2 shows the raw data of (x_1, x_2) plot and the capture efficiency of a 4 mm kick. The persistent displacement is about 1.2 mm with a 4 mm kick. This agrees with the capture efficiency of about 24%. Particles outside the island decohere in about 250 turns. Thus the phase space is a smeared donut ring with a concentrated small jumping raisin around five islands, which composed of about 24% of the total intensity.

For a single resonance at $m\nu_x = \ell$ with small island size, the Hamiltonian can be expressed as $H = \frac{1}{2}\alpha_{xx}(J - J_r)^2 + g\cos m\phi_x$. The island's tune and width are given by

$$\nu_{island} = m\sqrt{\alpha_{xx}g}; \quad \Delta J = 4\sqrt{\frac{g}{\alpha_{xx}}}.$$

Here $m = 5$ is the order of resonance and g is the resonance strength. Thus the island tune is related to the island width by

$$\nu_{island} = \frac{m}{4}\alpha_{xx}\Delta J = \frac{m}{4}\Delta\nu_r.$$

Here $\Delta\nu_r$ is the width of the capture efficiency shown on the right hand graph of Fig.2. One can conclude that the island tune is of the order of 0.006. Since the detuning parameter α_{xx} is large, the island width becomes small. Recently, an elaborated FNAL experiment reported[7] an island tune of 0.0063 by using external driven tune modulation to nudge out trapped particles.

The IUCF/SSC/FNAL/BNL group studied the fourth order resonance at the IUCF Cooler Ring. The IUCF Cooler is one of the new class of storage rings with electron cooling [see Table 1]. The rms emittance is smaller than 0.05 π mm-mrad and the rms momentum spread is about 4×10^{-5}. Due to the solenoidal field at the electron cooling section for the magnetized cooling, experiments show strong linear coupling. Fig. 3 shows the fourth order resonance data (inset). Note here that the winding motion around fixed points of a resonance island is due to linear coupling. Averaging the winding motion of linear coupling reveals an ellipse around an island's fixed point. FFT spectrum also shows the island tune[8]. Thus the Hamiltonian can be deduced.

Eliminating the linear coupling with skew quadrupoles, Fig. 4 shows the fourth order resonance Poincaré maps of the IUCF Cooler Ring[10]. The details of resonance island structure can be seen easily. Detail measurements give greater precision in predicting the nonlinear Hamiltonian of the synchrotron and a greater constraint in the theoretical modeling.

2.2. Single Resonance Dominant region in 2D (see also Appendix C)

The Hamiltonian for a single resonance, $m\nu_x + n\nu_z = \ell$, $m > 0$, in 2D is given by

$$H = H_0(J_x, J_z) + g J_x^{\frac{m}{2}} J_z^{\frac{n}{2}} \cos(m\phi_x + n\phi_z - \ell\theta + \chi) \tag{3}$$

Using a generating function, $F_2(\phi_x, \phi_z, J_1, J_2) = J_1(m\phi_x + n\phi_z - \ell\theta + \chi) + J_2\phi_z$, we obtain $J_x = mJ_1$; $J_z = nJ_1 + J_2$; $\phi_1 = (m\phi_x + n\phi_z - \ell\theta + \chi)$; $\phi_2 = \phi_z$. The new Hamiltonian is given by

$$\tilde{H} = \tilde{H}_0(J_1, J_2) + g(mJ_1)^{\frac{m}{2}}(nJ_1 + J_2)^{\frac{n}{2}} \cos\phi_1$$

Thus J_2 and $\tilde{H}$ are invariant of motion. For a given experiment, J_2 is a constant of motion given by $J_2 = J_{20} = J_{z0} - \frac{n}{m}J_{x0}$ with J_{x0}, J_{z0} as the initial kick amplitude. Poincaré maps in the J_1, ϕ_1 phase space should be similar to that of 1D motion.

Aladdin group recently reported its first 2D experiment at $\nu_x - 2\nu_z = -7$ ($\nu_x = 7.266, \nu_z = 7.135$) coupling resonance. Fig. 5 shows the x, z data vs turn number. The corresponding FFT spectrum should exhibit nonlinear coupling sidebands. Betatron amplitudes decoheres in about 800 turns due to a large tune spread within the beam. Since the betatron motion decoherent time is inversely proportional to the kicked amplitude, a smaller kick may help to minimize the effect. Sidebands in the FFT due to nonlinear coupling were washed away by the decoherent effect. The vertical betatron amplitude (see Fig.5) shows clearly the characteristic of nonlinear coupling. Based on the measured betatron detuning coefficients[5], i.e. $\alpha_{xx} = 296.5$ [$\frac{1}{\pi m}$], the beam should decohere in about 1800 turns. However in two dimensional motion, α_{xz}, α_{zz} are needed to understand the decoherence properly (see Appendix A).

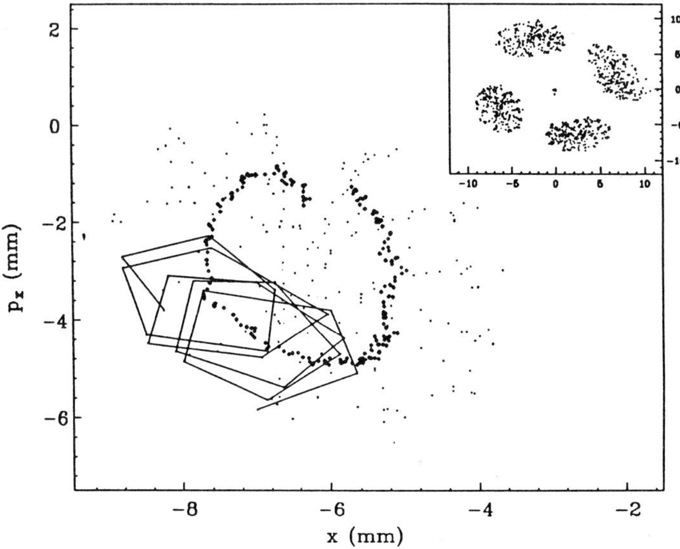

Figure 3. The Poincaré map at the fourth order resonance at IUCF Cooler Ring is shown in the inset. The effect of the linear coupling motion is shown as a winding motion around an island fixed point

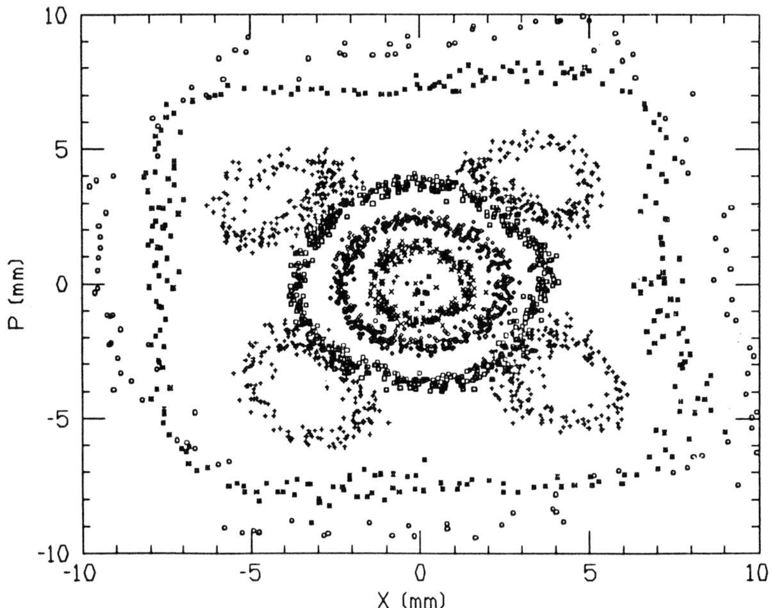

Figure 4. Nonlinear fourth order resonance at IUCF Cooler Ring after linear coupling correction

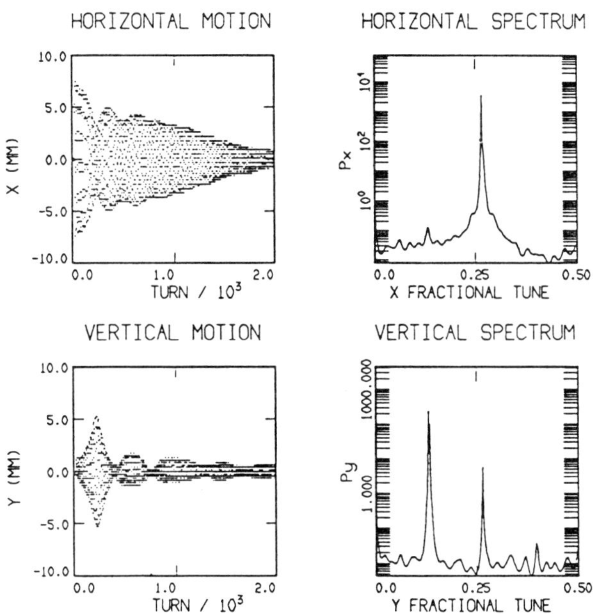

Figure 5. The measured x and z position vs turn number are showned for the Aladdin experiment at $\nu_x - 2\nu_z = -7$ resonance. The right frames show the corresponding FFT spectrum. The coupling sidebands are washed away by the betatron decoherent effect

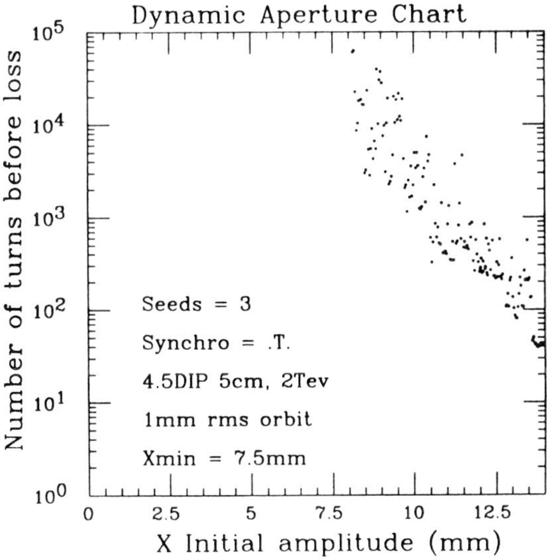

Figure 6. The survival plot of the SSC for the 5 cm coil i.d. magnets

2.3. Dynamical aperture studies

There are several dynamical aperture experiments at resonance conditions by kicking the beam and observing beam loss. Such experiments have been performed almost in all accelerators during the search of optimal operation condition. The procedure is routinely performed.

Another important dynamical aperture experiment is to carry out a test of the survival plot[11]. Assuming that we can prescribe a set of nonlinear elements to be installed in the Cooler ring, which simulates the nonlinear map of the superconducting super collider (SSC), the dynamical aperture experiments will verify the survival plot shown in Fig. 6[11]. Detailed study of the phase space evolution at the boundary of instability can be important in understanding chaos.

2.4. Tune Modulation

Stochasticity increases drastically with tune modulation. Tevatron group[7] studied recently the chaotic boundary of the tune modulation at the fifth order resonance island. A combined phase space evolution in such an experiment promises to be exciting. Such experiments are also planned at the IUCF Cooler Ring.

2.5. Longitudinal tracking

By shifting the rf phase nonadiabatically, the macroparticle can be set off to execute synchrotron motion. The synchrotron phase space coordinates are measured with a HBW wall gap monitor and a BPM at a high dispersion location. Fig. 7 shows an example of such a synchrotron tracking at the IUCF Cooler Ring[10]. Using such a technique, nonlinear beam dynamics studies can be performed for off-momentum particles.

Small amplitude harmonic modulation to the rf stable phase angle gives rise to the synchrotron equation of motion, $\ddot{\phi} + \omega_{syn}^2 \sin\phi = \omega_{syn}^2 g_m \sin\omega_m t$ which is equivalent to tune modulation on the particle motion in resonance islands. Detailed studies of this process will be carried out shortly at the IUCF Cooler Ring.

3. Conclusion

Detailed studies of nonlinear beam dynamics are important in understanding the particle motion in storage rings. Recent advances in the fast digitizing chips and also the availability of small emittance storage rings offer us the possibility of long term tracking of betatron motion. Combined with recent advances in theoretical nonlinear beam dynamics studies by using the Taylor map, Lie Algebraic and canonical perturbation techniques, nonlinear beam dynamics experiments is timely and important in supporting and verifying theories.

The limitations of nonlinear beam dynamics experiments rest on 1) finite beam size, 2) decoherence of betatron motion, 3) uncontrollable nonlinear resonances crossing. On the other hand, these limitations reflect a realistic storage environment. Theoretical calculation is usually limited by its possibility to predict the realistic situation. With a high quality beam, particle motion can be tracked to more than 10^5 turns. Such experiments have just begun to take place.

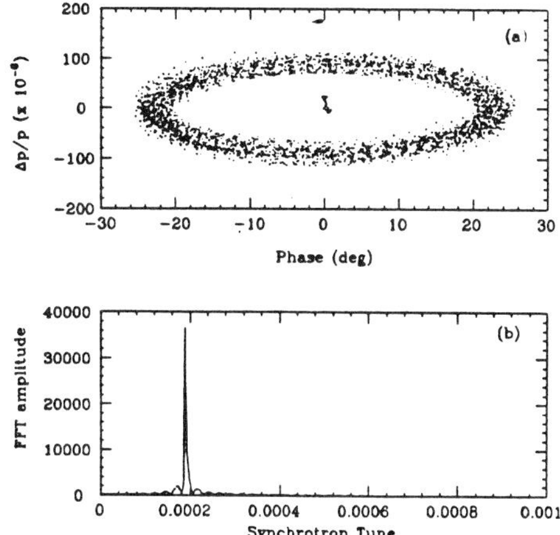

Figure 7. An example of Synchrotron phase space tracking at the IUCF Cooler

acknowledgement

I benefited greatly from discussions with the CE22 collaboration team, whose members are M. Ball, B. Brabson, D. D. Caussyn, J. Collins, S. Curtis, V. Derenchuck, D. Du-Plantis, G. East, M. Ellison, T. Ellison, D. Friesel, B. Hamilton, H. Huang W. P. Jones, W. Lamble, S.Y. Lee, D. Li, M.G. Minty, S. Nagaitsev, T. Sloan at IUCF, A. W. Chao, S. Dutt, M. Syphers, Y. Yan at the SSC, S. Tepikian at BNL, and W. Gabella, K.Y. Ng at Fermilab. I am also grateful to discussions with L. Teng, S. Peggs and T. satogata.

Table 1. Operation parameters in Nonlinear beam dynamics experiments

	SPEAR	Aladdin	TEVATRON	IUCF Cooler
E[GeV]	3.7	0.800	150	0.045
C[m]	234	88.9	6280	86.8
$\tau[\mu s]$	0.78	0.300	20.9	0.969
ν_x	5.3	7.14	19.4	3.7
ν_z	5.2	7.23	19.4	4.72
$\epsilon[\pi$ mm-mrad]		0.11	0.016	< 0.05
σ_x [mm]		0.85	1.4	< 0.7

References

[1] E.D. Courant and H.S. Snyder, Ann. Phys. (NY) **3**,1 (195 8).

[2] M. Cornacchia and L. Evans, Part. Accel. 19, 125(1986) ; L. Evans, et.al., Proc. 1st European Particle Accelerator Conference, Rome, p.619, (1988); L. Evans, J. Gareyte, A. Hilaire, and F. Schmidt, Proceedings, 1989 IEEE Particle Accelerator Conference, Chicago, p. 1376 (1989); L. Evans, et.al., 1403(1989); L. Evans, CERN SPS/83-38;

[3] P.L. Morton, et. al., IEEE Tran. Nucl. Sci. NS-32, 2291(1985).

[4] D.A. Edwards, R.P. Johnson, and F. Willeke, Part. Accel. 19, 145 (1986).

[5] J. Bridge, et al.,"Dynamic Aperture measurement on Aladdin", Part. Acc. 28,1 (1990). E. Crosbie, et al., IEEE PAC, p.1624 (1991); J. Liu, et al., Particle Accelerators, to be published; L. Teng, private communications.

[6] A. Chao, et al., Phys. Rev. Lett. 61, 2752 (1988); N. Merminga, et al., Proc. EPAC, p.791 (1988).

[7] T. Satogata, et al., Phys. Rev. Lett., 68, 1838 (1992) .

[8] S.Y. Lee, et al., Phys. Rev. Lett. 67, 2767 (1991)

[9] D.D Caussyn, et al., to be published

[10] M. Ellison, et al., to be published.

[11] Y. Yan, "Supercomputing for the Superconducting Super Collider", Energy Sciences Supercomputing 1990, pp. 9-13 (1990); DOE National Research Supercomputing Center, A. Mirin and G. Kaiper, eds.

[12] R.E. Meller, A.W. Chao, J.M. Peterson, S.G. Peggs, and M. Furman, "Decoherence of kicked beams", SSC-N-360 (1987).

[13] M. Abramowitz and I. Stegun, Eds," Handbook of Mathmatical Functions", (Dover, New York, 1965). I. S. Gradshteyn and I.M. Ryzhik, "Table of Integrals, series, and Products", Academic Press, NY (1980).

Appendix A. Two dimensional decoherence

From basic linear beam dynamics[1], a beam, kicked transversely from its closed orbit, will execute betatron oscillations. Betatron oscillations can be observed by a beam position monitor, which measures the centroid of particle distribution. Such procedures are commonly used to measure the betatron tunes.

This procedure is also useful in the nonlinear beam dynamics studies. A small emittance beam bunch is kicked transversely to set up betatron oscillations around the closed orbit. Nonlinear betatron motion is then observed through a detailed turn by turn phase space tracking. The tunes of the betatron motion can be obtained by Fourier analyzing data from beam position monitors. However, if beam particles have betatron tune spreads, then the observed centroid of the beam will decohere due to the accumulated betatron phase spread of particles. Knowledge of these decoherence effects can be used more reliably to obtain the betatron tunes and to understand nonlinear beam dynamics. When detailed phase space maps (Poincaré maps) are needed to study the nonlinear betatron motion, the decoherence process must be understood in order to deduce the effect of nonlinear motion.

A1. Decoherence due to Chromaticity

Due to the chromaticity, betatron tunes depend on momentum deviation as, $\Delta\nu = \xi\delta + \xi_2\delta^2$. where $\delta = \frac{\Delta p}{p}$ is the momentum deviation of the particle, ξ and ξ_2 are the linear and the second order nonlinear chromaticities respectively. Let us first assume a linear dependence of betatron tunes vs. momentum deviation, δ. The betatron tune shift of a particle at the nth turn is given by $\Delta\nu(n) = \xi\frac{\sigma_\delta}{\sigma_a}a\cos(2\pi\nu_s n + \phi_s)$, where ν_s is the synchrotron tune, a is the synchrotron oscillation amplitude of the particle, σ_a is the rms synchrotron amplitude of the distribution, σ_δ is the rms momentum deviation of the distribution. The accumulated betatron phase shift of the particle is then given by

$$\Delta\phi_\beta(a,\phi_s,n) = 2\pi\int_0^n \Delta\nu(m)dm = q\frac{a}{\sigma_a}\cos(\pi\nu_s n + \phi_s),$$

with

$$q = 2\xi\frac{\sigma_\delta\sin\pi\nu_s n}{\nu_s}.$$

Note here that the parameter $q = 0$ at $n = \text{integer}/\nu_s$. This means that the betatron phase spread vanishes every $\frac{1}{\nu_s}$ turns around the accelerator. The transverse position of a particle with synchrotron amplitude a and synchrotron phase ϕ_s at the nth turn is given by $x(n) = x_0\cos(2\pi\nu_\beta n + \phi_\beta + \Delta\phi_\beta(a,\phi_s,n))$. The centroid of the particle distribution is given by[12]

$$\langle x(n)\rangle = \int x_0\cos(2\pi\nu_\beta n + \phi_\beta + \Delta\phi_\beta(a,\phi_s,n))\rho(a,\phi_s)dad\phi_s = x_0 F_\delta\cos(2\pi\nu_\beta n + \phi_\beta), \quad (4)$$

with $F_\delta = \int_0^\infty J_0(q\frac{a}{\sigma_a})\rho(a)da$. where J_0 is the Bessel function[13]. Here x_0 is the initial kicked betatron amplitude. Eq.(4) shows that the measured centroid of a particle distribution executes betatron oscillations with a modulating form factor depending on particle distributions. The amplitude modulating factor recoheres every $\frac{1}{\nu_s}$ turns. The modulation does not affect tune measurement, i.e. there is no tune shift. The measured betatron tune is independent of particle distributions.

Using the Gaussian distribution function shown in Appendix B, one obtains easily[12]

$$F_\delta = \exp(-\frac{q^2}{2}).$$

The corresponding Fourier spectrum for a Gaussian bunch distribution is also Gaussian. Using the uniform particle distribution function, we obtain

$$F_\delta = \frac{2J_1(\sqrt{2}q)}{\sqrt{2}q} = {}_0F_1(2; -\frac{q^2}{2}) = \sum_0^\infty \frac{1}{m!\Gamma(m+2)}(-\frac{q^2}{2})^m,$$

where J_1 is the Bessel function and ${}_0F_1$ is the generalized hypergeometric function. The maximum of q is given by $q = 2\xi\frac{\sigma_\delta}{\nu_s} = 2\xi\frac{\sigma_t}{\tau_{rev}}$, where σ_t, τ_{rev} are bunch length and revolution time respectively.

A2. Higher Order Chromatic Effects

For a high energy particle accelerator, the linear chromaticity is normally corrected to zero value. Higher order chromatic effect may then become important. It is known that the second order chromaticity depends on the half integer stopband of accelerator.

Let us consider the tune spread due to the second order chromaticity, where the tune spread is given by $\Delta\nu(n) = \xi_2[\frac{a\sigma_\delta}{\sigma_a}\sin(2\pi\nu_s n + \phi_s)]^2$, where ξ_2 is the second order chromaticity. The betatron phase spread is then given by

$$\Delta\phi_\beta(a, \phi_s, n) = \pi\xi_2\frac{\sigma_\delta^2}{\sigma_a^2}a^2 n + 2\pi\xi_2\frac{\sigma_\delta^2}{\sigma_a^2}a^2 \sin(2\pi\nu_s n)\cos(2\pi\nu_s n + 2\phi_s).$$

Using a Gaussian distribution, we obtain easily the centroid of the beam distribution as $\langle x(n)\rangle = x_0 F_\delta \cos(2\pi\nu_\beta n + \phi_\beta + \Delta\phi)$, with

$$F_\delta = [(1 - 4\pi^2\xi_2^2\sigma_\delta^4 n^2 + 16\pi^2\xi_2^2\sigma_\delta^4\sin^2 2\pi\nu_s n)^2 + 16\pi^2\xi_2^2\sigma_\delta^4 n^2]^{-\frac{1}{2}},$$

$$\Delta\phi = \frac{1}{2}\arctan\frac{4\pi\xi_2\sigma_\delta^2 n}{1 - 4\pi^2\xi_2^2\sigma_\delta^4 n^2 + 16\pi^2\xi_2^2\sigma_\delta^4\sin^2 2\pi\nu_s n}.$$

The measured betatron amplitude decoheres with an amplitude function F_δ and that the phase is also modulating with the phase function $\Delta\phi$. The measured betatron tune is obtained from the fast Fourier transform (FFT) of $\langle x\rangle$, i.e. $\nu_\beta^{\text{measured}} \approx \nu_\beta + \xi_2\sigma_\delta^2$. We observe that the betatron tune is shifted by an amount of the rms tune spread . Since ν_s is a small number, we expect that $F_\delta \approx \frac{1}{1 + 4\pi^2\xi_2^2\sigma_\delta^4 n^2}$. The characteristic Fourier amplitude will be exponential.

When the linear chromaticity and the second order chromaticity are simultaneously present in the accelerator, then the ϕ_s integral in Eq.(4) can be performed to obtain a sum of Bessel functions. The radial integral becomes too complicated to be represented by special functions. However numerical solutions can be obtained easily.

A3. Decoherence due to the betatron tune spread

When the beam in an accelerator is kicked transversely by an amount $\Delta x'$ and/or $\Delta z'$, the particle distribution in the phase space coordinates is given by Eq.(5) in Appendix B. The kicked beam executes betatron motion around the closed orbit. The betatron tune of each particle depends on its betatron amplitude through the nonlinear multipoles. In lowest order approximation, we have $\Delta\nu_x = k_{xx}a_x^2$. Therefore betatron phase shift after n^{th} turns is given by $\Delta\phi_x = 2\pi k_{xx}a_x^2 n$. Due to betatron phase spread of particles in the beam, the betatron motion of the kicked beam will decohere. The decoherence can be calculated by the centroid of the kicked beam distribution as,

$$\begin{aligned}
\langle x(n)\rangle &= \int a_x \cos(2\pi\nu_x n + \phi_x + \Delta\phi_x)\rho_k(a_x, \phi_x)da_x d\phi_x \\
&= -\int a_x \sin(2\pi\nu_x n + \Delta\phi_x)I_1(\frac{a_x x_k}{\sigma_x^2})\rho_k(a_x, x_k)da_x, \\
&= -x_k F_x \sin(2\pi\nu_x n + \phi_{xx} + \frac{x_k^2}{2\sigma_x^2}\frac{\theta}{(1 + \theta^2)}),
\end{aligned} \tag{5}$$

where the Gaussian distribution function of Eq.(6) has been used in the integration. Here $x_k = \beta_x\Delta x'$; $\theta = 4\pi\Delta\nu_x n$; $\phi_{xx} = 2\arctan\theta$; and

$$F_x = \frac{1}{1 + \theta^2}\exp(-\frac{x_k^2}{2\sigma_x^2}\frac{\theta^2}{(1 + \theta^2)}).$$

Here x_k is the kicked amplitude, $\Delta\nu_x = k_{xx}\sigma_x^2$ is the rms tune spread in one degree of freedom. thus the measured betatron tune of the kicked beam is $\nu_x^{\text{measured}} \approx \nu_x + 4k_{xx}\sigma_x^2 + k_{xx}x_k^2$. The measured Fourier spectrum is initially a Gaussian and then evolves towards an expontial form.

A4. Decoherence in two degrees of freedom

The betatron motion of a kicked particle can be described by

$$y(n) = a_y \cos(2\pi\nu_y n + \phi_y + \Delta\phi_y); \quad \alpha_y y + \beta_y y' = -a_y \sin(2\pi\nu_y n + \phi_y + \Delta\phi_y),$$

where y stands for either x or z, a_x, a_z are the betatron amplitudes, ν_x, ν_z are the betatron tunes, ϕ_x, ϕ_z are betatron phases of the particle and $\Delta\phi_x, \Delta\phi_z$ are the betatron phase spread given by $\Delta\phi_x = 2\pi(k_{xx}a_x^2 + k_{xz}a_z^2)n; \quad \Delta\phi_z = 2\pi(k_{zx}a_x^2 + k_{zz}a_z^2)n$. The tune spread can arise from octupoles, higher order multipoles, or the second order effect of sextupoles. The betatron coordinates, measured from the beam position monitor, are then given by

$$\langle y(n) \rangle = \int a_y \cos(2\pi\nu_y n + \phi_y + \Delta\phi_y)\rho_k(a_x, \phi_x)\rho_k(a_y, \phi_y)da_x d\phi_x da_y d\phi_y.$$

Using the Gaussian distribution of Eq.(6), we obtain easily

$$\langle x(n) \rangle = -x_k F_{xx} F_{xz} \sin(2\pi\nu_x n + \phi_{xx} + \phi_{xz} + \frac{x_k^2}{2\sigma_x^2}\frac{\theta_{xx}}{1+\theta_{xx}^2} + \frac{z_k^2}{2\sigma_z^2}\frac{\theta_{xz}}{1+\theta_{xz}^2}),$$

$$\langle z(n) \rangle = -z_k F_{zx} F_{zz} \sin(2\pi\nu_z n + \phi_{zx} + \phi_{zz} + \frac{x_k^2}{2\sigma_x^2}\frac{\theta_{zx}}{1+\theta_{zx}^2} + \frac{z_k^2}{2\sigma_z^2}\frac{\theta_{zz}}{1+\theta_{zz}^2}),$$

where $x_k = \beta_x \Delta x'; \; z_k = \beta_z \Delta z'$, are kicked amplitudes with $\theta_{xx} = 4\pi k_{xx}\sigma_x^2 n; \theta_{xz} = 4\pi k_{xz}\sigma_z^2 n; \theta_{zx} = 4\pi k_{zx}\sigma_x^2 n; \theta_{zz} = 4\pi k_{zz}\sigma_z^2 n$, $\phi_{ij} = 2\arctan\theta_{ij}; \; (i,j = x,z)$ and

$$F_{xx} = \frac{1}{1+\theta_{xx}^2}\exp(-\frac{x_k^2}{2\sigma_x^2}\frac{\theta_{xx}^2}{1+\theta_{xx}^2}); \quad F_{xz} = \frac{1}{1+\theta_{xz}^2}\exp(-\frac{z_k^2}{2\sigma_z^2}\frac{\theta_{xz}^2}{1+\theta_{xz}^2}),$$

$$F_{zx} = \frac{1}{1+\theta_{zx}^2}\exp(-\frac{x_k^2}{2\sigma_x^2}\frac{\theta_{zx}^2}{1+\theta_{zx}^2}); \quad F_{zz} = \frac{1}{1+\theta_{zz}^2}\exp(-\frac{z_k^2}{2\sigma_z^2}\frac{\theta_{zz}^2}{1+\theta_{zz}^2}).$$

Within a short time interval after the kick, the measured betatron tunes, derived from the Fourier transform of $\langle x \rangle, \langle z \rangle$ are $\nu_x^{\text{measured}} \approx \nu_x + 4k_{xx}\sigma_x^2 + k_{xx}x_k^2 + 4k_{xz}\sigma_z^2 + k_{xz}z_k^2$, $\nu_z^{\text{measured}} \approx \nu_z + 4k_{xz}\sigma_x^2 + k_{xz}x_k^2 + 4k_{zz}\sigma_z^2 + k_{zz}z_k^2$. Note here that when the beam bunch is kicked in x direction, the decoherent amplitude is independent of the vertical tune spread. *However, the measured horizontal tune does depend on the vertical tune spread.* The amplitude modulation shortly after the kick is Gaussian. Thus the Fourier spectrum would also be Gaussian in shape. When the condition $|\theta_{ij}| \gg 1$ is met, the decoherence of the betatron oscillation obeys a power law. The corresponding Fourier spectrum is exponential.

Appendix B: Particle Distribution Functions

B1. Longitudinal phase space distribution

In the adiabatic region of the synchrotron motion, i.e. far away from the transition energy, particle orbit is an ellipse given by, $(\frac{\delta}{\sigma_\delta})^2 + (\frac{\phi}{\sigma_\phi})^2 = 1$, where $\delta = \Delta p/p$ and ϕ are conjugate variables of the synchrotron motion, σ_δ and σ_ϕ are the rms momentum deviation and rms phase angle of the bunch respectively. The rms phase space area of the bunch is given by $A = \pi\sigma_a^2 = \pi\sigma_\delta\sigma_\phi$, where σ_a is the rms synchrotron amplitude amplitude of the bunch. The phase space variables at the nth turn can then be described by $\delta = \frac{\sigma_\delta}{\sigma_a}a\cos(2\pi\nu_s n + \phi_s); \quad \phi = \frac{\sigma_\phi}{\sigma_a}a\sin(2\pi\nu_s n + \phi_s)$, where ν_s is the synchrotron tune and a is the synchrotron amplitude.

An equilibrium distribution function is a function of the invariant ellipse of the phase space coordinates, i.e. $\rho(\delta, \phi) = f([\frac{\delta^2}{\sigma_\delta^2} + \frac{\phi^2}{\sigma_\phi^2}])$. Transforming the phase space coordinates into amplitude and phase coordinates, the distribution function becomes $\rho(a, \phi_s) = \rho(a) = af([\frac{a^2}{\sigma_a^2}])$, where the Jacobian of the transformation is included in the definition of the distribution function. Some simple distributions are:

1. Gaussian distribution:
$$\rho(a, \phi_s) = \frac{a}{2\pi\sigma_a^2} \exp(-\frac{a^2}{2\sigma_a^2}).$$

2. Uniform distribution:
$$\rho(a, \phi_s) = \begin{cases} a, & a \le \sqrt{2}\sigma_a; \\ 0, & \text{otherwise.} \end{cases}$$

B2. Distribution function in betatron phase space.

Let (y, y') represent either of the transverse phase space conjugate variables (x, x') or (z, z'). In the absence of linear coupling, the action integral, or the Courant-Snyder invariant, of betatron motion is given by

$$J = \frac{1}{2\beta_y}[y^2 + (\alpha_y y + \beta_y y')^2].$$

Similarly, an equilibrium particle distribution must be a function of the Courant-Snyder invariant. Assuming a Gaussian distribution, the normalized distribution function is given by

$$\rho(y, y') = \frac{\beta_y}{2\pi\sigma_y^2} \exp[-\frac{(y^2 + (\alpha_y y + \beta_y y')^2)}{2\sigma_y^2}],$$

where $\sigma_y = \sqrt{2\beta_y J_0}$, where the rms beam emittance is given by $\epsilon = 2J_0$.

Let us now study the distribution function when the beam is kicked instantaneously at time zero by $\Delta y'$. Using the amplitude and phase coordinates (a_y, ϕ_y), i.e.

$$y(n) = a_y \cos(2\pi\nu_y n + \phi_y); \quad \alpha_y y + \beta_y y' = -a_y \sin(2\pi\nu_y n + \phi_y),$$

the distribution function of the kicked beam becomes,

$$\rho_k(a, \phi_y) = \frac{a_y}{2\pi\sigma_y^2} \exp(-\frac{a_y^2 + y_k^2}{2\sigma_y^2} + \frac{a_y y_k}{\sigma_y^2} \sin\phi_y), = \rho_k(a_y, y_k)\{\frac{1}{2\pi} \exp(\frac{a_y y_k}{\sigma_y^2} \sin\phi_y)\}, \quad (6)$$

where $y_k = \beta_y \Delta y'$ is the kicked amplitude. Note here that the Jacobian is also included in the definition of the distribution function. To simplify ϕ_y integration of the distribution function, the generating function of the Bessel functions becomes very useful, i.e.

$$\exp(u \sin\phi_y) = I_0(u) + 2\sum_{k=0}^{\infty}(-)^k I_{2k+1}(u) \sin[(2k+1)\phi_y] + 2\sum_{k=1}^{\infty}(-)^k I_{2k}(u) \cos[2k\phi_y].$$

Appendix C: Equation of motion for a single resonance in 2D

Let the unperturbed Hamiltonian H_0 of section 2.2 be

$$H_0(J_x, J_z) = \nu_{x0} J_x + \nu_{z0} J_z + \frac{1}{2}\alpha_{xx} J_x^2 + \alpha_{xz} J_x J_z + \frac{1}{2}\alpha_{zz} J_z^2.$$

We obtain then,

$$\tilde{H}(J_1, \phi_1, J_2) = \delta_1 J_1 + \frac{1}{2}\alpha_{11} J_1^2 + g(mJ_1)^{\frac{m}{2}}(nJ_1 + J_2)^{\frac{n}{2}}\cos\phi_1 + [\nu_{0z} J_2 + \frac{1}{2}\alpha_{zz} J_2^2] \quad (7)$$

with

$$\delta_1 = m\nu_{x0} + n\nu_{z0} - \ell + m\alpha_{xz} J_2 + n\alpha_{zz} J_2; \quad \alpha_{11} = m^2\alpha_{xx} + 2mn\alpha_{xz} + n^2\alpha_{zz}.$$

Two constants of motion, J_2, and $\tilde{H}$, determine the particle orbit completely, i.e.

$$\tilde{H}(J_1, \phi_1, J_2) = \tilde{H}(J_{10}, \phi_{10}, J_2). \quad (8)$$

As an example, let us consider $\nu_x \pm 2\nu_z = \ell$ resonance with initial horizontal kicks, i.e. $J_{10} = \mp\frac{1}{2}J_2$ The phase space evolution equation, Eq.(8), becomes,

$$(2J_1 \pm J_2)[\frac{\alpha_{11}}{8}(2J_1) \pm \frac{g}{\sqrt{2}}\sqrt{2J}\cos\phi_1 + \frac{\delta_1}{2} \mp \frac{\alpha_{11}}{8}J_2] = 0 \quad (9)$$

Thus the particle trjectory follows paths of two intersecting circles in the phase space map of $(\sqrt{J_1}\cos\phi_1, \sqrt{J_1}\sin\phi_1)$. The circle $2J_1 \pm J_2 = 0$ is called the *launching circle*, or more appropriately the Courant-Snyder invariant circle, while the circle

$$\frac{\alpha_{11}}{8}2J_1 \pm \frac{g}{\sqrt{2}}\sqrt{2J}\cos\phi_1 + \frac{\delta_1}{2} \mp \frac{\alpha_{11}}{8} = 0$$

is the nonlinear coupling circle. *The intersections of these two circles are unstable fixed points of the Hamiltonian.*

For a difference resonance, the launching circle is also the limiting circle[5]. For the sum resonance, the coupling circle can lead to instability depending on the detuning parameter α_{11}. When the detuning parameter is small, coupling circle becomes a straight line. Varying betatron tunes, the coupling circle scans through the launching circle for a given initial J_2 value. When these two circles overlap, the effect of resonance becomes strong. At the condition $\delta_1 + \alpha_{11}J_{10} = 0$, the coupling circle passes through the origin in the phase space. This is equivalent to full coupling.

For sum resonances, particles, started out from a launching circle, will reach outward in phase space along the coupling circle. For difference resonance, particle moves inward towards the origin on the coupling circle and reaches onto the other side of the launching circle. The left hand part of Fig. 8 shows a phase space map of $(\sqrt{J_1}\cos\phi_1, \sqrt{J_1}\sin\phi_1)$ for a simple linear lattice with a single sextupole at the $\nu_x - 2\nu_z = \ell$ resonance. Here the nonlinear coupling circle is a straight line due to $\alpha_{11} = 0$.

For $2\nu_x + 2\nu_z = \ell$ resonance with an initial horizontal kicks, similar equation of motion can be derived, i.e.

$$(2J_1 + J_2)[\frac{\alpha_{11}}{8}(2J_1) + g \cdot (2J_1)\cos\phi_1 + \frac{\delta_1}{2} - \frac{\alpha_{11}}{8}J_2] = 0. \quad (10)$$

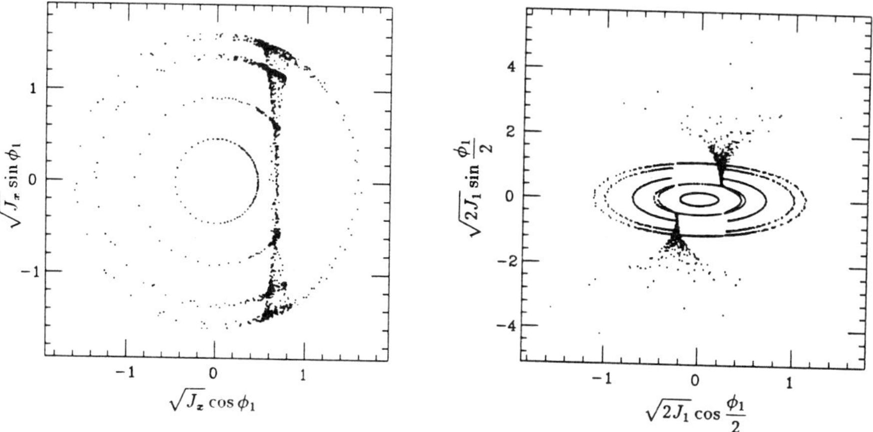

Figure 8. The Poincaré maps (see text for explanation) are shown for a Simple tracking calculation with a single sextupole (left) at a $\nu_x - 2\nu_z = \ell$ resonance, and a single octupole (right) at a $2\nu_x + 2\nu_z = \ell$ resonance.

Let us now define the phase space as $(\sqrt{2J_1}\cos\frac{\phi_1}{2}, \sqrt{2J_x}\sin\frac{\psi_x}{2})$. Eq.(10) is composed of a launching circle $2J_1 + J_2 = 0$ and *an ellipse or a hyperbola* described by

$$\frac{\alpha_{11}}{8}(2J_1) + g \cdot (2J_1)\cos\phi_1 + \frac{\delta_1}{2} - \frac{\alpha_{11}}{8}J_2 = 0,$$

for nonlinear resonance. Locations where the nonlinear resonance curve intercepts the launching circle are *unstable fixed points of the Hamiltonian*. Particles follow the Courant-Snyder invariant circle until they encounter the unstable fixed point, which leads to unstable motion afterwards. The right hand side of Fig. 8 shows a phase space plot of $(\sqrt{2J_1}\cos\frac{\phi_1}{2}, \sqrt{2J_1}\sin\frac{\phi_1}{2})$ for a simple tracking with linear lattice and a single octupole at $2\nu_x + 2\nu_z = \ell$ resonance. Clearly, particles follow the Courant-Snyder invariant curve until it encounter the unstable fixed point. Particle motion appears chaotic afterwards. *Particle loss is located at a unique phase space curve in the* (J_1, ϕ_1).

When the the betatron tunes are such that the nonlinear resonance coupling curves does not intercept the launching circle, particle motion is stable (see a small circle on the right hand side of Fig. 8).

Author Index